LEGAL DISCLOSURE

AUTHORS

Professor Dr. Martin S. Fischer
Friedrich-Schiller-Universität Jena
Institute of Systematic Zoology and Evolutionary Biology with Phyletic Museum
Erbertstrasse 1
D-07743 Jena
Germany

Dr. Karin E. Lilje
Kistlerstrasse 18
CH-3065 Bolligen
Switzerland

DESIGN

Jonas Lauströer | www.jonas-laustroeer.de
Amir Andikfar | www.andikfar.de

TRANSLATION

Lucy Cathrow

ISBN 978-3-9814339-2-0

Printed by Bonifatius GmbH, Paderborn
Printed in Germany

Print product with financial
climate contribution
ClimatePartner.com/53323-2405-1008

MIX
Paper from responsible sources
FSC® C011558
FSC
www.fsc.org

DOGS IN MOTION

Martin S. Fischer and Karin E. Lilje
with illustrations from **Jonas Lauströer and Amir Andikfar**

CONTENTS

THE STUDY WAS ENABLED BY THE FINANCIAL AND THE ORGANIZATIONAL SUPPORT OF THE FOLLOWING ASSOCIATIONS:

Gesellschaft zur Förderung Kynologischer Forschung (gkf) www.gkf-bonn.de

Members of the German Kennel Club (VDH)

1. Deutscher Shar-Pei-Club 1985 e.V. I www.1-dspc.de
Akita Club e.V. I www.akita.de
Bundesverein für Weiße Schweizer Schäferhunde in Deutschland e.V. I www.bvws.de
Club Berger des Pyrénées 1983 e.V. I www.cbp-online.de
Club für Britische Hütehunde e.V. I www.cfbrh.de
Deutscher Doggen-Club 1888 e.V. I www.doggen.de
Deutscher Pudel-Klub e.V. I www.pudel-klub.de
Deutscher Retriever Club e.V. I www.drc.de
Deutscher Teckelklub e.V. I www.teckelklub.de
Deutscher Windhundzucht- und Rennverband (DWZRV) e.V. I www.dwzrv.com
Eurasier-Klub e.V. I www.eurasier-klub.de
Internationaler Klub für Französische Bulldoggen e.V. I www.ikfb.de
Internationaler Klub für Tibetische Hunderassen e.V. I www.tibethunde-ktr.de
Jagdspaniel-Klub e.V. I www.jagdspaniel-klub.de
Klub für Terrier e.V. I www.kft-online.de
Old English Mastiff Club Deutschland e.V. I www.mastiff.de
Pinscher-Schnauzer-Klub e.V. I www.psk-pinscher-schnauzer.de
PON-Club e.V. I www.pon-club-vdh.de
Rassezuchtverein für Hovawart-Hunde e.V. I www.hovawart.org
Schweizer Sennenhund-Verein für Deutschland e.V. I www.ssv-ev.de
Verband für Kleine Münsterländer Vorstehhunde e.V. I www.kleine-muensterlaender.org
Verband Große Münsterländer e.V. I www.grossemuensterlaender.de
Verein für Deutsche Schäferhunde (SV) e.V. I www.schaeferhunde.de

PREFACE BY PETER FRIEDRICH, PRESIDENT OF THE GERMAN KENNEL CLUB (VERBAND FÜR DAS DEUTSCHE HUNDEWESEN - VDH)

Dog lovers often have a high level of specialist knowledge and very clear ideals. As a result, they sometimes tend to regard dogs almost as a puzzle solved. Reams and reams of breed standards, commentaries, photographic works, club rules and text books would appear to back up this claim. Does that mean, though, that we can afford to dispense with the virtue of openness, that quality which makes it possible to challenge, think through, learn and relearn everything there is to know about dogs? Or do we need, over and over again, new questions, materials and findings to spark productive discussion, make the acquisition of new knowledge a pleasure and expand the factual basis which feeds our actions and increases our competence? Of course we do! And sometimes we are lucky enough to be presented with exactly what we need – this time by Martin Fischer and Karin Lilje in the form of their book "Dogs in Motion". The authors provide us with exciting information, captivate us with stunning visual material, challenge us in a deeply constructive way with arguments and ideas and point the way to significant new lines of investigation. In short, it is astonishing how convincingly they present important new insights into what has traditionally been one of the core areas of cynological research. The locomotion of dogs has preoccupied us for years. Continuous, long-term commitment to the topic has given us an idea about what invisible processes may be at work when a dog is walking, trotting or galloping, what relationships of cause and effect exist, how an individual's gait can fittingly be described in words, and what impact all this needs to have on breeding decisions. Many of our previous beliefs are shown in this book to be right, but by no means all of them stand the test of this new research: the results of the Jena Study challenge many an established view. May I arouse your curiosity by presenting three of the most surprising aspects of the book as a taste of what awaits you? One of the most widely propagated explanations for short-strided, ineffective, restricted seeming forelimb movement in the trot is the assumption that the upper arm is disproportionately short. You will discover why this is false and what has been proven instead. Dogs of one and the same breed often move very differently. Often decorated champions are no exception. Breeds which appear to be as different as chalk and cheese sometimes move in a surprisingly similar way. Here too, this is not always purely the result of comparing top animals with individuals which fail to conform to the breed standard. On the basis of video recordings, you will be able to put together the reasons for this phenomenon. Can we draw reliable conclusions about motion parameters from the angulation of a standing dog? The text explains clearly and convincingly why this is only likely to be the case under very specific conditions.

The Jena Locomotion Study is a huge project, and projects of this nature are, inevitably, incredibly expensive. Happily, the research program was supported by a whole range of institutions, not least because the results will be helpful in improving the quality of life and healthcare of dogs. Its sponsors include the VDH-organized breed clubs for those dogs which took part, the Society for the Promotion of Cynological Research (GFK) and the umbrella organiza-

tion Verband für das Deutsche Hundewesen (VDH/ German Kennel Club).
How can it be that previously unheard of yet utterly conclusive arguments can suddenly be put forward on a topic that has been so thoroughly investigated already? The answer is simple: state-of-the-art methods of data collection and processing that no one would have dared to dream of just a few years ago now bring objectively to light what previously we could only speculate on. Professor Fischer and his team took high-speed video recordings, conducted marker-based movement analyses, and made use of X-ray videography. Specialized analysis software enabled them to evaluate the data scientifically and produce movie animations in various forms – the quality of which is unparalleled. For anyone seriously interested in the locomotion of dogs, the video material which accompanies the book is of inestimable value. Everyone who purchases a copy of "Dogs in Motion" has access to the entire film archive. I personally cannot wait to get down to discussing film sequences in detail with colleagues, friends and other experts. Not only will it be fun, it will be useful, because this kind of discussion will provide a better basis for the judging decisions made at shows and breeding inspections. In this context, the huge variety of breeds included in the study is a definite advantage. Ten representatives each of 32 breeds of varying sizes and functional groups, from the Whippet to the Great Dane and from the Dachshund to the Golden Retriever were investigated in minute detail. The practical implications of this are manifold, and will be even more fruitful in the context of breeding and selection than I know they will be in the show ring.
Even the most modern technology and the greatest expense would be little more than smoke and mirrors in the absence of a decent analysis of the results. I raise my hat to the authors, who consulted over four thousand scientific publications – hundreds of which are cited in the book – in order to produce this scholarly work. The figures and illustrations by Jonas Lauströer and Amir Andikfar are another absolute treasure trove. The anatomical figures and their explanations alone elevate the book to a standard work, and the documentation of the insights into motion sequences is simply unique and – finally a justified use of the word – truly innovative. Another aspect which deserves special praise is the logical way in which the results are presented. At the beginning of the book, foundations are laid and terms explained, be they anatomical, evolutionary or from other essential fields, and only then, on the basis of this, are the insights into the walk, the trot and the gallop introduced and discussed. In my view, this makes this valuable contribution to cynology suitable for a particularly wide audience. Breeders, breed judges, trainers, dog owners, students of veterinary medicine, established veterinarians and biologists alike will all find something of interest here, as will numerous other groups of readers. I am an imaginative enough person, but even I cannot imagine that any dog lover will be able to study this book and not learn a lot, and that in the most pleasurable of ways.

Professor Dr. Peter Friedrich

DOGS WILL BE DOGS

The aim of this book is to help dog lovers better understand their dog. Much has been written in recent decades about canine behavior, nutrition and health, and, encouragingly, dogs are increasingly viewed as living beings worthy of respect. By dealing with the particular aspect of canine biology to which this book is dedicated, we draw renewed attention to something that every dog owner knows, but which is often still ignored, and that is the immense significance of movement for dogs.

When zoologists write about dogs, they are normally interested in the biology of the dog. Like all dedicated dog breeders nowadays, zoologists see dogs as domesticated creatures, but not as pawns to be subject to the mercy of human whims or the dubious human desire to control and subjugate nature. Zoologists recognize the wolf in dogs, the wolf that is necessarily there thanks to dogs biological descent, and urge dog owners to accept their dogs as they are. Many aspects of the canine anatomy have not changed or have changed astonishingly little vis-à-vis the wolf, despite millennia of domestication and centuries of pedigree dog breeding. This book is intended to help readers to find all this out for themselves.

Another, no less important, aim of the book is to examine in a critical light the "knowledge" that has found its way into many popular and pseudo-scientific books on dogs. Where not enough facts are available, and even where they are, opinions are formed. The world of dog breeding is currently swamped with opinions on how dogs run, how they should run and how they must never run. In the few years we spent among dog enthusiasts, we have been confronted at the talks we give by a huge variety of opinions: on body shape, on the "right" position of the shoulder blade, the "right" length of the forearm and – very importantly – the "right" angulation. Curtis M. Brown put his finger on it 25 years ago when he wrote in his then-representative book that "Some of the locomotion concepts presented here are in direct conflict with accepted tradition, which means that some fanciers will object vehemently. It is time to re-examine tradition and differentiate between what is fact and what is fetish." (Brown 1986, p. 17-18).

In order to understand the locomotion of dogs in general and of the different breeds in particular, a fair amount of expertise is required, from knowledge of the locomotor habits of wolves and of canine anatomy, including the way canine muscles are structured, to the ability to describe the kinematic behavior of joints, extremities and the rump during the different gaits. We need to look at the forces at play during locomotion, and ultimately to explore the minor differences and many areas of common ground between the different breeds.

The idea of writing a book on canine locomotion came about as a result of the enthusiasm with which the study we have been carrying out in Jena into canine locomotion over the last few years was received by dog lovers. In the final section of the book, all 32 of the breeds investigated are profiled individually. The initiative for the study and the initial idea for the book came from Dr. Helga Eichelberg, who in her compellingly charming way gave us (next to) no choice but to agree to the project. The study was funded by the German Research Council (Deutsche Forschungsgemeinschaft DFG). But more important, together with the German Kennel Club, Dr. Eichelberg developed a financing plan, which for the first time saw a major scientific study jointly funded by the

23 breeders' associations involved and the Gesellschaft für kynologische Forschung (GKF, Society for Cynological Research).
Our investigation into canine locomotion involved three different high-frequency measuring techniques: videography, marker-based movement analysis and biplanar X-ray videography. Over 300 representatives of 32 breeds were filmed using high-speed cameras and their movement sequences while walking, trotting and galloping recorded using a marker-based measuring system. Some breeds of dog were also filmed using X-ray technology, and these films became the basis for the first ever whole-body skeleton animations – a very complex process but one which we hope will help readers to understand the movements of a healthy dog. A study on this scale was only possible thanks to the untiring commitment of the dog owners involved. Many thanks again to everyone who made the journey to Jena and Bad Langensalza with their dogs.
One important aim of our book is to present the scientific literature available on the subject of locomotion and anatomy in a way which the "normal" reader can understand. There are countless veterinary studies on the disorders which can afflict the musculoskeletal system of dogs and many works available on the anatomy of the dog, but too few works on the locomotion of healthy dogs. The main results of over 300 scientific studies have gone into our book. As scientists we cite these sources in the customary way, by indicating the name or names of the author(s) and the date of publication. In the case of publications by more than one author, it is customary not to name all the contributors but to use the Latin abbreviation "et al." (which stands for et alii or et aliae depending on gender). Full bibliographic details, arranged in alphabetical order based on the name of the first author, are contained in the chapter "Literature cited".
Our mission to write a scientific book for non-scientists was only made possible by two years of cooperation with the illustrators Jonas Lauströer and Amir Andikfar, during which we were able to develop a visual language especially for this book. As authors, we learned to convert as many words as possible into as few pictures as possible. It's not just the illustrations in this book which are so unique: Jonas and Amir also managed to create the first 3D animations ever of a dog based on X-ray films and marker-based movement data. These animations make the movement of the skeleton of a walking, trotting and galloping dog clearer and more easily visualizable than ever before. The high point of the media library is the 3D animation of the musculature of the limbs. The modeling of each individual muscle with its connection to the virtual skeleton took months of work, but these muscles can now be viewed in action for the first time. The media library which contains over 300 high-speed video films, X-ray films, animations and more, was produced by William Takashi Ahrend, who sifted through enormous amounts of data to produce a fascinating visual document.
But what would a book be without a publisher? We would like thank the German Kennel Club (VDH), and in particular Bernhard Meyer and Leif Kopernik for such fruitful collaboration. Not many scientific authors are as heartily encouraged to pass on their knowledge as we have been over the past few years by those responsible at the VDH. We also thank the President of the VDH, Professor Peter Friedrich, for his preface and a chapter on the practical consequences of the results generated by this study.

WOLVES ARE THE ANCESTORS OF ALL TODAY'S DOMESTIC DOGS. TO PROPERLY UNDERSTAND THE LOCOMOTION OF DOGS, IT IS ESSENTIAL TO KNOW SOMETHING ABOUT THE HABITS AND LIFESTYLE OF WOLVES.

THE LIFESTYLE AND LOCOMOTION OF WOLVES

The natural range of the modern wolf (*Canis lupus*) extends across Eurasia and North America. Within it, the animals face extremely varied living conditions. Climate, food supply and, not least, the proximity of the wolves' habitat to that of humans demand place-specific behavior patterns. The various populations and subspecies of wolf thus differ accordingly in size, body shape, color and lifestyle. A population is a group of individuals of one species in a particular area which form a common gene pool. A wolf from Italy looks and behaves differently to one from North America. Most of what is known about wolves today stems from observations of individual populations from North America, Eastern Europe and Italy – just small pockets of the enormous distribution area.

Wolf distribution

Recent natural range
Former natural range

The Arctic Wolf (*Canis lupus arctos*) still lives on the Canadian Arctic Archipelago and on the north and east coasts of Greenland. Withers height: 65-80 cm, weight: 50-80kg.

The Mackenzie Valley Wolf (*Canis lupus occidentalis*) was once found in Western USA and Canada up to Alaska. In Western USA it today just lives in Yellowstone National Park and reintroduced in Idaho. Withers height: up to 90 cm, weight: 45-65 kg.

The Italian Wolf (*Canis lupus italicus*) used to be found all over Italy and Sicily. Today it only occurs in the Apennines from Calabria to Liguria and in the western Alps. Withers height: up to 70 cm, weight: up to 35 kg

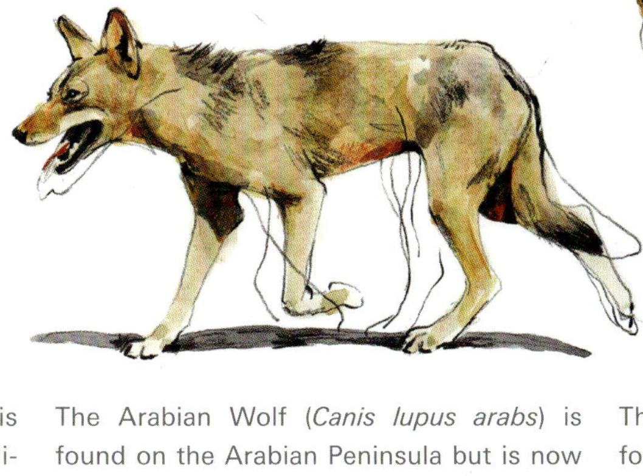

The Eurasian Wolf (*Canis lupus lupus*) is found in parts of Central Europe, Scandinavia, Russia, China, Mongolia and the Himalayas. Withers height: up to 60-80 cm, weight: 30-70 kg.

The Arabian Wolf (*Canis lupus arabs*) is found on the Arabian Peninsula but is now facing extinction. Withers height: up to 65 cm, weight: up to 20 kg.

The Indian Wolf (*Canis lupus pallipes*) is found in India, Afghanistan, Pakistan and Iran and is extremely endangered. Withers height: up to 65 cm, weight: up to 20 kg.

A WOLF'S TERRITORY CAN COVER AN AREA AS LARGE AS THE ISLE OF ELBA

The main purposes of wolf locomotion are finding food and territory patrol. Territory size in Europe ranges from 82 to 243 square kilometers (Okarma et al. 1998, Kusak et al. 2005, Ansorge et al. 2010), meaning that it can take an animal around 10 days to cover the entire area. Territory boundaries are set by chorus howling and scent marking. This means that wolves need to show their presence throughout the territory at regular intervals, which forces them to move around continuously. Wolves travel at an average speed of 6 – 13 kilometers per hour (Mech 1994, Jedrzejewski 2001). This enables them to travel distances of 400 – 600 km per month, though some wolves have been observed to travel as far as 800 kilometers in one month (Fritts 1983). The distance travelled per day can vary widely (0.4 – 64 km). On average, wolves cover between 22 and 27 km per day, but as the crow flies only move around 4 km away from the previous day's locality. In Dalmatia, the distances covered are smaller (nine kilometers in total, leaving the wolves 2.5 km from where they were the day before). The reason for this discrepancy is unclear and there does not seem to be any connection to territory size. In general, male wolves cover larger distances than females (28 km as opposed to 22 km). If the bitch is suckling young, the distance she covers in a day is reduced to 15 km. Locomotor activity within territories was studied in wolf populations in Poland (Jedrzejewski 2001), Alaska (Burkholder 1959), Minnesota (Mech et al. 1971, Fuller 1991), Italy (Ciucci et al. 1997) and Dalmatia (Kusak et al. 2005). Wolves do not move around all day long. Half of the wolf populations investigated so far are mainly active at dusk and at dawn (Theuerkauf 2009). As these are the times at which wolves can see best (Kavanau & Ramos 1975) and at which most of the animals which make up their prey have their main activity phase, it is assumed that dawn and dusk are the original periods of activity for wolves. In Alaska, however, wolf activity depends on the time of year. In winter they are active during the day, in summer they are nocturnal. Italian wolves, on the other hand, are nocturnal all year round. This change in activity rhythms is thought to be a result of the danger posed by humans, as Italian wolves live in an area which is densely inhabited by humans (Merrill & Mech 2003, Theuerkauf et al. 2007).

Both territory marking and finding food are usually carried out as a pack activity. Wolf packs generally consist of one set of parents and their two or three years-old offspring, but the groups are fairly fluid and flexible. In a study carried out over almost two years, the members of one pack were only observed all together in a third of sightings (Mallonée 2008).

Generally, packs hunt and kill a large animal every one to four days. Although wolves do not eat regularly (Jedrzejewski et al. 2002), the amount of food that the northern breeds take in averages out at five to six kilograms per day. In the north, wolves usually prey on large ungulates (hoofed animals) such as red deer, white-tailed deer, moose, reindeer, roe deer and wild boar. Mountain populations also eat mouflon, wild goats and chamois. In the southern distribution areas, which are often more densely populated by humans than those in the north, wolves also prey on domestic animals.

Hunting methods can vary widely from pack to pack. One thing which all hunting strategies for large prey have in common is that the pack will hound the animal in question before dragging it to the ground together and killing it with a series of quick bites. The danger of being injured by a large prey animal is smaller for each individual wolf if the animal is already tired and weak. Hounding is only possible when the wolves act as a pack and requires a huge amount of stamina.

The way in which large prey animals are caught differs fundamentally between canines and felines. Unlike cats, which pounce on their prey with their front limbs and hold onto it with their claws, dogs make almost exclusive use of their teeth. As well as differences in the actual snatching of the prey, hunting strategies also differ between the two groups. As a rule, canines hunt large prey in packs, while most felines hunt alone. Van Valkenburgh & Ruff (1987) draw a connection between the different hunting strategies and differences in the teeth of the two families. Canine teeth are, on the whole, much less specialized than feline teeth. In canines, the eyetooth (also known as the canine!) is elliptical in diameter, while in felines it is round. This difference in shape causes a difference in the stability of the tooth. Because its bending strength is much lower, the elliptical canine tooth would break much more easily than the round feline tooth if plunged into the throat or neck of a struggling animal. This means that when it comes to prey which is larger than the hunters themselves, canines are not able to deliver such powerful, targeted and rapidly fatal bites as felines.

Hunting in packs has disadvantages, however, as each individual ends up with proportionally less food (Schmidt & Mech 1997). A wolf which hunts as part of a pack of six to twelve has to make do with a maximum of 60 – 70 % of the amount available to a lone hunter or to animals which hunt in pairs (Thurber & Peterson 1993, Hayes 1995). Not only does hunting in packs result in less food for each animal, it also has a lower success rate than hunting in pairs. There must, then, be other reasons for the joint hunting strategy. It is likely that the amount of energy required for hunting and the risk to each individual is lower in large packs.

Because a pack is usually made up of family members, whatever food is caught benefits the wolves' own offspring. This principle is known in evolutionary biology as kin selection and is seen in many animal species.

The daily routine of today's domestic dogs differs greatly from that of the wolf in many ways. Two of the main driving forces behind wolf locomotion – the perennial need to hunt and the patrolling of the territory – simply do not exist for domestic dogs. This gives rise to the question of whether and how the structure of the organs related to locomotion might have changed in the course of domestication and across thousands of generations of dogs.

THE DOMESTICATION OF THE WOLF

In the course of history, man has often removed animals from wild populations and familiarized them with human presence – or, in other words, tamed them. The difference between taming and domestication is that in the latter, animals taken from the wild are gradually prevented from reproducing naturally with wild-living animals of the same species. In the domestication process, the mating habits of animals are directed by humans; domestication is achieved through selective breeding. As such, it is irrelevant whether the domestication of wolves came about through the removal of pups from the wild or through wolves actually choosing to remain in the presence of humans (Budiansky 1992). Domestication began in both cases with the interruption of the gene flow to and from the natural population, and not before. Initially, domestication is not geared to any clear and specifically defined goal, as it is not clear what parameters can actually be changed. Man did not know at the beginning of the wolf domestication process, for instance, that body size would be variable. It is important to remember the basic premise that in the thousands of years over which domestication has taken place, only those characteristics have been changed that it is possible to alter.

Konrad Lorenz still believed that particularly affectionate and sociable dogs stemmed from wolves, and that less sociable dogs were descended from the jackal. Even Charles Darwin believed that various species of wild canine had been tamed and their descendants crossed with each other. A species is a natural group of individuals whose members can interbreed to produce fertile offspring (biological definition of species as proposed by Ernst Mayr). The results of research in biological disciplines such as morphology, archeozoology, behavioral science and molecular genetics, however, now show clearly that all modern dogs are descended from wolves. The scientific name for the wolf is *Canis lupus*; that of the dog *Canis lupus* forma familiaris or *Canis lupus* f. familiaris. The point in time in which domestication began is still a matter of debate, as is the question of the number of different locations in which it took place and the number of individual wolves it involved.

Just a few thousand years ago, the natural range of the wolf was the largest of all living land mammals. Wolves, split into various subspecies, inhabited the entire northern hemisphere, from Greenland to the Arabian Peninsula, Mexico and India. Subspecies are geographical variants of a species which can be distinguished from one another on the basis of differences in certain characters. Body size and weight vary widely between subspecies: the Arctic Wolf has a withers height of up to 80 cm and can weigh up to 80 km; the European Wolf has a withers height of up to 90 cm and weighs up to 70 kg, while the Arabian Wolf weighs less than 20 kg. These differences show that variation in body size is inherent in the wolf. The enormous variety of body shapes seen in today's dogs thus stems essentially from variability already present in wolves. The differences in appearance between dog breeds are based on "just" sixty or so characters, the great majority of which are influenced by a maximum of three genes.

THE ORIGIN OF DOMESTICATION

At the beginning of 2009, something happened to revive the discussion surrounding the starting point of wolf domestication. Until then it had been assumed that the domestication process began around 14,000 years ago, but when the Bel gian researcher Germonpré and her team (2009) re-examined a skull that had been found in the 1860s in the Goyet Cave in the valley of the Samson river (a tributary of the Maas in what is today Belgium), they determined that it was actually an astonishing 31,700 years old. Since the publication of this result, scientists have been discussing whether domestication may actually have begun much earlier than previously thought. The next oldest bone finds to have been discovered were unearthed in Russia and Oberkassel (Nobis 1979) and are less than half as old. The skull from the Goyet Cave shows changes that can be interpreted as the result of domestication, including a shortening of the muzzle region and changes in the cranium and the teeth. However, doubts remain about whether the Goyet skull really came from a dog. The latest publication by Napierla & Uerpmann (2010) challenges the idea that the Goyet and Eliseevichi finds can be classed as the earliest dogs. The authors claim that the remains of a tooth they describe found in the Kesslerloch cave near Schaffhausen in Switzerland constitutes the oldest evidence of dog domestication.

Inevitably, finds dating from the period when dogs split off from wolves will always be ambiguous as the two were still very similar at this point. The Goyet find cannot be unambiguously allocated either to a wolf or to later Paleolithic dogs. Other indications that domestication began earlier than previously thought include the preserved footprints of a child and what appears to be a dog which were found together in the Grotte Chauvet in the South of France (Garcia 2005).

Lindblad-Toh et al. (2005) use computer programs to simulate the population history of the dog on the basis of data obtained from the genetic material of modern dogs. They calculate the separation between wolves and dogs to have taken place over 27,000 years ago. The Goyet skull, then, is the first archeozoological find so far to date from the period that has been mathematically calculated to be that in which dogs split off from their ancestors the wolves.

The oldest evidence of a human buried together with a dog, Ein Mallah, Israel, 12,000 years old. (Reprinted by permission from Macmillan Publishers Ltd: Nature 276: 608-610 Davies & Valla 1978).

Other skeletal remains of dogs dating from between 14,000 and 10,000 years ago have been found in western and southern Europe, Russia, Kamchatka, the Middle East and North America. This shows that at the end of the Paleolithic, dogs were already widely distributed. The finds also testify to great variation in body size. The smallest, a 10,000 year old dog found in Saint-Thibaud (France), measured just under 40 cm, while other dog skeletons from the same period are the size of Great Danes.

Paleolithic dogs have a shorter and broader muzzle than wolves, and a broader cranium. They were probably broad and stocky in appearance, like most of the Nordic breeds of dog (Müller 2010). They were fed on horsemeat and the meat of large cloven-hoofed animals (Germonpré 2009). Large dogs were probably used as beasts of burden, as guard dogs and as food, as is still the case today in various Native American tribes. However, even in the Paleolithic, dogs were probably more than just working animals, with psychological interaction between them and humans more than likely (Müller 2010). In a 12,000- year old tomb in Ein Mallah, Israel (Tomb H. 104), archaeologists found that the hand of the deceased had been placed on a four to five- month old puppy (Davis & Valla 1978). This tomb is no exception: Morey (2005) claims that dog burials were prolific in the late Upper Paleolithic. The first Paleolithic domestication of the wolf, then, took place before the advent of sedentism, during the age of hunter-gatherer societies (Clutton-Brock 1995).

It is possible, however, that domestication took place a second time, independently, in the "Fertile Crescent" region at the beginning of the Neolithic between 12,000 and 8,000 years ago (Driscoll & MacDonald 2010). Dogs became smaller in connection with human sedentism and may have been selected on the basis of new criteria. The research of the next few years will hopefully show the extent to which today's dogs in all their variety are the result of this second domestication and whether or not the influence of Paleolithic dogs can still be traced in their genome.

THE MOLECULAR CLOCK

For over 50 years, scientists have been using the "molecular clock" technique to determine the separation points of populations and species. Zuckerkandl & Pauling (1962) discovered that the genetic material DNA changes within species over time. As they assumed that the rate of change had remained constant over the course of evolution, they argued that two species with only minor differences in DNA must have separated later than two species whose DNA differed more significantly. The number of changes would, theoretically, be an indication of the point in time in which the species had separated. In practice, however, it quickly became clear that different sections of DNA change at different rates. Some regions are very conservative, others change more quickly. The speed at which changes appear is not the same for all DNA. The change rates of the various regions depend on a number of factors such as the size of the population and the length of a generation and, as a result, are hard to work out. Another factor which plays a role is whether or not a change has phenotypic implications and is thus subject to selective pressure. We are only just beginning to understand which regions may change at what speeds. Furthermore, molecular clocks need to be calibrated with the help of archeological and paleontological finds and thus depend on the accuracy of the dating of such finds. Although the molecular clock is used to determine the starting point of the domestication of animals, even modern approaches can only calculate the degree of probability that a species or population separation took place within a particular period of time. The periods of time in question, however, are still extremely large, making it impossible to make a reliable statement about the point at which wolf domestication began (Ho et al. 2005, Ho et al. 2007). Archeozoology and molecular genetics are developing quickly, however, making it likely that a precise answer will be available in the near future (see Zeder 2006).

THE GENETIC VARIETY OF THE INITIAL POPULATION DETERMINES THE VARIETY IN FUTURE BREEDS.

NUMBER OF PREHISTORIC DOMESTICATION EVENTS

Genetic material from fossilized and prehistoric dogs indicates that wolves were domesticated a number of times in different places in Eurasia (see Verginelli 2005, Vilà & Leonard 2006). If domestication only took place once, several hundred wolves would need to have been domesticated to account for the genetic variety of today's dogs. If we assume, however, that various wolves from various different populations were involved, far fewer individuals would have been necessary. The investigations carried out by Vilà & Leonard (2006) showed that at least six different domestication localities are required to explain the number of different alleles found, for example, in the MHC, a genomic region which plays an important role in immune defense.

To account for the variety indicated by bone finds in the body shapes and sizes of Paleolithic dogs, population geneticists proceed on the basis of there having been six different founder populations of 60 wolves per population (Vilà & Leonard 2007). However, if wolves and dogs were crossed again at a later date, the calculations are wrong and far fewer wolves would have been necessary in the founder populations. Wolf-dog hybrids can and do exist, as evidenced by molecular genetic data (Andersone et al. 2002, Vilà et al. 2003, Verardi et al. 2006).

The North American wolf was never domesticated (Leonard et al. 2002). When North America was colonized via the Bering Straits from Alaska around 14,000 years ago, the immigrants brought dogs with them – something which provides further, indirect information about the age of the wolf domestication process. The oldest skeletal remains of a dog from North America are just under 10,000 years old and were found in Danger Cave in Utah (Crockford 2005).

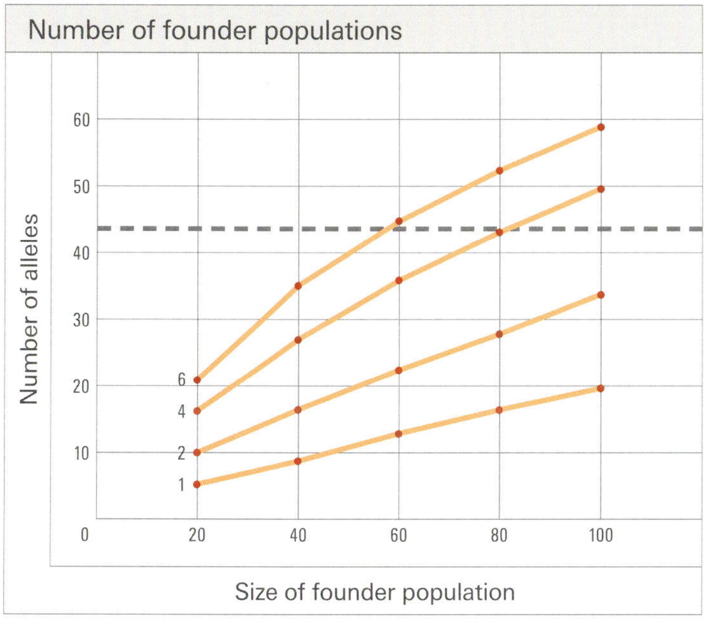

The level of variety which exists today in the alleles of the MHC-complex (dashed horizontal line) requires either 80 wolves from four founding populations or 60 wolves from six founding populations. If fewer founding populations were involved, they must have been very large, as many, many individuals would have been required.

As many finds from the same period show, the dogs brought to North America varied greatly in appearance, which points to a large and varied dog population in Eurasia before the colonization of North America. Whether this variety was the result of the multiple domestications of various different wolf populations or already a product of targeted breeding is a question we cannot answer yet. Interestingly, North American wolves appear to have interbred with the imported dogs, as dog genes have been found in the DNA of American wolf populations (Anderson 2009).

DNA

The genetic material DNA is found in every cell of every living being. 95% is found in the nucleus, and around 5 % in the mitochondria, the powerhouse of the cell. Offspring receive the genetic information housed in the cell nuclei from both their parents, while that housed in the mitochondria is only passed down from the mother. Paternal mitochondrial DNA is not passed on. Molecular genetic analyses which only take mitochondrial DNA into account cannot detect the influence of male wolves. Maternal lines can be traced back, however, as this genetic information is not diluted.

GENETIC BOTTLENECKS

When a small group of individuals is separated from the rest of their species, either as a result of natural circumstances or human intervention, genetic information changes much more quickly than it would in a larger population only subject to external selective pressure. This is due to the reduction in the reproductive community – or in other words, incest. This reduction and concentration of genetic information is also termed a "genetic bottleneck". The population of a few individuals is known as the founder population. If the genetic bottleneck is not followed by selective pressure (e.g. because there are no predators in the area, or because the animals are fed by humans), broad genetic variability quickly develops. This effect is known as the founder effect.

© SPL / Agentur Focus

SILVER FOX REACTIONS TO CONTACT WITH HUMANS

AGGRESSIVE

FRIENDLY

FEARFUL

THE SILVER FOX AS AN EXAMPLE OF DOMESTICATION

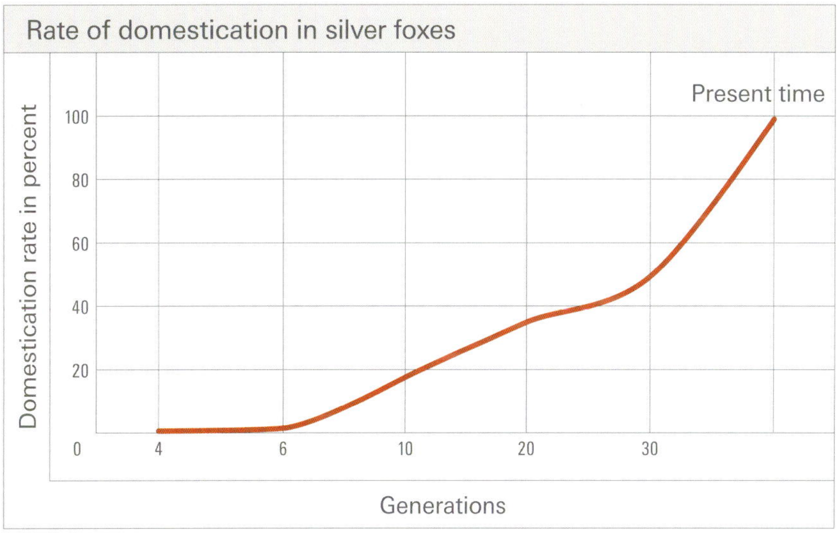

The Russian geneticist Dmitry Konstantinovich Belyaev worked out the principle by which selective breeding brings about anatomical changes and identified the speed at which traits can change. He assumed that the key criterion in the selection of dogs for breeding was not an anatomical trait but a behavioral one: tameability (Belyaev 1979, Trut 1999, Trut et al. 2004, Trut et al. 2009). In 1959 he began, at the Russian Academy of Sciences in Novosibirsk, to tame and domesticate silver foxes. After his death, his research was carried on by Lyudmila Trut.

In a breeding programme spanning 50 years and involving over 50,000 silver foxes, the team selected those individuals whose behavior around humans was the most companionable and sociable. Selection began when the foxes were still cubs and carried on until sexual maturity was reached. Only those animals which actively sought human company and which greeted humans in a similar way to dogs were selected to breed.

Within just four generations, cubs were produced which greeted humans by wagging their tails. In the sixth generation, just under 2 per cent of the cubs (4 of 213) displayed this behavior. By the tenth generation, 18 per cent of the cubs were tail waggers, after 20 generations 35%, after 30 generations 49 % (804 of 1641 cubs) and by 2005/2006 almost all the foxes at the breeding station wagged their tails in greeting. This proves clearly that predisposition to the behavioral trait "tamability" is inherited.

Over the course of the domestication project, changes in various anatomical traits were also observed in the foxes. Firstly, coats became dappled, then floppy ears and curly tails appeared. Skulls became flatter and snouts shorter. After 20 generations, shorter tails and shorter limbs were observed, as was the appearance of overbites and underbites. This shows that even when the sole selection criterion is a behavioral trait, in this case "tamability", changes also occur in anatomical characters such as coat color, skull shape, dentition and body shape. What is still not completely clear is whether these anatomical changes are directly related to the trait of tamability or whether the differences arise as the result of the lack of natural selective pressure.

As well as anatomical changes, changes also occurred in the foxes' hormonal ba-

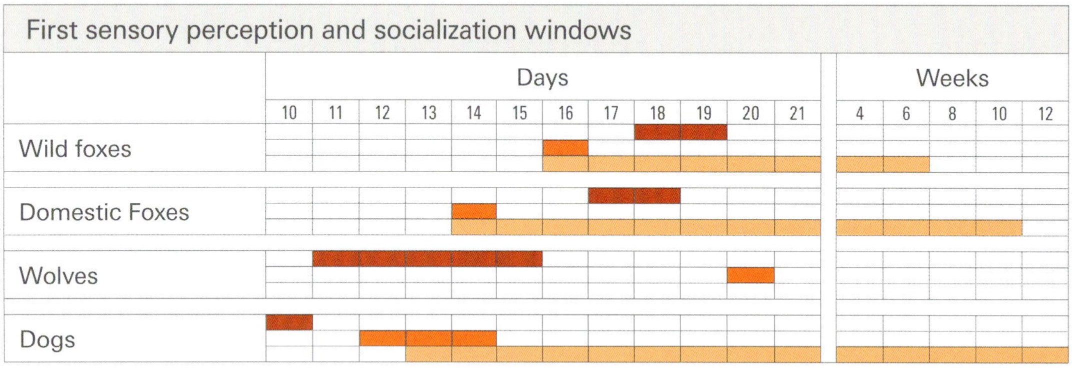

Eyes fully open
Response to sound
Window of socialization

In domestic animals, sensory perception begins earlier and the window of socialization is larger.

lance. The adrenal cortex produced less and less of the stress hormone corticosteroid, and after 30 generations only a quarter of the level present in wild silver foxes was present in the blood of the domesticated animals. The "feel-good" hormone serotonin, on the other hand, which plays a role in the suppression of aggressive behavior, increased. An overall reduction in the hormones responsible for the regulation of stress and the immune system was observed.

To be confronted by humans is normally an extremely stressful situation for a wild animal. Selecting the silver foxes on the basis of docility and tolerance of humans makes a reduction in aggressive behavior both the precondition for (those animals are selected which produce less of the stress hormone) and the consequence of selection (their descendants, in turn, release even less of the stress hormone). This reduction in stress hormone has also been observed in dogs. The genetic process which controls this hormone in the hypothalamus region of the brain differs significantly in dogs from that in wolves (Saetre et al. 2004).

Domestication also has an effect on reproduction. All domestic animals go into heat much more frequently than their wild relatives, and almost independently from the time of year. They also reach sexual maturity earlier and have more young. Estrus no longer depends on the lengthening of daylight in the spring. After birth, the young develop more quickly and the period in which they familiarize themselves with their environment, the "socialization window", is longer.

In the course of the domestication process, silver foxes came to understand more and more human signals. Interspecies communication developed, resembling that between dogs and dog owners. The ability to read behavior probably came about as a result of the social interaction to which the foxes were exposed. It may also have been strengthened by the reduction in fearful and aggressive behavior caused by artificial selection (Hare et al. 2002, Hare & Tomasello 2005, Hare et al. 2005, Horowitz 2009).

The explanation for the various phenomena which occurred over the course of the domestication of the silver fox is to be sought in the gradual shifts observed in the animals' development "schedule". Traits which stem from the period before sexual maturity, i.e. juvenile traits, are retained and carried over into later developmental stages. Hilzheimer (1926) termed this phenomenon juvenilization; in science it is known as pedomorphosis. Pedomorphosis does not just affect physiological and anatomical traits (one example of anatomical pedomorphosis is the broader skull found in dogs), it also extends to behavior. Certain gestures of submissiveness and the barking of adult dogs can be regarded as pedomorphic traits, as in wolves, only cubs "bark".

The experiment with the silver foxes demonstrates clearly that domestication can be achieved within a relatively short timeframe. If the process of selection begins while the animals are still cubs, it is not necessary to think in terms of millennia. The decisive criterion in cubs is behavior, as the other traits have not yet become evident. Domestication can progress so fast that within just a few human generations, wolves can be bred which, in terms of their attitude to humans, can actually be described as dogs. Domestication, then, does not necessarily have to be preceded by a long phase of wolf taming (Benecke 1994). Selection originally took place on the basis of behavioral characteristics, and the alteration of physical characteristics was an unintended consequence. Selection on the basis of size or other physical traits is something which did not occur until much, much later.

THE ORIGIN OF DOG BREEDS

The term dog "breed" has only been in use since the mid-19th century. It originates in England, where dog breeding as we know it today (with stud books and pedigrees) began. The first stud book was submitted in 1874 and covers the years 1859 to 1873 (Sampson & Binns 2006). The first breeders' associations were founded around 150 years ago, and as time went by they joined up to form various umbrella associations. One of the largest such associations is the FCI (Fédération Cynologique Internationale), formed in 1911, which now counts breeders' clubs from over 80 countries among its members. Under the leadership of the umbrella associations, it is the breeders' associations which decide what constitutes a breed of dog. The very formation of a breeders' association means that the members have agreed on certain traits which a dog has to have in order to belong to a breed. These traits, which mainly concern aspects of physical appearance, are set down in writing in the breed standard.

Breed standards may lay down specific requirements with regard to traits such as coat color (this applies to certain French Hounds) or body size (as in the case of the Standard Schnauzer), while in other breeds, such as Poodles and Pomeranians, coat color and size are insignificant: the various "varieties" all belong to the same breed. Dog breeds are thus defined not by biological characters, but by humans. In some cases, dogs can look extremely similar to each other yet belong to two different breeds, while in other cases, two dogs of the same breed can look very different from each other. In most cases, the standards leave room for interpretation. Differences in the way they are interpreted mean that breeds change over time. Changes of this nature are well documented for the St. Bernard (Räber 2010) and the Bull Terrier (Fondon III & Garner 2004).

The latest molecular genetic investigations make it clear that the extraordinary degree to which appearance varies across dog breeds is governed by simple genetic mechanisms, even where the variation affects complex traits (Boyko et al. 2010, Shearin & Ostrander 2010). Furthermore,

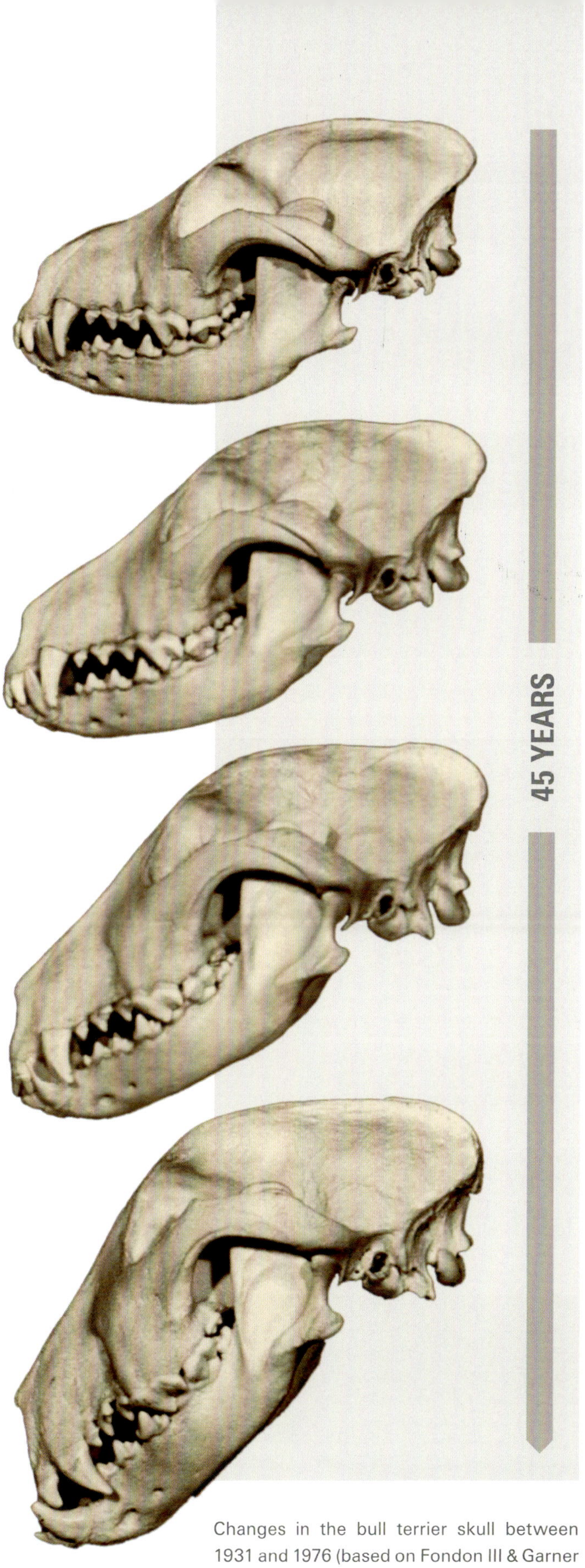

Changes in the bull terrier skull between 1931 and 1976 (based on Fondon III & Garner 2004). With kind permission of Passionate Productions London.

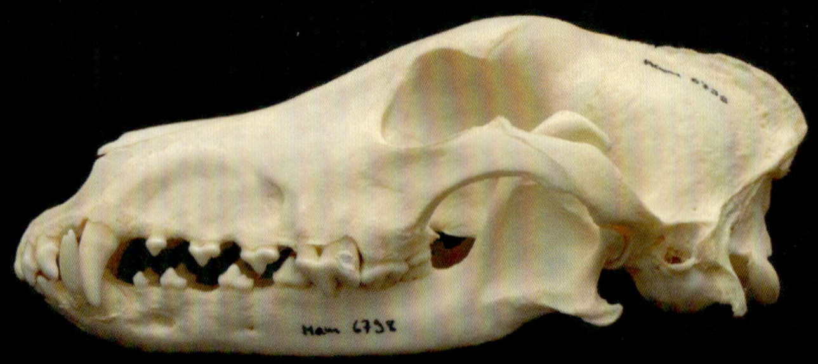

BORZOI

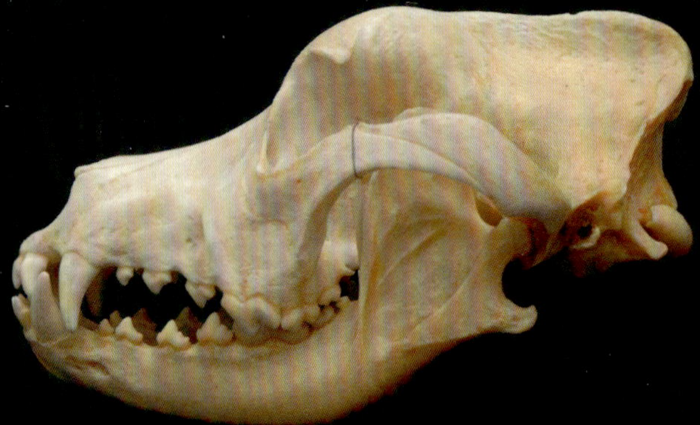

ST. BERNARD

The diversity of dog breeds is seen in the differences in their skulls.

variation is limited to relatively few traits. Bone finds from the Paleolithic show that dogs of different sizes existed even then. Different types of dogs which were used for different purposes are known to have been around for 4,000 years. A variety of types has also been proven to have existed in the Roman Empire (De Grossi Mazzorin & Tagliacozzo 2000). For thousands of years, dogs were selected according to the purpose they were to serve. The formulation of breed standards marked a break with selection for purely functional purposes and saw, for the first time, the focus move to appearance. In an era in which the sole purpose of most dogs is to entertain their owners or act as a status symbol, some breeds are now selected for "aesthetic" traits (and as we know, beauty lies in the eye of the beholder).

In recent years, scientists have attempted, using molecular genetic techniques, to distinguish the breeds from each other on the basis of DNA (Parker et al. 2004, Lindblad-Toh et al. 2005, Sundqvist et al. 2006, Streitberger et al. 2008). As the majority of today's dog breeds are fairly new, however, with their beginnings coinciding with the establishment of the breeders' associations, no clear genetic differentiation is yet seen (Parker et al. 2004, Lindblad-Toh et al. 2005). Only two clearly distinct groups exist alongside the numerous "young" breeds: firstly, a group of Asian dog breeds (Shar Pei, Shiba Inu, Chow Chow, Akita), and secondly the Basenji.

The way in which breeds are categorized by the various umbrella associations is based on pragmatic criteria. The FCI, for example, bases its classification scheme on the use and supposed geographic origin of breeds. The American Kennel Club's classification system, on the other hand, is based almost solely on function and recognizes fewer groups. Neither system reflects the temporal sequence in which the breeds emerged or the actual relationships between them.

GREYHOUND

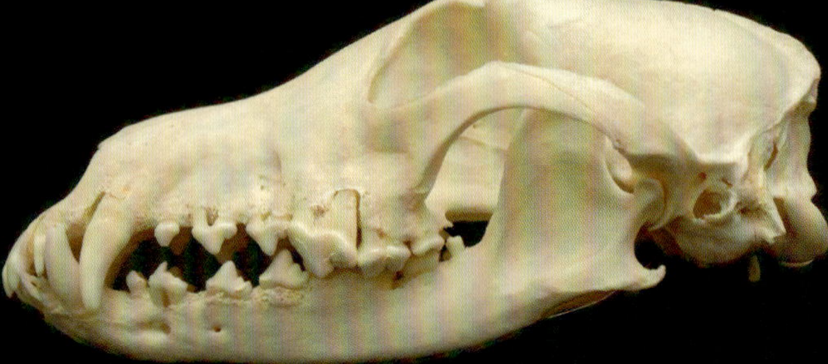

YORKSHIRE TERRIER

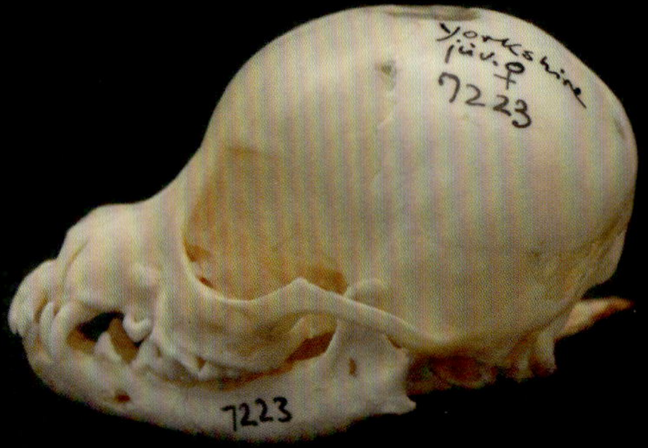

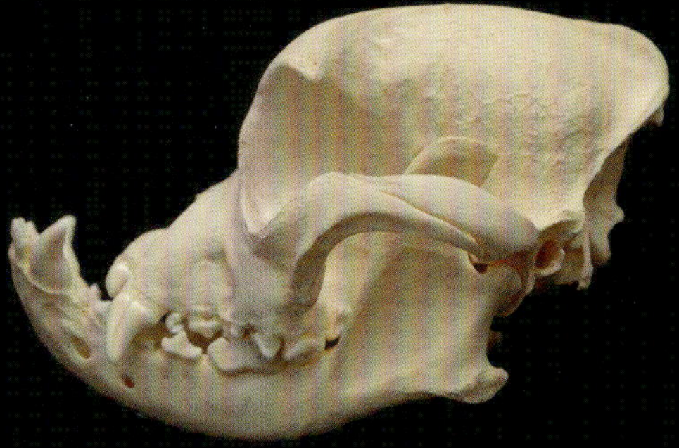

ENGLISH BULLDOG

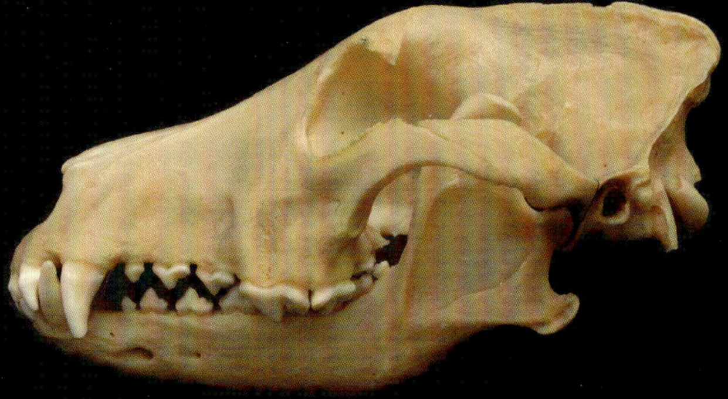

GERMAN SHEPHERD DOG

Phylogenetic tree of dog breeds

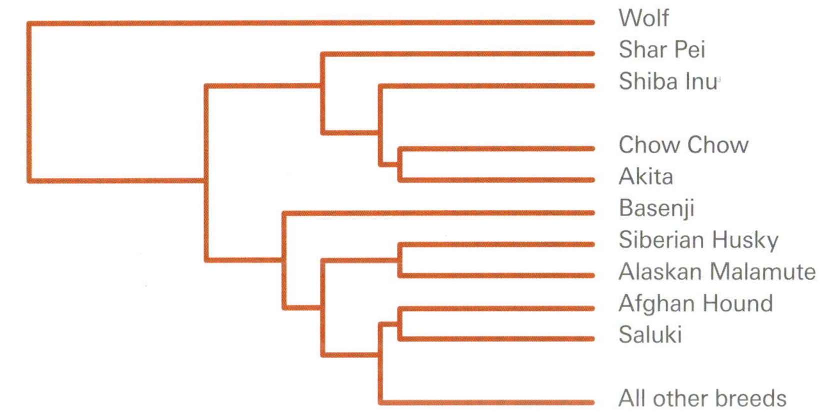

Of the 85 breeds investigated, 76 were barely distinguishable (= all other breeds). Just nine breeds form the first branches in the phylogenetic tree (based on Parker et al. 2004).

HUNGARIAN WOLFHOUND

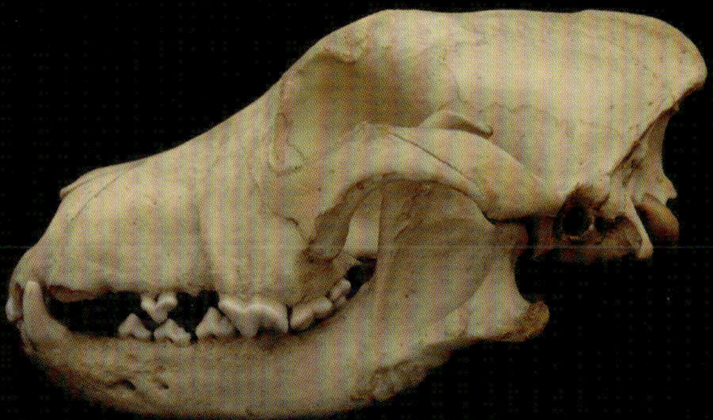

PUG

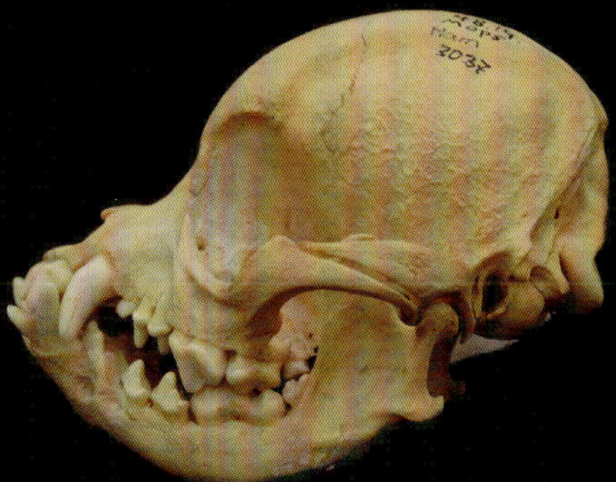

FCI	
Fédération Cynologique Internationale	
Group 1	Bearded Collie
	German Shepherd
	Pembroke Welsh Corgi
	Polish Lowland Sheepdog PON
	Pyrenean Shepherd
	White Swiss Shepherd
Group 2	Appenzell Mountain Dog
	Bernese Mountain Dog
	Entlebucher Mountain Dog
	Great Dane
	Hovawart
	Miniature Schnauzer
	Mastiff
	Shar Pei
	Standard Schnauzer
Group 3	Airedale Terrier
	Bedlington Terrier
	Scottish Terrier
Group 4	Dachshund
Group 5	Akita
	Eurasier
Group 6	-
Group 7	Large Münsterländer
	Small Münsterländer
Group 8	English Cocker Spaniel
	Golden Retriever
Group 9	Chihuahua
	French Bulldog
	Tibetan Terrier
Group 10	Borzoi
	Whippet

KC	
English Kennel Club	
Pastoral	Bearded Collie
	German Shepherd
	Pembroke Welsh Corgi
	Polish Lowland Sheepdog PON
	Pyrenean Shepherd
Hound	Borzoi
	Dachshund
	Whippet
Utility	Akita
	Eurasier
	French Bulldog
	Miniature Schnauzer
	Shar Pei
	Standard Schnauzer
	Tibetan Terrier
Gundog	English Cocker Spaniel
	Golden Retriever
	Small Münsterländer
	Large Münsterländer
Terrier	Airedale Terrier
	Bedlington Terrier
	Scottish Terrier
Toy	Chihuahua
Working	Bernese Mountain Dog
	Entlebucher Mountain Dog
	Great Dane
	Hovawart
	Mastiff
Miscellaneous	Appenzell Mountain Dog
	White Swiss Shepherd

AKC	
American Kennel Club	
Herding	Bearded Collie
	Entlebucher Mountain Dog
	German Shepherd
	Pembroke Welsh Corgi
	Polish Lowland Sheepdog PON
	Pyrenean Shepherd
Hound	Borzoi
	Dachshund
	Whippet
Non-Sporting	French Bulldog
	Shar Pei
	Tibetan Terrier
Sporting	English Cocker Spaniel
	Golden Retriever
Terrier	Airedale Terrier
	Bedlington Terrier
	Scottish Terrier
	Miniature Schnauzer
Toy	Chihuahua
Working	Akita
	Bernese Mountain Dog
	Great Dane
	Mastiff
	Standard Schnauzer

International Kennel Club classifications of the 30 dog breeds involved in the Jena Dog Study

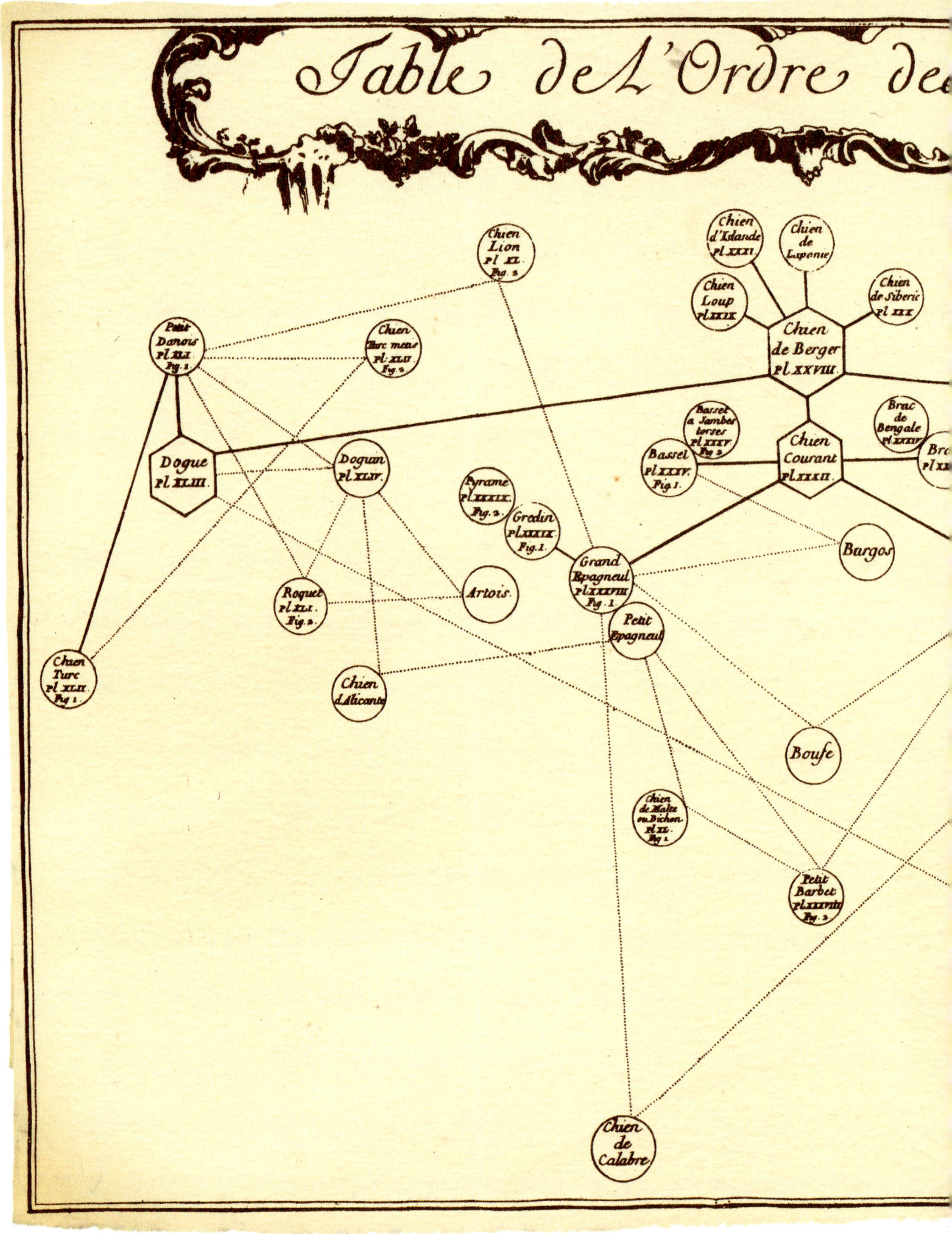
Table de L'Ordre de
Chien Lion Pl XL Fig. 2
Chien d'Islande Pl XXXI
Chien de Laponie
Chien Loup Pl XXIX
Chien de Siberie Pl XXX
Chien de Berger Pl XXVIII
Petit Danois Pl XLI Fig. 1
Chien Turc metis Pl XLII Fig. 2
Dogue Pl XLIII
Doguin Pl XLIV
Basset a Jambes torses Pl XXXV Fig 2
Brac de Bengale Pl XXXIV
Chien Courant Pl XXXII
Basset Pl XXXV Fig. 1
Pyrame Pl XXXIX Fig. 2
Gredin Pl XXXIX Fig. 1
Grand Epagneul Pl XXXVIII Fig. 1
Burgos
Artois
Roquet Pl XLI Fig. 2
Petit Epagneul
Chien Turc Pl XLII Fig 1
Chien d'Alicante
Bouſe
Chien de Malte ou Bichon Pl XL Fig 1
Petit Barbet Pl XXXVIII Fig. 2
Chien de Calabre

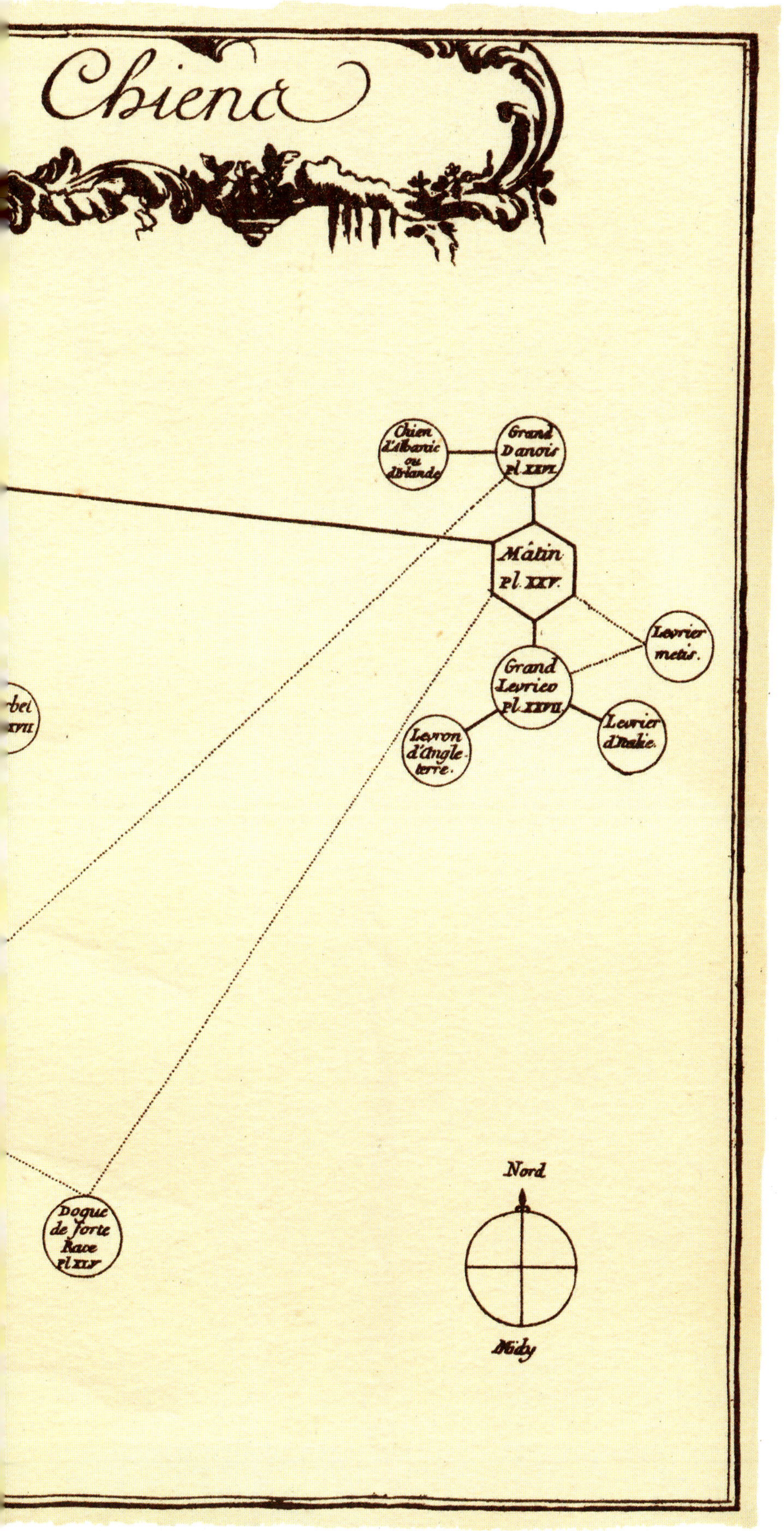

In the 18th century, Buffon & Daubenton (1755) produced the first classification of dog varieties, working on the assumption that the varieties around at the time had arisen mainly as a result of climatic differences. The four main categories into which dogs are divided are the mastiffs (dogue), shepherd dogs (chien de berger), scent hounds (chien courant) and guard dogs (mâtin).

DOGS AND THEIR BODY SHAPE

Starting from the body shape of a wolf, a unique variety of builds has been achieved over the course of dog evolution, all within one biological species. Differences in today's dogs relate to both size and shape. When two dogs have the same ratio of withers height to body length, i.e. the same shape, their relationship is one of geometric similitude. The smaller can be transformed into the larger via simple linear enlargement, a geometric relationship that has been recognized as a rule of mathematics since Galileo.

However, canine history has brought forth completely new body shapes, too, as a comparison of Airedale Terriers and Dachshunds shows. Did selective breeding change one trait at a time, or are some traits genetically coupled to form inheritable complexes of traits? This question is of immense importance for breeders, for it determines how dependently or independently physical traits are passed on.

THE CONSEQUENCES OF CHANGES IN BODY SIZE

According to the principle of geometric similitude, the shape of a body can only be transformed when its length ratios change. When this happens, the surface area and volume of the body change at different rates. Surface area scales as length dimensions squared and volume scales as length cubed. A dog which is twice the size of another has, therefore, a proportionally greater volume^3 but a proportionally smaller surface area^2 than the smaller dog. This relationship is important in determining the upper limit for body sizes, as the cross-sectional area of the bones can only increase by a power of two, while the dog's mass, which constitutes a volume, will increase by a power of three. The bones of large dogs, then, are proportionally narrower than those of small dogs. There would come a point, once a dog had reached a certain size, that the bones would no longer be able to carry the weight of the body. The relative smallness of the surface area in larger animals leads to a lower level of heat loss via the body's surface: the animal does not get as cold as quickly. Rubner (1893) worked out a "surface rule" according to which the metabolic rate, or the rate of energy consumption per kilogram of body weight, increases as body size decreases. As maintaining body temperature requires a lot of energy, larger dogs use less energy, relatively speaking, than smaller dogs. However, the decrease in energy consumption is exponential, not steady.

CHANGING TRAITS

The father of modern genetics, Gregor Mendel, proposed at the end of the 19th century that every trait corresponded to one gene (or "hereditary factor" as he called it) and that changes in the trait were caused by changes in the gene. We know nowadays that this simplistic view of hereditary transmission, the "one gene one phene" model, only holds true in a few very exceptional cases. The majority of traits develop as the result of interaction between various genes (polygenic inheritance), and, in numerous cases, a change in one gene affects several traits at once (pleiotropy = polyphenism). Hip dysplasia, for instance, is a polygenic disorder, and the gene for the coloring "merle" is a good example of pleiotropy as it influences not just coat color but the development of the eyes and ears too. Homozygotic merle dogs are, in some breeds, relatively often also blind and/or deaf. To complicate things further, there are also master genes which are responsible for the regulation of other genes.

In the fruit fly Drosophila melanogaster, geneticists have discovered a master gene which regulates 2,000 other loci. Master genes are central information points which reduce the amount of information that needs to be present in the genome. A mutation in one of the master genes usually has drastic consequences, for it affects not just one gene but a whole gene cascade. Another important genetic phenomenon is that of coupled genes. In this case, although one gene is responsible for the development of one trait, it is "coupled" with one or more other gene and cannot be passed on in isolation. The same applies, of course, to the traits determined by the coupled genes. If one trait changes, the coupled trait changes too, without fail. In this way, the amount of information necessary to produce a trait can be reduced: trait A automatically appears in conjunction with trait B. At the same time, the coupling of traits also limits their scope for change. Coupled inheritance can apply to both qualitative traits (the trait is present or not present) and quantitative traits (the trait is present to differing degrees). The gene sequences responsible for the formation of quantitative traits are known as QTLs (quantitative trait loci). Various teams of scientists (Chase et al. 2002, Carrier et al. 2005, Parker & Ostrander 2005, Lark et al. 2006) have explored the connection between the occurrence of such traits, kinship, and correspondence between gene sequences. Experiments were carried out on 330 Portuguese Water Dogs, as the standard for this breed tolerates a high level of variation with regard to size and build. First of all, pelvis width, skull shape, the thickness of the limb bones and other skeletal structures were measured using X-ray. The statistical procedure known as principal component analysis (PCA) was then carried out to reveal which of these traits, if any, were connected, and which were independent from each other. Predictably, almost all the anatomical traits of the Portuguese Water Dogs were found to depend on body size and sex, but the study also revealed some unexpected connections, such as that between pelvic shape, limb bones and skull shape. Two developmental trends are evident in Portuguese Water Dogs: some individuals are stocky and strong, others are narrow and slim, and it is possible to interpret the latter as an adaptation to running at higher speeds. Blood samples were taken from all the dogs involved in the study and subjected to genetic analyses. It was shown that the two different tradeoffs are associated with genetic changes in various specific chromosome sequences. Changes in pelvic shape, limb bones and skull shape are

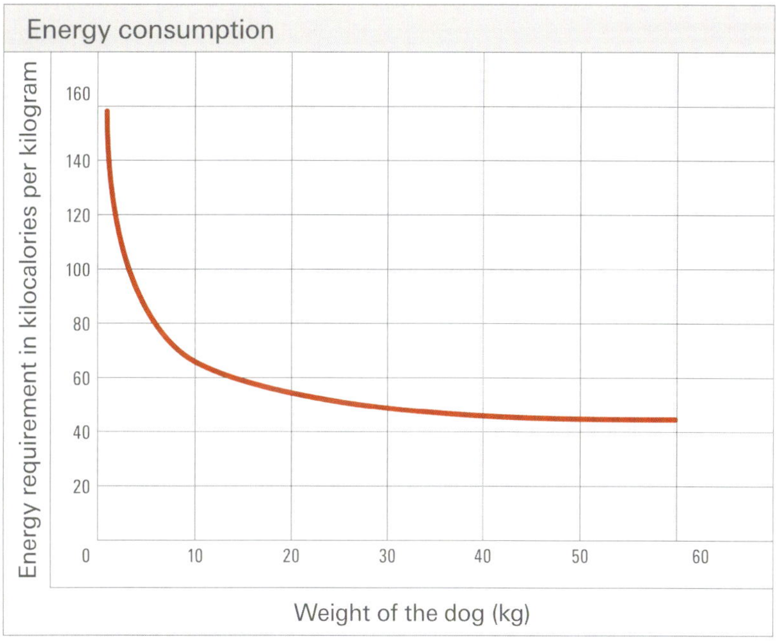

The volume of a body increases faster by a power of one than its surface area.

genetically coupled. The same connection was found to exist in domesticated silver foxes (Trut et al. 2006). It appears, then, at least in canines, that there is a basic regulatory mechanism which selects for either "strength" or "speed", although of course neither of these trends has to manifest itself in an extreme form.
These results show that no matter how long the attempt continued, it would never be possible to breed a dog with the body of a Greyhound and the skull of a Molosser.

SPEED OR STRENGTH?
Speed types and strength types differ in skull and pelvis shape and in the cross-sectional profile of their long bones. A difference in ribcage shape is also observed (either slim or round in cross-section), and the position of the limbs differs accordingly. The distribution of body weight across the fore- and hindlimbs is also different in the two different types. Dogs built for speed have a long, slim body which does not interfere with freedom of leg movement (legginess index >1.2), and when they trot they place their paws under their body in a straight line. The legginess index was introduced by Brown (1986) to describe the ratio of trunk height to free leg length below the trunk.
The shape and thickness of their long bones (femur and lower leg, upper arm and forearm) are not just adapted to locomotion but serve other important purposes. Kemp et al. (2005) showed that there can be a real conflict between the two functions of running and fighting and based their conclusions on two specialists in the respective disciplines, Greyhounds and Pit Bull Terriers. Fighting, unlike running, necessitates the rapid mobilization of large forces to accelerate the body, and the directions of force are varied. As expected, Pit Bull Terriers have sturdy bones, and the middle of the upper arm shaft in particular

SPEED TYPE | STRENGTH TYPE

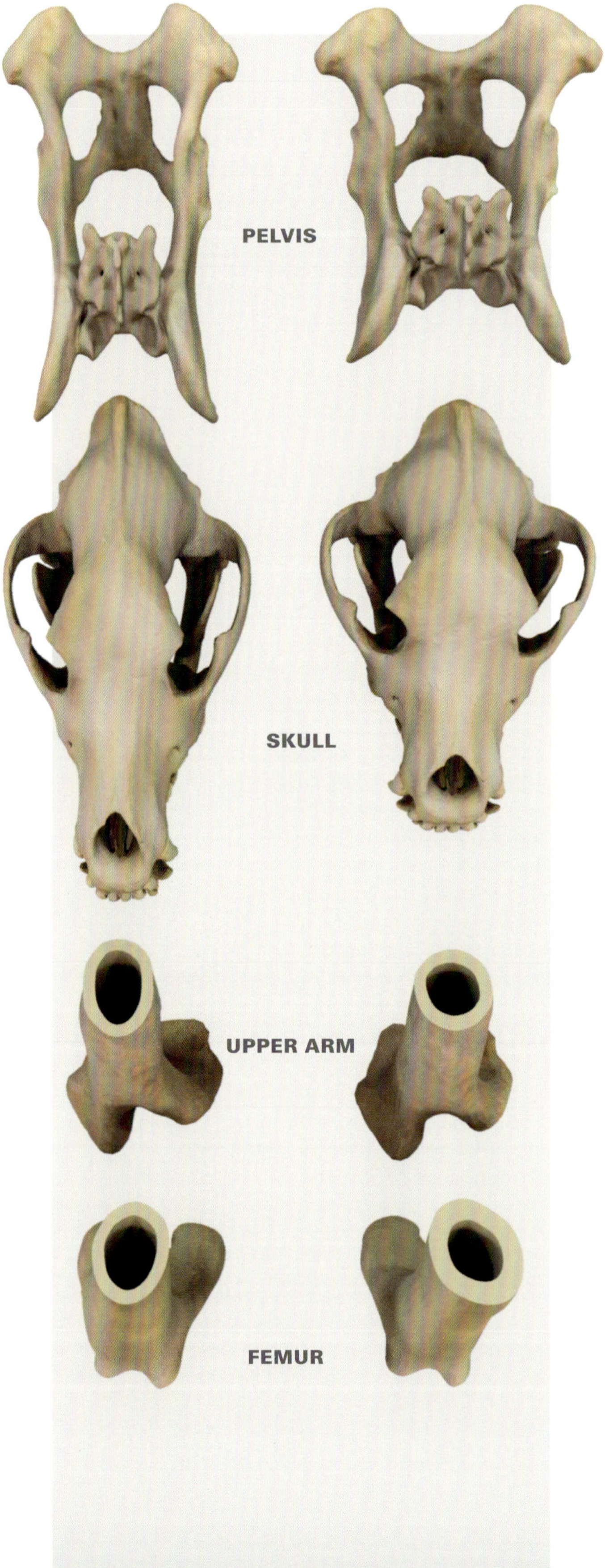

Built for speed and built for strength (reproduced with the kind permission of Prof. David Carrier).

is round in cross-section. In Greyhounds it is elliptical. This corresponds to the difference in the canine teeth of cats and dogs described in the first chapter. Resistance to breaking was two-fold greater in the pit bulls, whereas the stiffness of the bones was 60% higher in the greyhounds.

THE GENETIC REGULATION OF DOG GROWTH

As well as genes responsible for the inheritance of qualitative traits such as coat color, there are also gene regions which influence quantitative, phenotypic traits such as body size. Sutter et al. (2007) discovered that a quantitative trait locus on chromosome 15 is responsible for the size differences observed within the Portuguese Water Dog breed. An investigation of the DNA of a total of 3200 dogs from another 143 breeds showed that the same QTL that is responsible for size differences within Portuguese Water Dogs also determines the size differences between breeds.

Chromosome 15 carries a gene called IGF1 (insulin-like growth factor) which plays an important role in growth, not just in dogs but in mice and humans too. This gene triggers the production of a growth-stimulating hormone. Next to this gene in dogs is a sequence of regulatory DNA which acts like a dimmer switch, not only turning the growth hormone gene on or off, but also controlling the amount of growth factor it stimulates the body to produce. In small dogs, this regulatory DNA ensures that the neighboring gene triggers the production of less growth-stimulating hormone. Large dogs, on the other hand, generally do not possess the sequence at all. The exceptions are Rottweilers, Mastiffs and Bullmastiffs (Gray et al. 2010), in which breeds other, as yet unknown factors seem to turn off the effect of the regulatory DNA.

THE JENA STUDY ON DOG LOCOMOTION

The Chihuahuas which took part in the Jena Study.

The question at the heart of the Jena Study was what similarities and differences exist in the locomotion of different breeds of dog. This question is not only of interest to dog owners but to all interested in locomotion, especially as dogs are the only one species in which it is possible to study the influence of body weight in a more the 25-fold weight spectrum and also of body shape.

In order to answer this question, gait analyses were carried out on over 300 dogs. The Jena Study is thus the most extensive study into dog locomotion to have taken place anywhere in the world to date. Various breeders' associations played an important role in the financing of this study. The overall sponsor was the "Gesellschaft zur Förderung Kynologischer Forschung - GKF" (Society of Cynological Research), and also the German Research Council.

To record locomotion, three different high-frequency techniques were used: videography, marker-based movement analysis (Qualisys®) and biplanar X-ray videography. The video films taken at 500 frames-per-second capture the full beauty of the dogs' movements. Where technically possible, dogs were filmed moving on a treadmill from the side and from the front using two cameras simultaneously. A selection taken from over 300 of these films is included in the media library. Furthermore, using X-ray films and marker-based data, dog movement was animated for the first time.

HIGH-FREQUENCY VIDEO FILMS AND MARKER-BASED MOVEMENT ANALYSES

32 different breeds were involved in the Jena Study. Ten representatives of each breed were analyzed performing the gaits walk, trot and gallop. Attention was paid in selecting the breeds to ensure that all the build types seen in dogs were represented. The spectrum ranged from the smallest breeds to the largest, and included breeds which were small and light (Chihuahua), small and heavy (French Bulldog), large and light (Borzoi) and large and heavy (Mastiff). Square builds were represented (Miniature Schnauzer, Airedale Terrier), as were short-legged, long-backed breeds (Dachshund, Pembroke

Welsh Corgi). The dogs selected for the study were chosen by the associations responsible for each breed. The Chihuahuas took part as part of a privately organized initiative.
The logistics and organization of the project, along with the film recordings and the evaluation of over 1,000 films kept one scientist (Dr. K. E. Lilje) and numerous student assistants fully occupied for over four years. The study was only possible thanks to the help and support of the breeders' associations involved and the willingness and dedication of the dog owners who allowed their pets to participate. Most of the breeds (30) were investigated in Bad Langensalza, Germany, while the two smallest, the Chihuahuas and the French Bulldogs, were analyzed in Jena, Germany.
fzmb GmbH, a medical and biotechnology company with associated large animal hospital in Bad Langensalza provided a large treadmill set into the floor with continuously adjustable speed settings from a minimum speed of 0.7 m/second (Mustang 2200®). The treadmill was 4.25 meters long and 1.10 meters wide. Its breadth meant that the dogs and their owners could warm up together, allowing the animals to acclimatize very quickly to movement on the treadmill. Previous studies have shown that locomotion on a treadmill does not differ from movement on the ground, except for the fact that the stance phase is slightly longer
When the dog was moving confidently on the treadmill, it was secured to a leash hanging loosely from above. The dog's owner would stand in front of the treadmill encouraging the dog to run towards him/her. Filming did not depend on the reaching of a specific speed, but started as soon as the dog was moving freely in the desired gait. The preferred speed could differ even between dogs of the same breed, according to individual fitness levels.
The treadmill in Jena (ADAL3D-WR®) was slightly raised and, at 2.10 meters long and 60 cm wide, significantly smaller than that in Bad Langensalza.
During this part of the study, the dogs were measured using a high-speed video system (Photron®, Camsys®) and the movement analysis system Qualisys®. Reflective markers were glued to the dogs' joints, and six cameras simultaneously emitting infra-red flashes registered the reflection and transmitted the information via internet. This too was a high-frequency system, recording at 280 frames per second. Using specifically designed software, the 3D coordinates of the markers were calculated, making it possible to work out the angular displacement of the joints as they moved.

HIGH-SPEED VIDEO RECORDING
(Mikromak Camsys® and Photron® Fastcam-X 1024PCI)

High-frequency video systems made it possible to film the dogs continuously at a frequency of 500 frames per second. The recordings which result appear to be in slow motion, revealing every detail of the animals' movements. One second of "real" time is extended to 20 seconds of film time. Unlike many slow motion recordings seen on television, the effect is not achieved by stretching out a small number of frames to fill a longer stretch of time – what we see actually are twenty different frames. In the first year of the study, the films were recorded using the black and white system Mikromak Camsys®. The first camera filmed the dogs moving on the treadmill from the side, the second from the front. The cameras were connected to a computer which transmitted the pictures to two different monitors. From the monitors, the films were recorded onto video.
Thanks to a grant from the German Research Foundation (Deutsche Forschungsgemeinschaft), the second year of the study saw the introduction of a Photron® Fastcam-X 1024PCI (Model 100KC). This system is exclusively digital and boasts excellent color rendering (3200 ISO/ASA). Both cameras are connected to a computer and the recordings are saved directly to the hard drive. Resolution is 1024 x 1024 pixels at 1,000 color frames per second.
Both systems save the film sequences to a frame buffer so that they can be viewed before being definitively selected.

To enable us to describe the differences between the motion sequences performed by the different breeds in the three gaits walk, trot and gallop, it is first necessary to lay down parameters which can then be subject to a comparison. These parameters are the angular displacement of the joints, the stride length of the fore- and hindlimbs and the excursion of the fore- and hindlimbs at touchdown and liftoff. In this study they were all calculated on the basis of the Qualisys® films. To obtain the three-dimensional coordinates which permit the calculation of angular progression, the individual markers first need to be identified and named. Only when each marker has been clearly identified in all 2400 frames/pictures/images per gait and dog can its three-dimensional coordinates be exported out of the program. Following this, the frame/image which captures the precise moment of touchdown and lift off has to be identified for each of the four limbs. Finally, the raw data are combined with these data and processed using a specially developed analysis program (AiM: Animal in Motion, Hackert 2005).

The test setup at the Forschungszentrum für Medizintechnik und Biotechnologie mit Großtierklinik (fzmb GmbH) in Bad Langensalza (With kind permission of J. Kehrl).

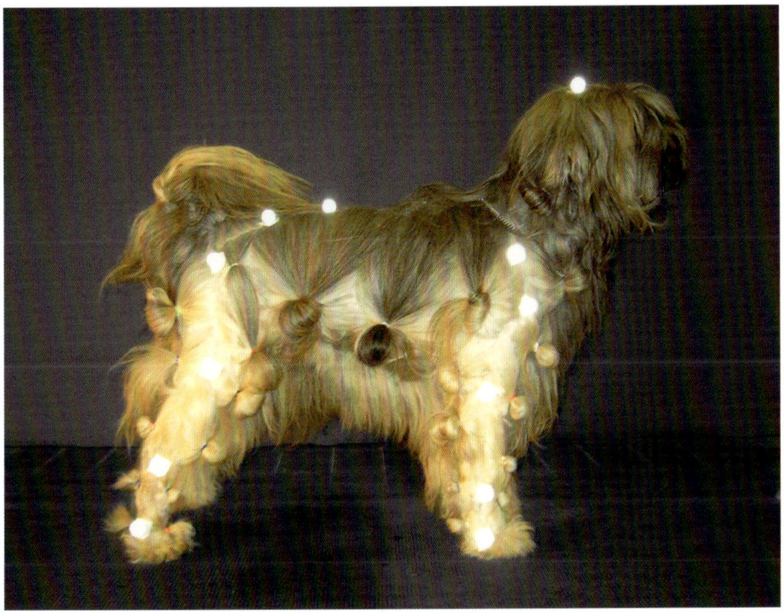

A Polish Lowland Sheepdog being prepped to take part in the marker-based movement analysis.

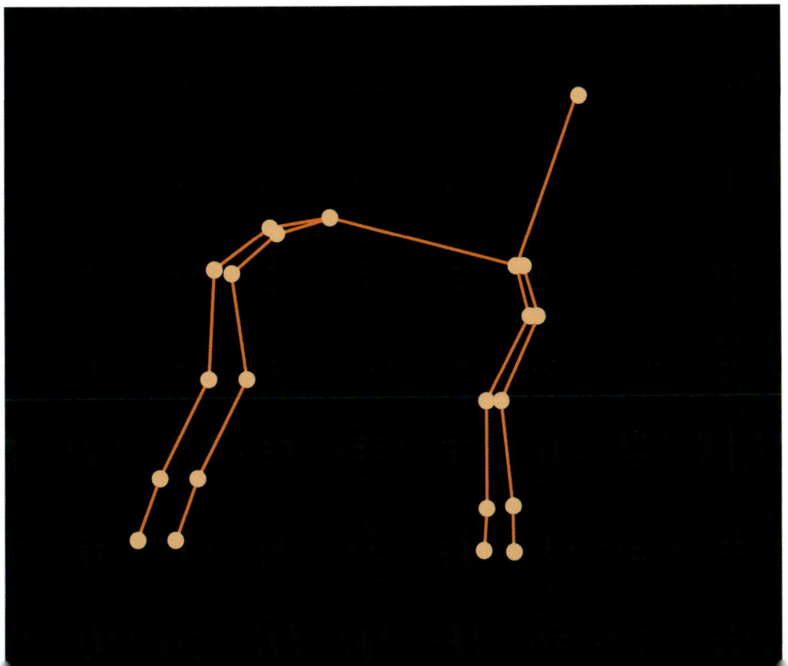

HIGH-SPEED MOVEMENT ANALYSIS SYSTEM (Qualisys®)

This system was developed for the analysis of human movement and works using reflective skin markers which look like miniature table tennis balls that have been flattened on one side. The markers are covered with a reflective layer. Six cameras which emit infra-red flashes are connected via a network to a computer. The same cameras record the infra-red light that is then reflected back by the markers. When the reflection from one marker is captured by at least two cameras, a special computer program is able to calculate the three-dimensional position of the marker in space. To make this calculation possible, the system is calibrated at the beginning of every measuring session using a metal grid in which reflective markers are positioned at predefined intervals.

Using double-sided tape, markers were attached to the joints of the fore- and hindlimbs, the spine of the scapula, the iliac blades, the paws, the head and along the back of each dog. The level of precision with which the markers are placed strongly affects the quality of the measurements. Depending on the type of coat in question, different methods were used to ensure that the markers could be positioned in exactly the right place. In the simplest cases (short, smooth coats), the markers could simply be stuck on. In most cases, however, some hair had to be cut, and in addition to this, the long-haired dogs even had to have their coats tied into pigtails.

The dogs were filmed from both sides for ten seconds at 280 frames per second. The result was a sequence of the markers moving in space.

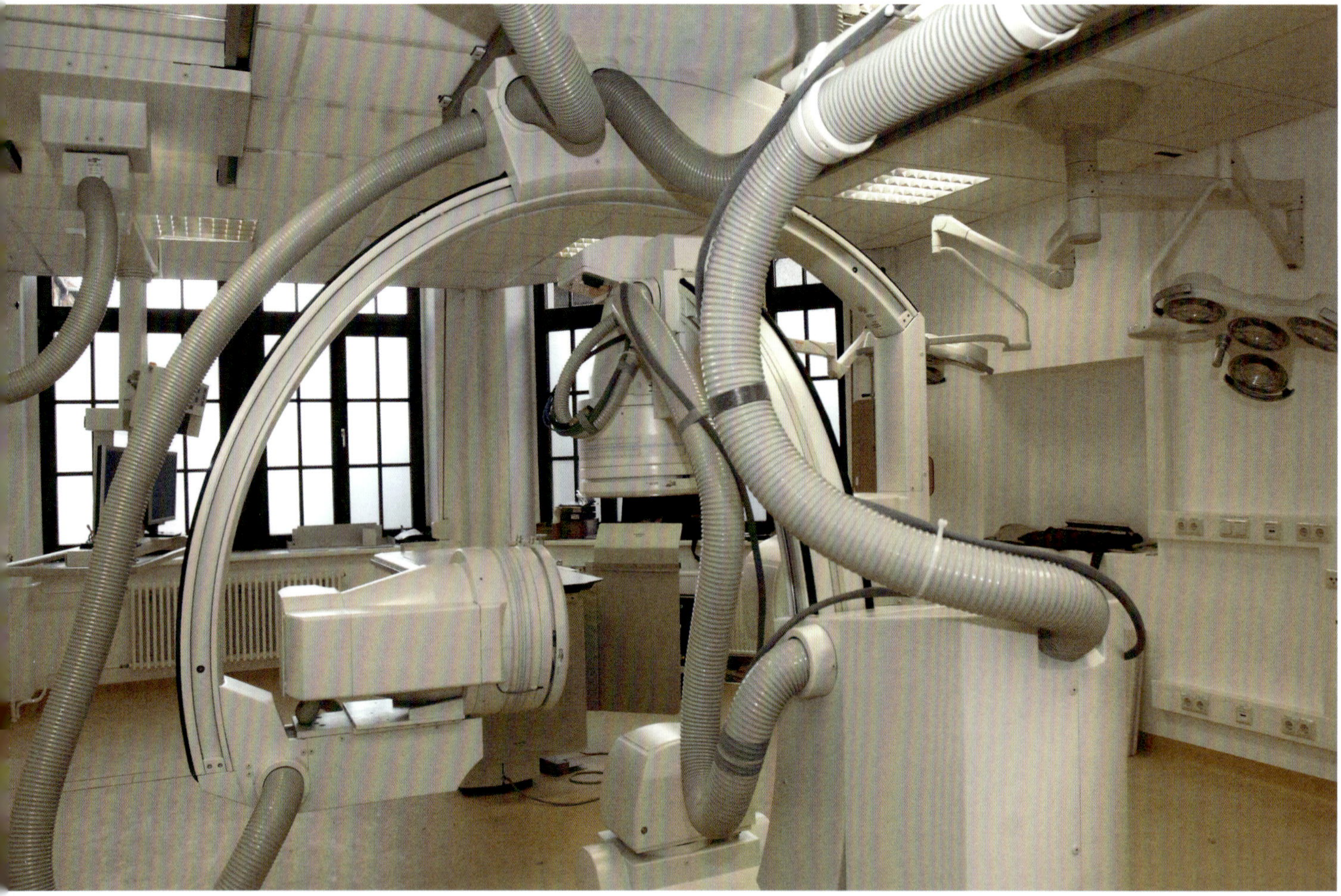

The biplanar, high-frequency video radiography system in Jena.

X-RAY VIDEOGRAPHY

Our Institute in Jena boasts what is currently the most powerful X-ray videography system for animal research in the world. Recording at rates of up to 2000 frames per second, the equipment makes it possible to capture the movements of animals such as small dogs at the highest possible resolution from two different directions in space. This complex technology enabled us, for the first time ever, to take X-ray recordings of the entire front half or entire rear half of some of the smaller dogs. In dogs with a withers height of over 35 cm, only isolated regions of the body, such as the shoulder area, can be filmed. Eight breeds of dog have been filmed using X-ray videography so far, all at 500 frames per second.

Analyzing the X-ray films is even more complex and time-consuming than analyzing the Qualisys® data, as each frame has to be imported into a special software package (SimiMotion®), where every joint then has to be individually identified via mouse click. Doing this for tens of thousands of X-ray films is a job for an entire army of students!

Only when all three investigative techniques are used together can dog locomotion be described with any precision. Every technique has its advantages and disadvantages. X-ray videography captures the movement of the individual bones with unparalleled precision, but the 40 cm diameter of the image intensifier, which is the largest currently available for clinical systems, limits the dogs which can be displayed in full height to small breeds. Moreover, analyzing the data currently takes such a long time that we would need around 30 years to study 300 dogs.

X-RAY VIDEOGRAPHY SYSTEM
(Siemens)

The X-ray videography system is based on the Siemens Neurostar ®, a high-end neuroradiology system. Two X-ray sources and two X-ray image converters 40 cm in diameter make it possible to take synchronous recordings of movement at any level. Depending on the frame rate, two high-performance video cameras (Weinberger SpeedCam Visario g2 ®) record anywhere up to the last 16 seconds of a movement. The decision regarding the selection of the film is thus not taken until the desired movement has been recorded. The video cameras used here combine high light sensitivity with very high resolution and high image frequencies. Using a matrix of 1536x1024 pixels, a maximum frame rate of 1000 fps can be obtained. As well as the two X-ray video cameras, the system features two conventional video cameras which record the movements synchronously. All the cameras are controlled externally using the software VISART. The pictures from all the active cameras are shown simultaneously on monitors. The image data are transmitted via a Gigalink network to a powerful PC equipped with a RAID 0 and a Dual Core E 6400 processor.

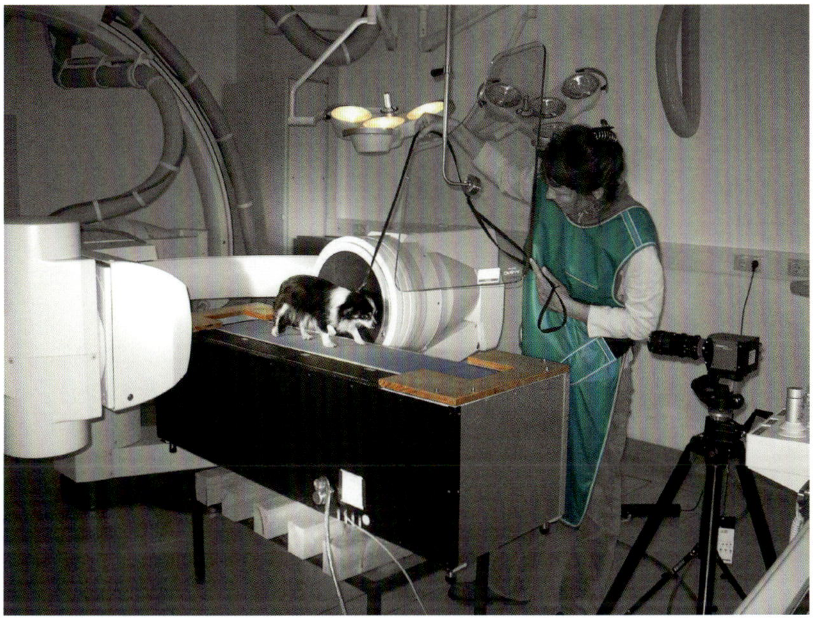

Recording the video film of the Chihuahua depicted on the next page.

The markers used in the Qualisys® movement analysis system depict the movements of a dog in a way to which humans are used: from the outside. Analyzing 300 dogs only took four years and the result is a three-dimensional description of the limb movements of a complete dog. The disadvantage of this technique is the fact that the markers have to be placed on the skin in an attempt to gauge the movement of the bones. The inevitable shifting of the skin relative to the body leads to systematic error (i.e. discrepancies which are inherent in the technique), especially where there is a significant distance between the bone and the marker thanks to a layer of connective tissue or muscle. In some dogs, we actually calculated the deviation between the marker measurements and the X-ray measurements. Markers can measure the movement of joints such as the ankle and elbow very precisely, but that of joints such as the hip or shoulder only approximately. The anterior position of the rotating shoulder blade, for instance, always appeared around 10° smaller on the basis of the markers than it did on X-ray. Every measuring technique has its uses and its limitations, and these must be recognized and made transparent. The values given here in the descriptions of the breeds (see p. 138) were all calculated exclusively using the movement analysis system Qualisys®. This is the only way to ensure that the data pertaining to the various breeds are comparable. We corrected some of the systematic error and marked the discrepancies in a italics.

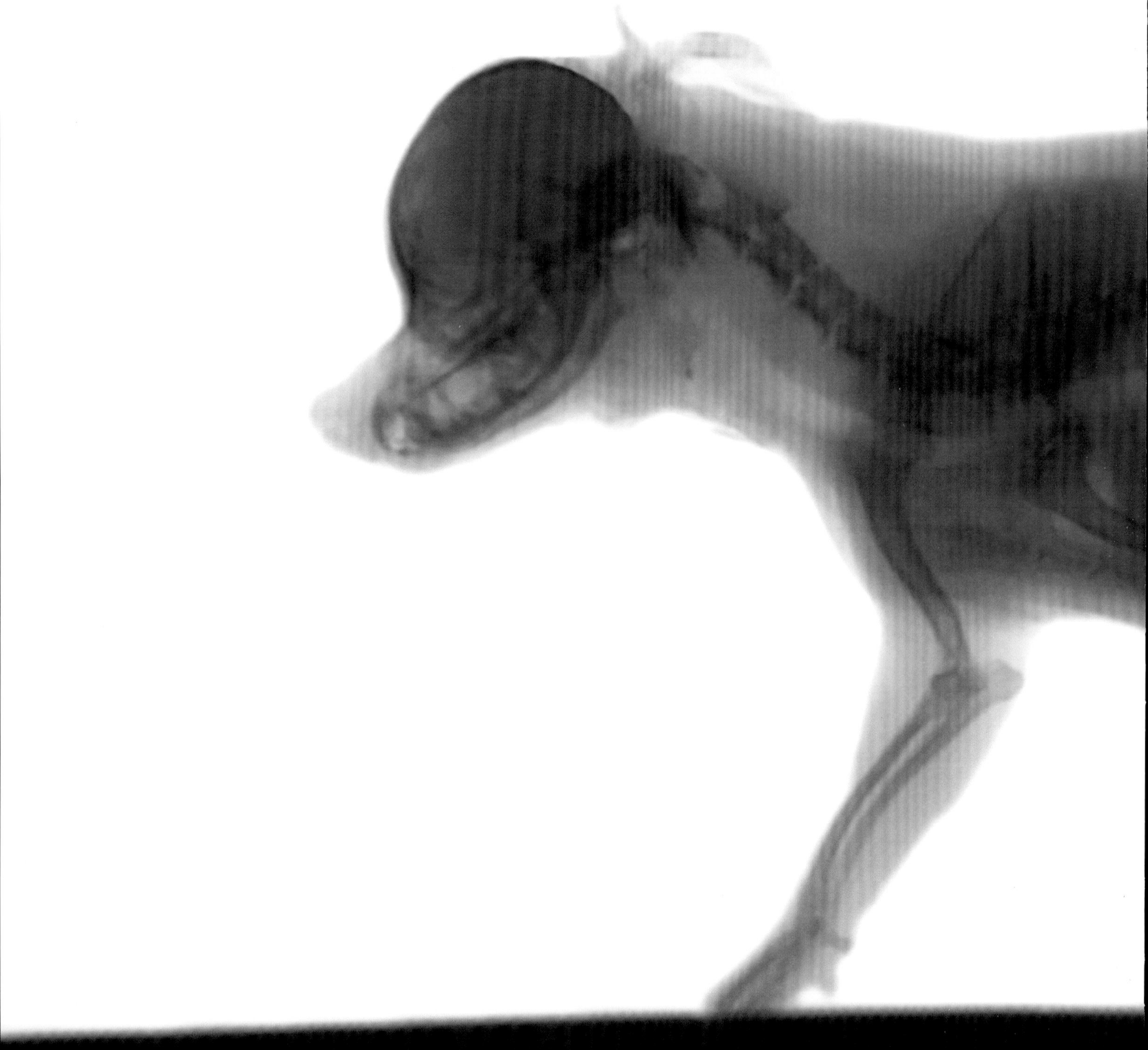

A CHIHUAHUA WALKING. THIS IMAGE WAS PUT TOGETHER FROM TWO SEQUENCES OF A 500 FPS X-RAY FILM.

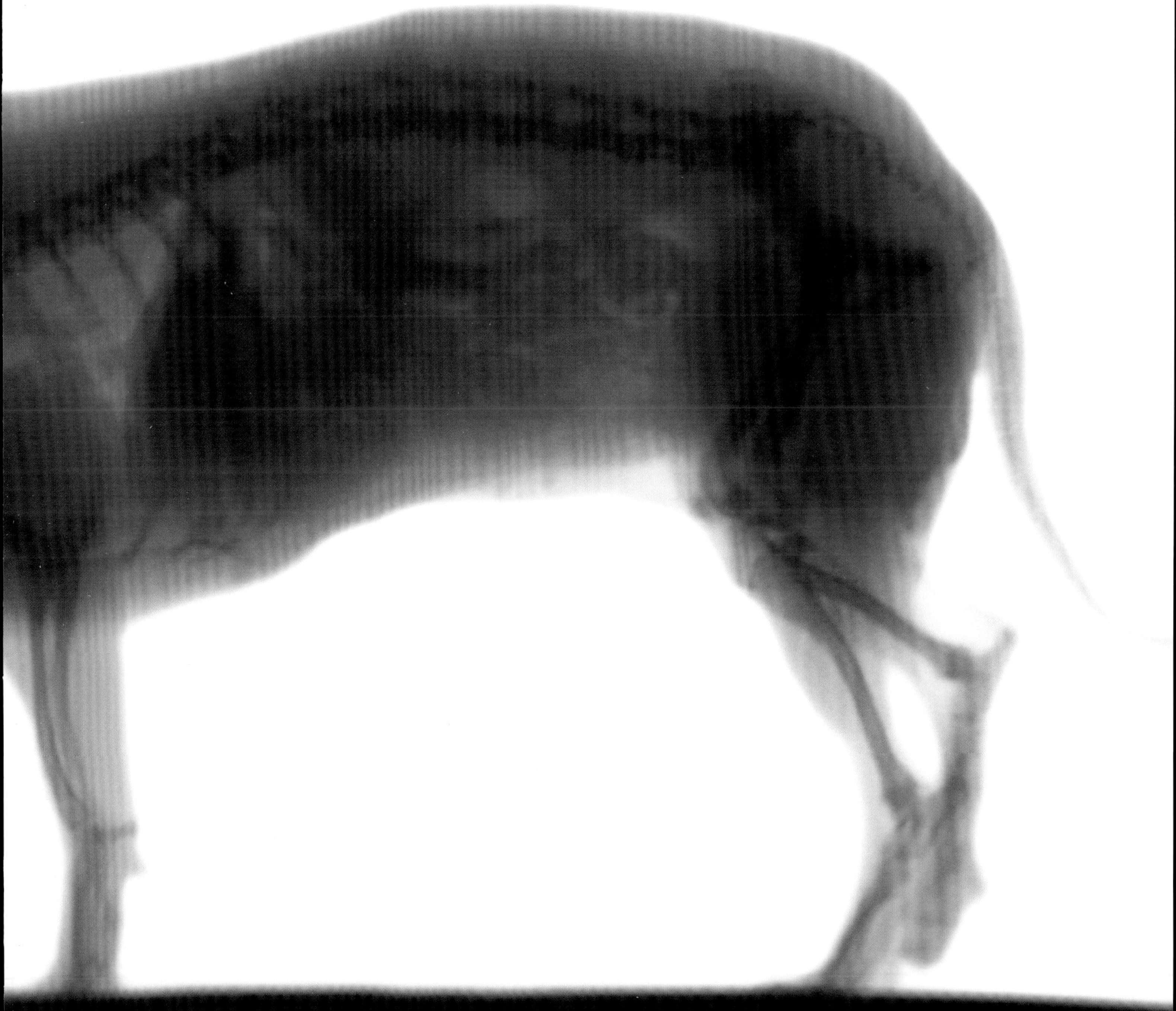

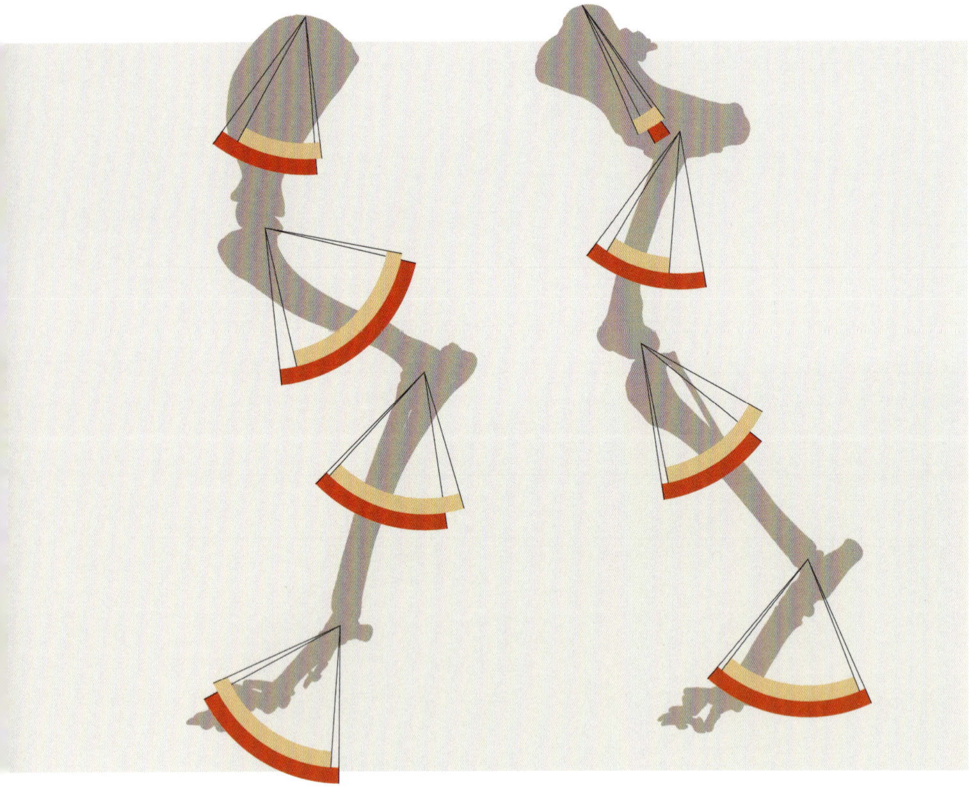

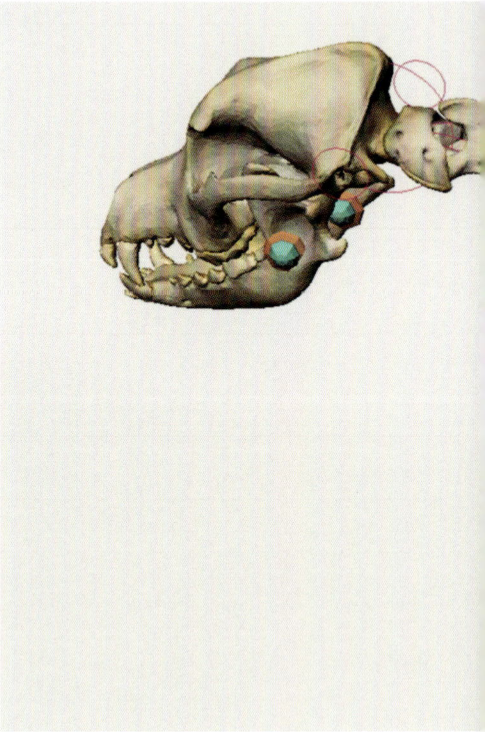

By comparing the X-ray films and the video recordings of the skin markers taken at the same time, it was possible to calculate the level of discrepancy between the two measuring techniques. The angle shown in red is the real angle measured using X-ray. The angle shown in light brown is the one measured using skin markers.

Marker-based angle X-ray-based angle

COMPUTER ANIMATION

Animating the skeleton

Animation involves creating, from a number of still images, a series of images which, in the eye of the beholder, appear to move. The basis of our computer animation of the locomotion of a large and a small dog in the gaits walk, trot and gallop were X-ray films and marker-based motion analyses. Replaying X-ray images one after the other does not reveal any more or any less than an X-ray film would. Visualizing the motion sequence of a skeleton using X-ray images involves a complex, controlled process. First of all, the bodies of several dead dogs were imaged using computer tomography (CT) at the Institute for Diagnostic and Interventional Radiology (IDIR) in Jena. Both whole corpses and macerated skeletons were reproduced in this way. Although computer tomography has a very high polygon density, it fails to completely resolve certain structures, even when they are actually isolated. For this reason the CT data were then imported into the 3D program Autodesk Maya®. The skeletal elements were then separated on the basis of the dog's actual anatomy and missing surfaces were digitally re-modeled. After this, the number of polygons was reduced to diminish the volume of data and thus permit it to be used for animation purposes in the first place. After optimization, the surfaces were smoothed and the polygon raster projected like a sewing pattern onto a two-dimensional plane in a way similar to the projection of the globe in an atlas so that we could create textures and surfaces for the 3D skeleton. The bones "edited" in this way were then put together on the basis of X-ray images taken from various perspectives to form a virtual dog skeleton. To give the whole thing a more realistic and lifelike character, the surfaces of the material were actually painted to lend them more structure. The manual application of paint is still the best way of tempering the stiff "computer graphics"

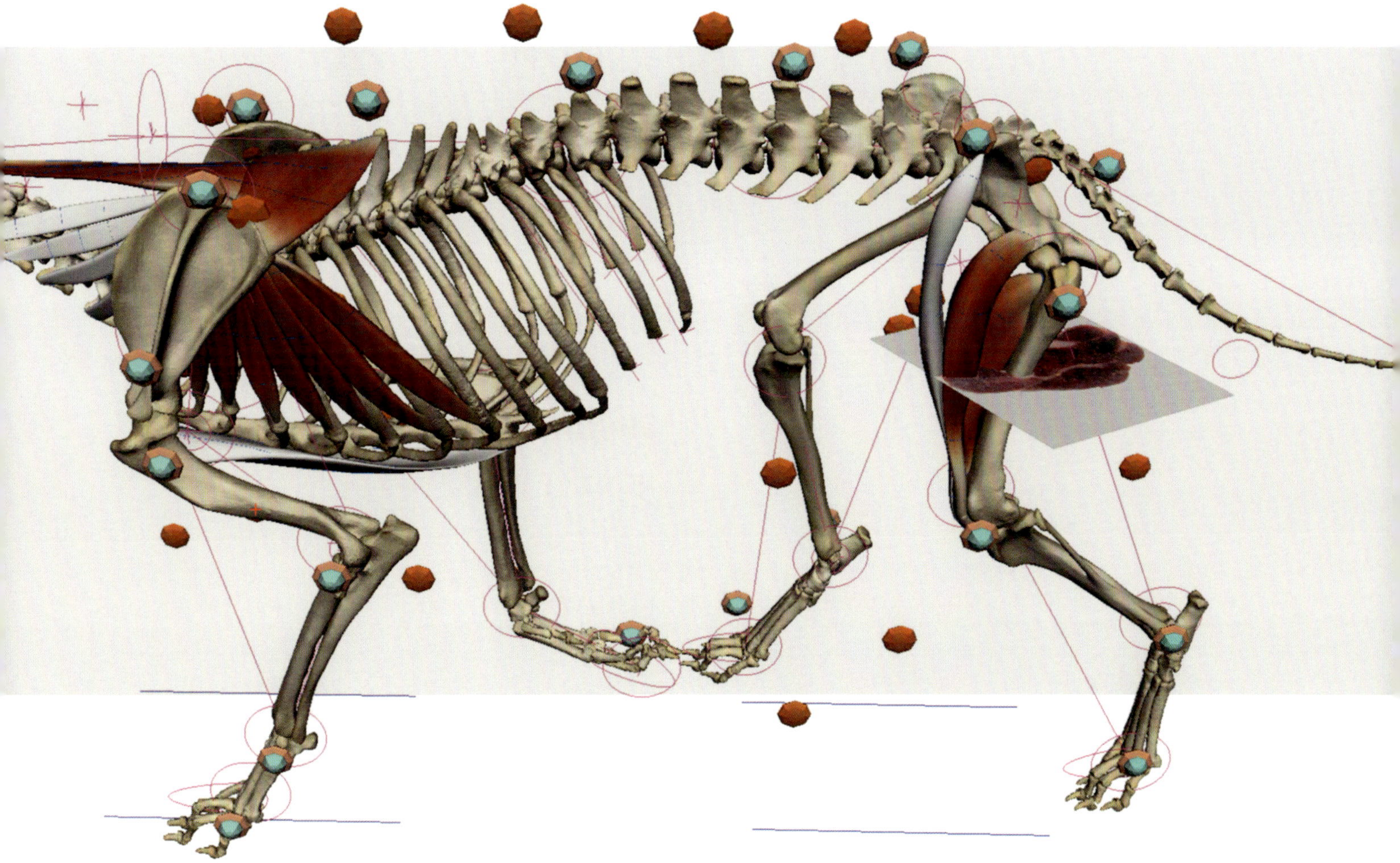

look and in this case allowed us to unite the precision afforded by the computer with the lively quality of freehand art. The axes of movement were determined after an analysis of the X-ray films. The kinematic dependencies observed in the motion analyses, such as the rotation of the knee cap as a function of the angulation of the knee joint, were taken into account. Using the marker-based data, a decision was then taken regarding the hierarchical order of the joints. The knee joint, whose movement can only be described approximately using markers, was, for example, made dependent on/treated as a function of the hip and ankle joints. To enable us to use the Qualisys data in Autodesk Maya®, a program was written to convert the CT Qualisys data into 3D-compatible files.

Animating the muscles

As CT does not provide a particularly clear rendering of muscle tissue, the frozen left leg of a dog was cut into regular sections. These sections were photographed and used as the basis for the cross-sectional curves of the muscles. In addition, the right side of the same dog was sectioned to allow us to study its anatomy. The result is the most precise reproduction of the arrangement of the muscles of a dog ever achieved in a virtual model. This model is the basis of all the muscle diagrams in this book. The animation of the muscle model is based on published data stemming from electromyographic studies on dogs (figs. 120-123). Using different colors, the duration of activity of the muscles involved in locomotion was visualized for the different gaits. The animation thus combines the findings of electromyographic investigations with the anatomically precise topography of the model and is the first film ever to accurately visualize muscle activity. Art and science meet to bring the results of various studies and a vast amount of knowledge together in one fascinating animation.

An object's center of mass is the point at which the total mass of the object is calculated to form the angle of attack for gravity. If the object were to be suspended from this center of gravity, it would remain in static equilibrium. The center of mass is not a fixed point but can move around the body.

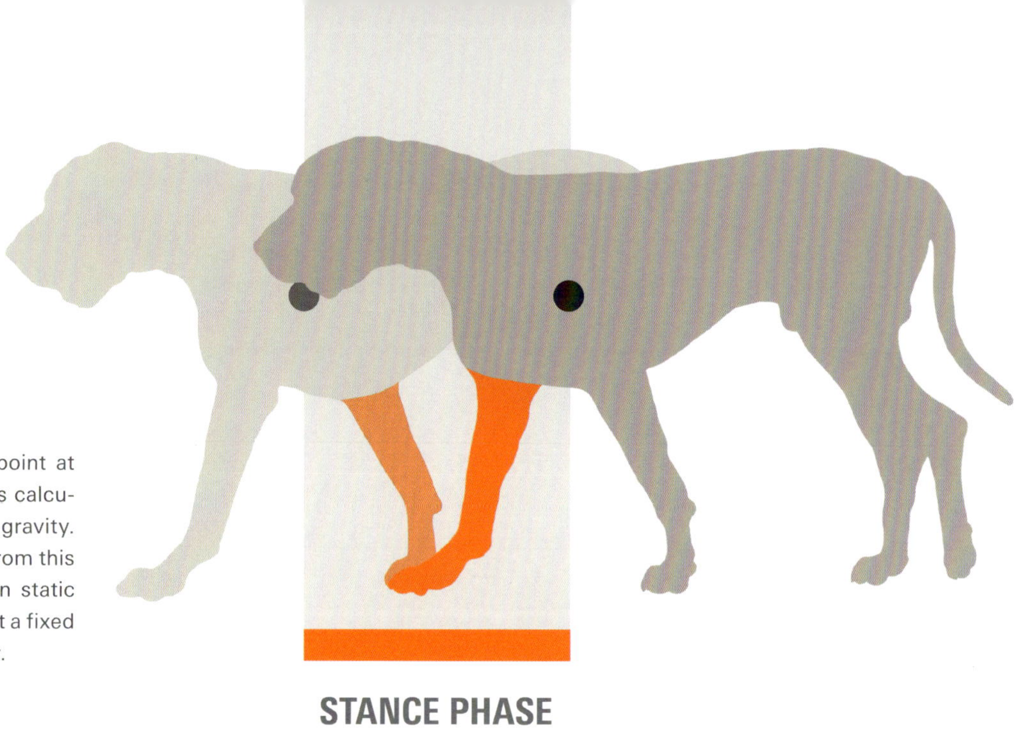

SPEED
is distance / time, so the distance a dog covers within a particular time.

TOUCHDOWN
is the point in time at which the paw of a limb makes contact with the ground.

LIFTOFF
is the point in time at which the paw of a limb breaks contact with the ground.

STRIDE DURATION
is the length of time between two consecutive touchdowns of the same limb.

STANCE PHASE DURATION
is the length of time between a limb touching down and lifting off again.

SWING PHASE DURATION
is the length of time between a limb lifting off and touching down again.

DUTY FACTOR
is the percentage of stride duration taken up by the stance phase: a duty factor of 50 means that stance phase duration and swing phase duration are identical; a duty factor above 50 means that the stance phase lasts longer than the swing phase.

GAITS AND GAIT PARAMETERS

GAITS

All dog owners know that dogs can move using different gaits. A sniffing dog moves its limbs in a different rhythm from a dog running alongside a bicycle or one that just happens to be chasing a cat. As the rhythm and frequency of limb movements change, so does the pattern in which the fore- and hindpaws are placed on the ground and the amount of time they remain there. This is known as the footfall pattern. A gait is defined as a regularly repeated sequence of limb movements. The following double page provides a visual guide to the various gaits. The walk, the trot and the pace are classed as symmetrical gaits, as both in terms of rhythm and in terms of movement the limbs of one side of the body do the same as the limbs on the other side, although not at the same time. Within the symmetrical gaits, the gait itself is determined by the amount of time by which the movements of the fore and hindlimb on the same side of the body are staggered. During the pace, the fore and hindlimb touch down at the same time; during the walk touch down is staggered between them by around one quarter of the stride cycle; during the trot this increases to half a stride cycle. The symmetrical gaits form a continuum, which means that the pace-like walk (also known as the fast walk) and the trot-like walk are equally possible. Because of this, a dog can never go straight from the pace into the trot or vice versa, but must always gear up or down via the walk.
Another definiton distinguishes trot and walk by the presenece or absence of an aerial phase.

Gallop is actually just an umbrella term for various galloping gaits. The breeds involved in this study were observed to perform the slow gallop, also known as the canter, and the two rapid gallops known as the diagonal gallop and the rotary gallop. During the canter, the flight phase is either very short or non-existent. During the diagonal and rotary gallops at least one flight phase occurs, and sometimes two when the dog is travelling at very high speeds. When two flight phases occur, the gait is described as the racing gallop. This gait is observed in various breeds of sighthound, and in the Doberman, Basenji, Pointer, Corgi and Dachshund (Brown, 1986). The gallop in all its variety is described as an asymmetric gait, as it is the limb pairs (the two hindlimbs and the two forelimbs) which touch down one after another rather than the limbs on the same side of the body. Neither the forelimbs nor the hindlimbs touch down at exactly the same time, however. The one which touches down first is described as the trailing limb and the one which touches down second is described as the leading limb.

This "collaboration" between the fore- and the hindlimbs makes it possible for the back to contribute to stride length, too. When the forelimb pair is on the ground and the hindlimb pair in the air, the lumbar spine flexes so that the hindlimbs can be brought down further forward. When the hindlimbs make contact with the ground, the lumbar spine starts to be stretched. Acceleration is only possible during the gallop at those points in the stride cycle when the limbs are in contact with the ground and the dog can thus transfer forces. During the flight phase, the animal's speed decreases.

When transitioning between gaits, dogs do not have to move via the walk and the trot before they can gallop or vice versa before they can stop. Unlike cars, dogs are able to go straight into top gear. Having said this, no mammal has ever been observed to go from motionless directly into a typical gallop footfall pattern: although in a sprint start the forelimbs move as they do in the gallop, the hindlimbs touch down simultaneously. This is known as a "half-bound", and only after a few strides of this does the dog begin to gallop. Blaszcyk (2001) found that the transition between gaits in dogs is usually made within just two strides.

The walk, trot, pace and gallop can all be performed at a variety of different speeds: one dog may be trotting while another of the same breed walks quickly alongside it. For this reason, the dogs in this study were not filmed once they had reached particular, pre-defined speeds, but simply at moments when they were observed to be moving easily in a particular gait. The range of speeds at which gaits can be performed can be very large, as the occurrence of the racing trot and the racing pace illustrates. The gait with the smallest speed range is the walk – a dog walking fast soon transitions into a trot or a pace. The trot and the pace can be performed at a greater range of speeds as they only require the coordination of the frequency of two limbs.

STRIDE LENGTH
is the distance covered by the dog's trunk between two consecutive touchdowns of the same limb. As the dogs being filmed on the treadmill were running on the spot, stride length was not measured but calculated from stance phase length, stance phase duration, swing phase duration and treadmill speed.

STANCE PHASE LENGTH
is the distance which the dog's trunk covers between the touchdown and takeoff of a limb. On the treadmill of course, the dog's movement takes place on the spot and distance is marked by the treadmill itself. This particular situation calls for a change in the point of reference, so stance phase length is the distance covered on the treadmill in the period between a paw touching down and the same paw lifting off again.

SWING PHASE LENGTH
is the distance covered by the dog's trunk between the liftoff and touchdown of a limb. On the treadmill, with the dog moving on the spot, the swing phase cannot be measured but must be calculated. Swing phase length is the difference between stride length and stance phase length.

RELATIVE STRIDE LENGTH
refers to the relationship between stride length and functional limb length. It is the quotient of withers or hip joint height and stride length.

HORIZONTAL EXCURSION
describes the forward and backward excursion of the limb as a whole. It corresponds to the distance between the point at which a limb touches down and the perpendicular of its pivot point (shoulder blade or hip joint), or to the distance between the perpendicular of a pivot joint and the point at which a limb lifts off.

RELATIVE HORIZONTAL EXCURSION
refers to the relationship between excursion and functional limb length. It is the quotient of withers or hip height and horizontal excursion.

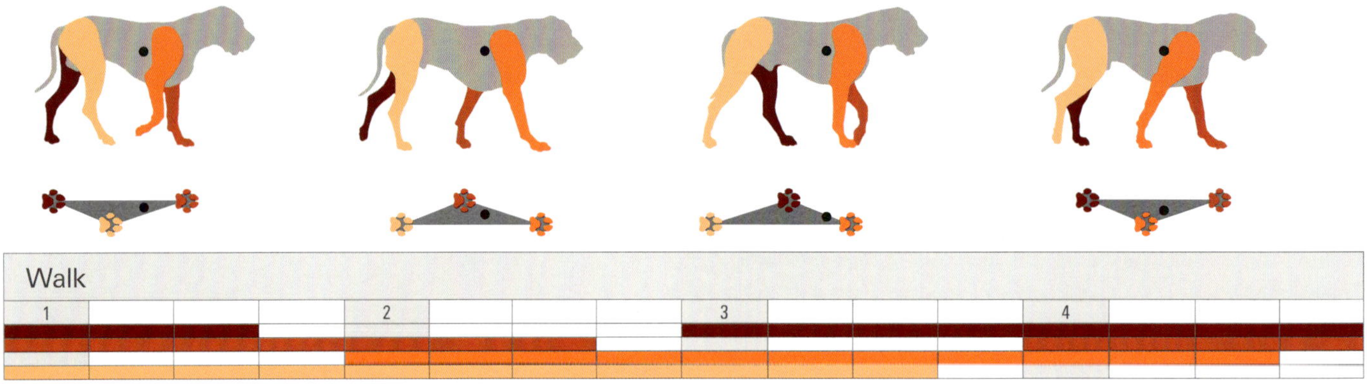

During the walk, phases of two-legged support alternate with phases of three-legged support, though the phases of two-legged support are very short. The advantage of this is that the body's center of mass is almost always located within the support triangle on the ground.

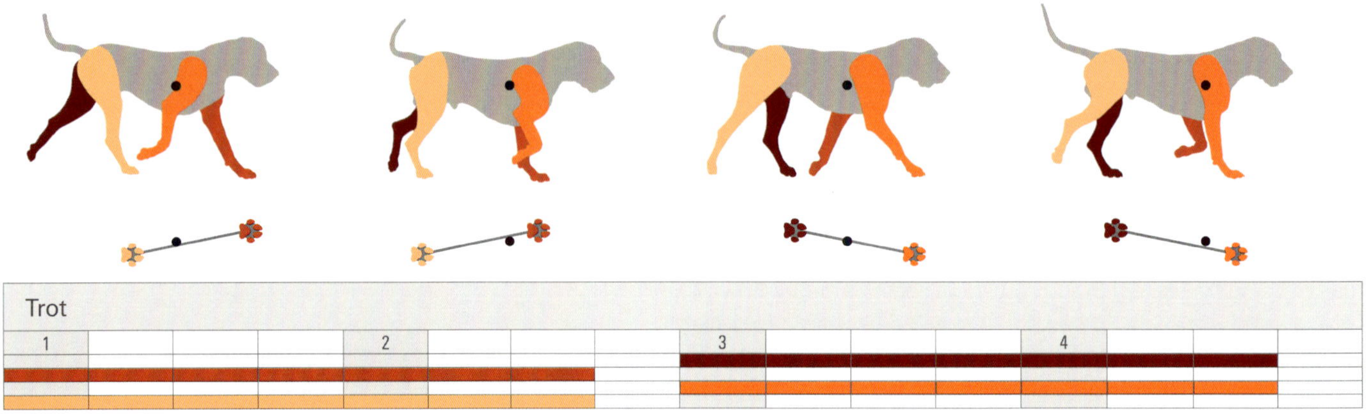

During the trot, the body is supported by one diagonal limb pair at a time. The stance phases of the right forelimb and left hindlimb alternate with those of the left forelimb and right hindlimb, sometimes with a flight phase in between. Ideally, the body's center of mass during the trot will remain at all times on the support line formed by the two limbs in contact with the ground. The "follow-the-leader-principle" guarantees that the hindlimb touches down on safe terrain.

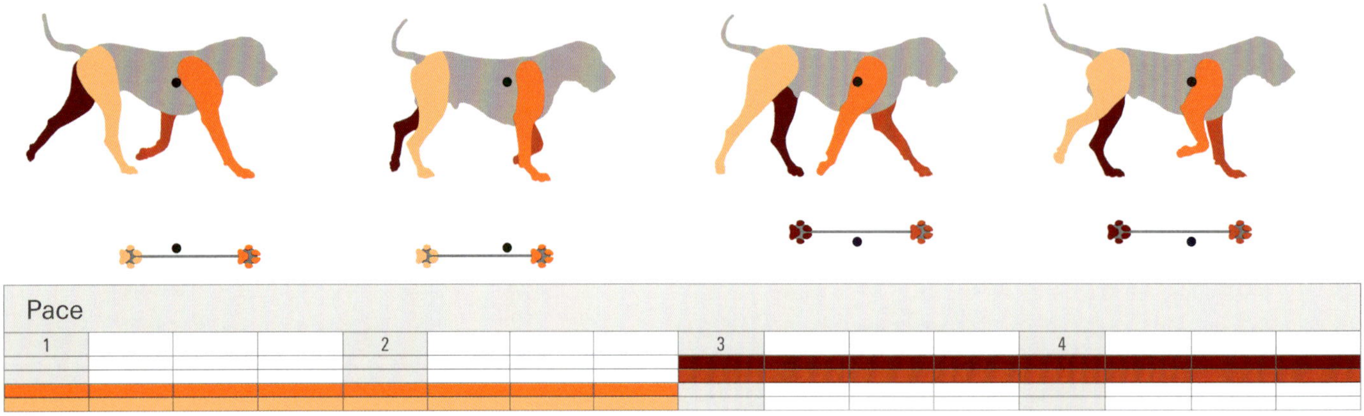

During the pace, the body is supported by the fore- and hindlimbs of the same side of the body. The center of gravity shifts from one side to the other and is usually not located above the line of support. It is this which causes the slightly rolling air associated with this gait.

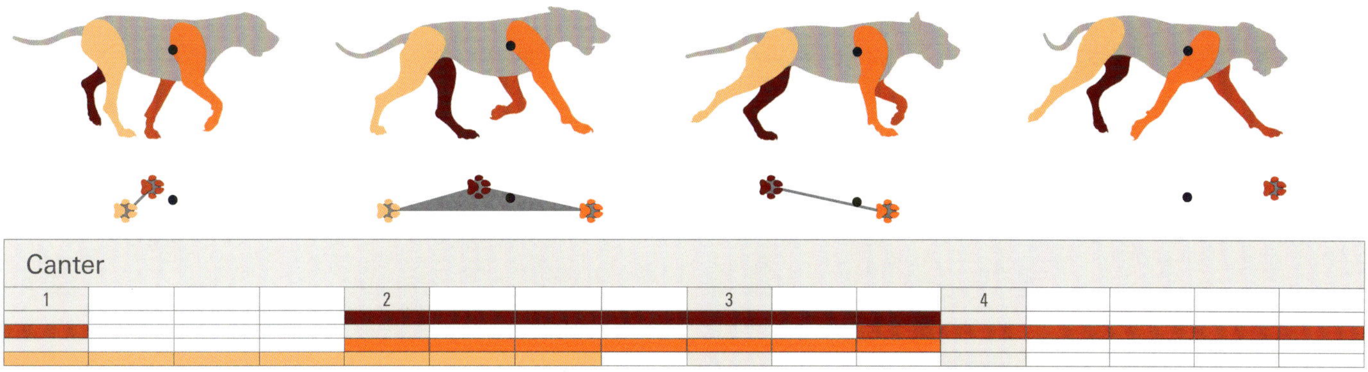

The canter is the slowest form of gallop and does not involve a flight phase. One-legged support alternates with three-legged support.

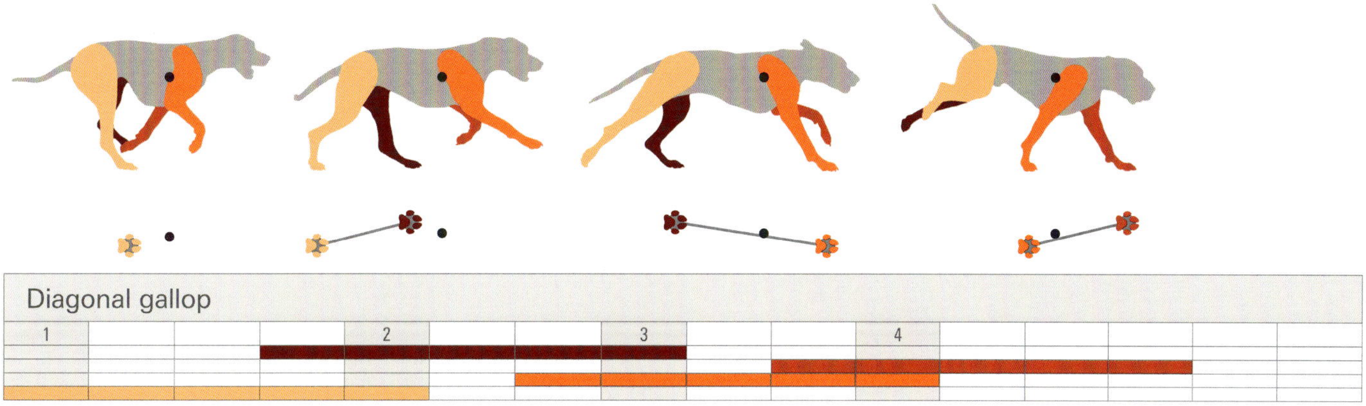

During the diagonal gallop the touchdown of the leading hindlimb is followed by the touchdown of the forelimb diagonal to it. The trailing fore- and hindlimbs are located on the same side of the body.

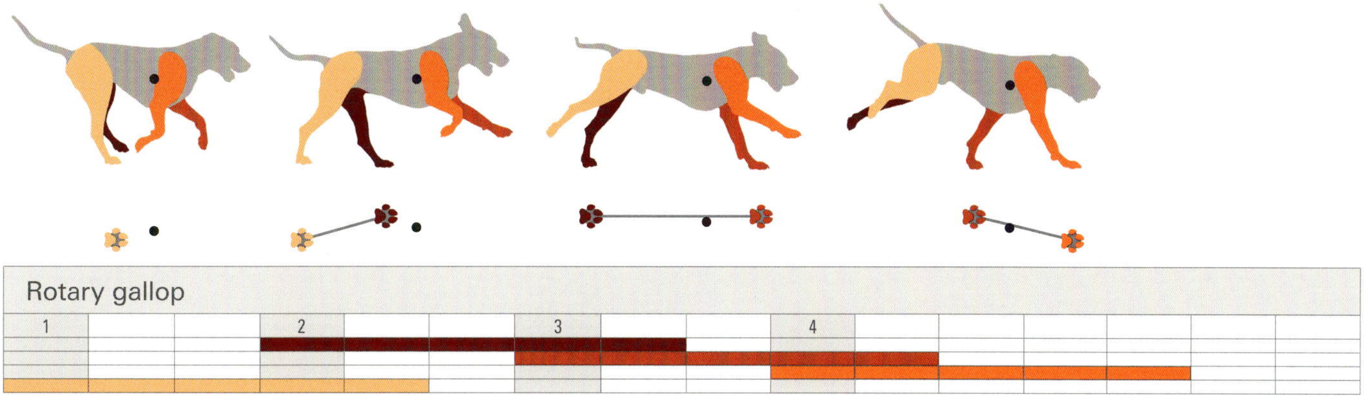

During the rotary gallop the limbs are placed on the ground either in a clockwise or an anticlockwise pattern. Touchdown of the leading hindlimb is followed by that of the trailing forelimb on the same side.

THE GENERATION OF PROPULSION IN LOCOMOTION

FORCE MEASUREMENT

During locomotion, dogs must exert forces on the ground in order to bear the weight of the body and propel it forwards. Forces applied by the dog are "ground forces" and the force applied by the ground are the "ground reaction forces". They can be measured using a force plate. Depending on the size of the plate, it is possible to measure the ground reaction forces generated by one limb, and those generated by more than one limb, e.g. a diagonal pair of limbs when the animal is trotting. These forces have been being measured in dogs for over 40 years (Hutton et al. 1969). Most of the studies focus on the Greyhound trot (Jevens et al. 1993, Rumph et al. 1993, 1997, 1999, McLaughlin et al. 94, Roush & McLaughlin 1994, Bertram et al. 2000, Besancon et al. 2003). Labradors have been the subject of force analyses four times, and other breeds that have been investigated include Beagles, Beaucerons, German Shorthaired Pointers, English Setters, Flat Coated Retrievers, Golden Retrievers, Huskies, Malinois, Irish Wolfhounds, Burmese Mountain Dogs, Weimaraners, 43 "mesomorphic dogs" (Rumph et al. 1994) and over 40 mongrels (Hutton et al. 1969, Budsberg et al. 1987, Franchen et al. 2006, Bockstahler et al. 2007, Kapatkin et al. 2007, Madore et al. 2007, Walter & Carrier 2007). This puts us in a position to be able to make general statements about the medium-sized to large breeds at least, and gives us specific information about the breeds investigated. More research is needed into small and long-backed breeds, however, as they have never been investigated with regard to ground reaction forces.

In the experiments conducted so far, dogs were either kept on a lead and led over a force plate in the ground, or called from the end of a 10 – 40 meter-long track way. The number of measurements required was obtained through frequent repetition. In the early experiments, only one force plate was used, which meant that at best, only the ground reaction forces of the fore- and hindlimbs of one side of the body could be measured (e.g. Jayes & Alexander 1978, Budsberg et al. 1987, Jevens et al. 1993, Gurevich et al. 1994, Renberg et al. 1999, Rumph et al. 1997, 1999). Later, at least four plates tended to be used, and in some cases whole measuring tracks (e.g. Bertram et al. 1997, 2000, Lee et al. 1999, McLaughlin 2001). This meant that the ground reaction forces of all four limbs could be measured at the same time, moving in different gaits and at different speeds. Since the beginning of the 1990s, treadmills have been available with ground reaction force measurement systems integrated into them, and have been used successfully in experiments involving dogs (Brebner et al. 2006, Fanchon et al. 2006, Bockstahler 2007). The advantage of this kind of system is that the treadmill can be set to run at a specific speed. Essentially, the technically complex studies conducted using this equipment confirmed the results of earlier studies regarding the distribution of the vertical forces exerted by the fore- and hindlimbs and the relationship between vertical and horizontal forces.

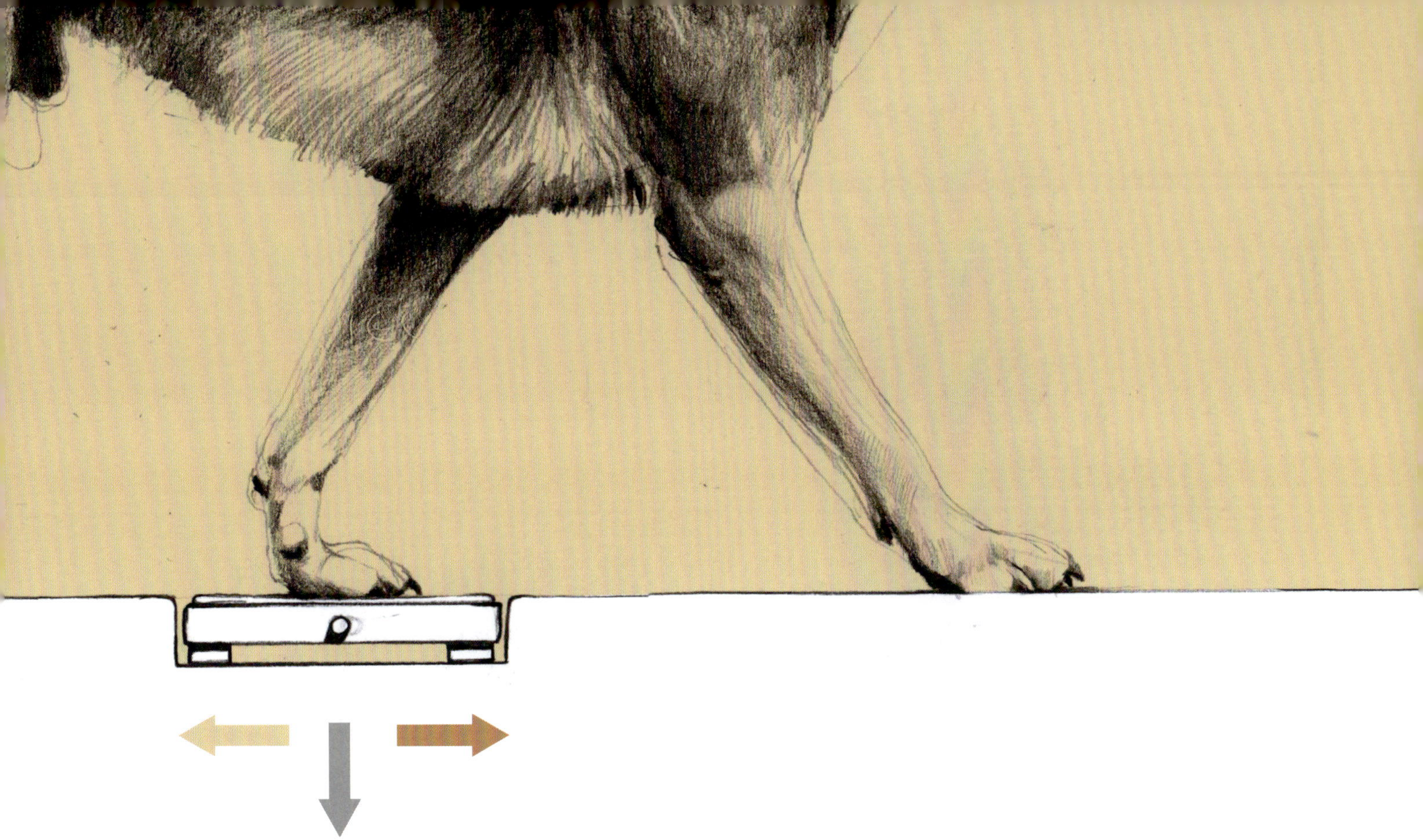

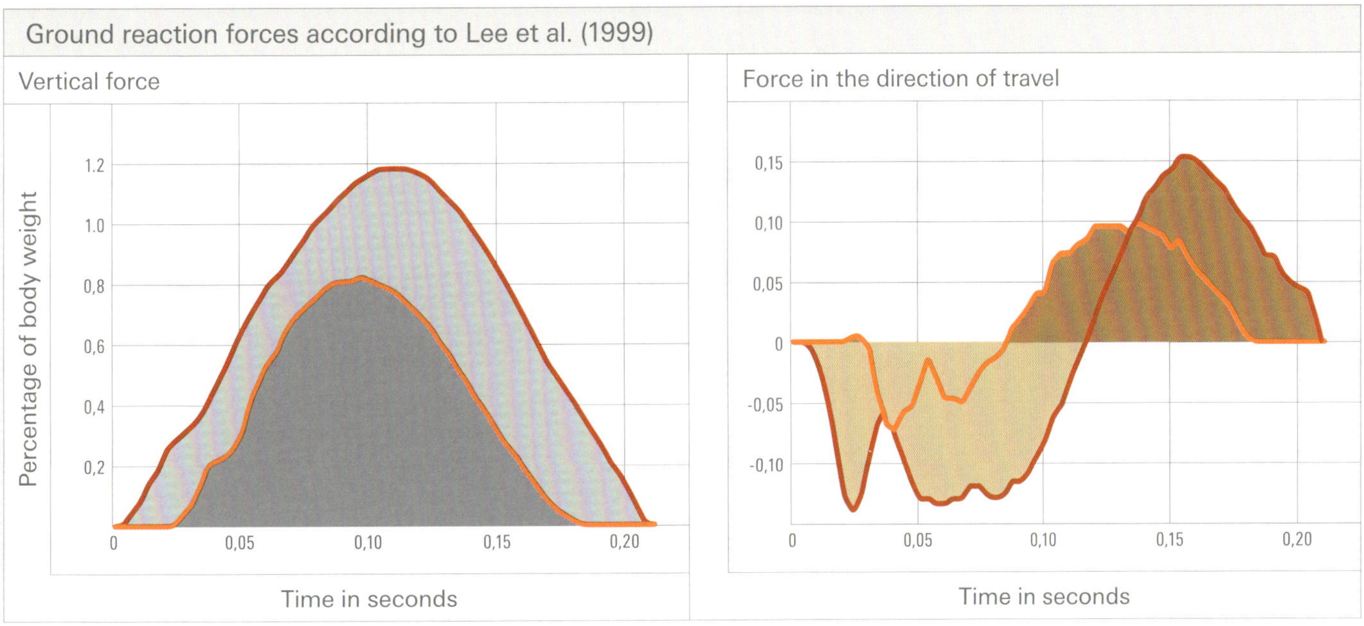

- Forelimbs
- Hindlimbs
- Decelerating force
- Accelerating force
- Vertical force

The vertical ground reaction force is around 50% higher in the forelimbs than the hindlimbs. The peak vertical force is reached by the forelimbs in the middle of the stance phase and by the hindlimbs much earlier. The forelimbs decelerate more sharply and for longer than the hindlimbs, which begin to generate propulsion after just a third of the stance phase.

Ground reaction forces are exerted in all three directions in space. The vertical force is equal to the body weight transferred. Forces in the direction of movement indicate deceleration or acceleration. If acceleration over the course of a step is zero, i.e. there is no change in the overall forward speed, the dog is moving at a constant rate described as "steady state". Mediolateral forces, i.e. those forces directed inwards or outwards perpendicular to the direction of travel show whether and, if so, to what extent the paw is pushed outwards or inwards. A vertical ground reaction force curve shows the load borne by a limb at a specific moment in the stance phase up to the peak vertical force, while the slope of the curve indicates the speed of force development. The way the force changes in the direction of travel over time shows that each step begins with a deceleration event (negative values) during which the fore- and hindlimbs begin, at different times and to different degrees, to accelerate (positive values). The ground reaction forces are dependent on various general factors:

- the distribution of the body mass over the fore- and hindlimbs: this depends on the distribution of mass in the dog (e.g. large or small head) and the distance between foot placement point and the center of mass of the body;
- speed and thus the duration of the stance phase: when the same amount of weight is placed on the paw for a shorter amount of time, the maximal vertical force is higher;
- the stiffness or flexibility of a limb;
- the gait: the fewer the limbs on the ground, the greater the load borne by each limb and, should a flight phase occur, the greater the gravitational acceleration at touchdown;
- the acceleration or deceleration of the movement.

However, other specific factors related to the experiment situation also play a role, including whether or not the dog is used to running on the treadmill (habituation), which is the case after a few minutes. Even the shape the dog is in on any particular day is measurable: there can be significant day-to-day variations in the size of the forces generated by any given dog (Rumph et al. 1999). The influence of the dog handler is very small.

The transmission of forces in dogs is distributed almost evenly over the paw. The pressure point only shifts marginally from the back of the paw to the front and from the inner edge to the outer (Budsberg & Verstraete 1993). Forces are measured in newtons (N). Ten newtons correspond approximately to one kilogram of mass on the ground. In order to be able to draw a comparison between dogs of different masses, forces are given as a percentage of body mass; they are "normalized". In everyday language no distinction is made between mass and weight. Physically speaking, weight is a force (force = mass x gravitational acceleration) which used to be measured in kiloponds and today is given in newtons.

GROUND REACTION FORCES DURING THE WALK AND TROT

Whether a dog is standing or moving, its body weight is unevenly distributed across the fore- and hindlimbs. Even in very derived shaped breeds such as the Basset Hound, this distribution of weight is maintained (Hitz 1973). In the German Shepherd, each forelimb bears 31 percent and each hindlimb 19 percent of the body weight when the dog is standing still. When it is walking, the maximum vertical force exerted on the forelimbs amounts to approximately 60 percent of its body weight and that exerted on the hindlimbs to about 40 percent. This corresponds approximately to the static mass distribution across the front and rear part of the body. Accordingly, the front paws are bigger than the hindpaws. Unkel-Mohrmann (1999) calculated in a walking German Shepherd that 63 percent of the total force was generated by the forelimbs and 37 percent by the hindlimbs. Walking, the values observed in breeds with similar body shapes are almost identical (Roush & McLaughlin 1994, Besancon et al. 2003, Page et al. 1993).

When the animal is trotting, the forces the limbs apply to the ground must increase to accelerate the body. At times, more than the total body weight is borne by a single forelimb. During the gallop, this increases to over twice the body weight. The increase is due to the accelerations of the body in the earth's gravitational field at the end of the flight phase observed in the trot and the gallop. The highest value recorded in this context was observed by Bryant et al. (1987), who found that a Labrador Retriever travelling at 27 km/h bore 2.6 times its body weight on one forelimb.

Of the breeds investigated to date, the Greyhounds and, probably, other sighthounds stand out as an exception with over 10% more weight borne by the hindlimbs than the forelimbs as compared to other breeds during the trot (Griffon et al. 1994, Bertram et al. 2000, Lee et al. 1999, Williams et al. 2008a). It is, then, not true that in Greyhounds and other sighthounds 70 – 80 percent of the body weight is borne by the forelimbs as Off & Matis (1997) claim.

The maximum vertical ground reaction forces are closely connected to the duration of the stance phase, which in turn depends on the speed of movement. Contact with the ground is longer in large dogs than in small dogs travelling at the same speed. This means that they have longer to distribute their body mass, which means that the maximum force is proportionally smaller. Peak vertical forces thus decrease, relatively speaking, as body size increases. As speed increases, which is achieved by a combination of increasing step length and shortening ground contact time, the peak vertical forces increase. In the walk and in the gallop, this correlation is stronger for the forelimbs than for the hindlimbs (Roush & McLaughlin 1994, McLaughlin et al. 1994). As the stance phase of the fore- and hindlimbs are almost equal in length, but the amount of weight borne by the forelimbs is greater, the vertical force exerted by the front limbs after touchdown increases significantly quicker and reaches its maximum earlier in the stance phase than in the hindlimbs (Budsberg et al. 1995).

In a comparison between Labrador Retrievers and Greyhounds, Bertram et al. (2000) discovered the difference between the two breeds. When running at the same speed, there is a difference in the stance duration of the two breeds, but step duration and step length are the same. In Greyhounds, the stance is 18 percent shorter in the fo-

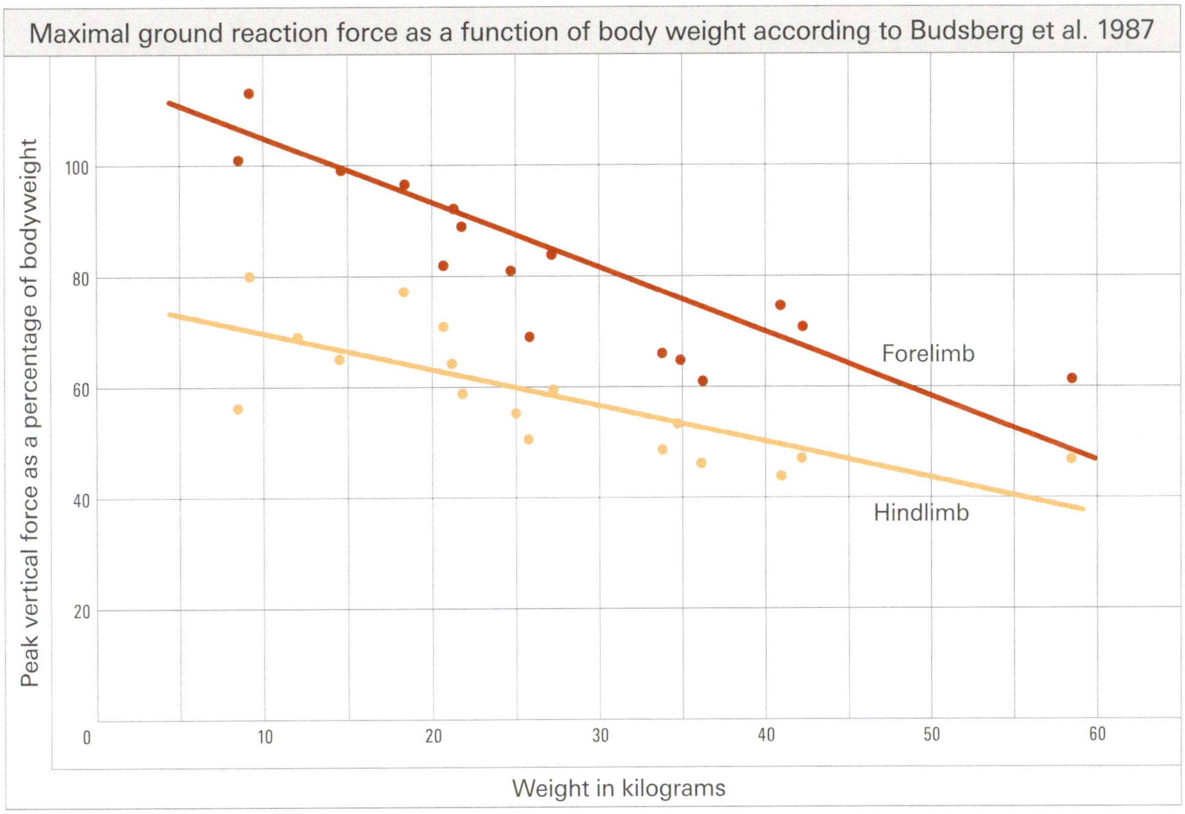

relimbs and 6 percent shorter in the hindlimbs than is the case in the Labrador Retriever. The normalized peak vertical force of the forelimbs is 117 percent in the Greyhound and 117 percent in the Labrador Retriever, while that exerted on the hindlimbs is 107 percent in the Greyhound and just 76 percent in the Labrador Retriever. A Greyhound thus needs to muster more strength in its hindlimbs within a shorter time in order to shift a greater proportion of body mass. This is solved as follows: in Labradors, the hindlimbs touch the ground after and leave the ground before the forelimbs, which means that only the stance phase of the forelimbs is relevant to the size of the forces. In the Greyhound, the hindlimbs touch down and take off before the forelimbs, which means that the total stance phase duration is made up of that of both the fore- and hindlimbs together. In the Greyhound, both limb pairs are on the ground for a shorter time and thus tread more powerfully. The short, powerful tread gives the Greyhound its characteristic bouncing gait.

The forces produced in or counter to the direction of travel are accelerating and decelerating forces. These too differ greatly between the fore- and the hindlimbs (Bertram et al. 1997, McLaughlin et al. 1994, 1995, Riggs et al.1993, Rumph et al. 1997). After touchdown, both limb pairs initially exert a decelerating force. During the walk, the front limbs decelerate for more than half the stance phase and their decelerating force impulse is greater than the accelerating impulse. In the hindlimbs, which have to compensate for this, the switch from deceleration to acceleration comes after the first third of the contact phase and the net force impulse is one of propulsion (Budsberg et al. 1987). The term force impulse designates the total force generated within a specific period. It is calculated by integrating the area under the force curve.

The point at which a limb begins to exert a braking or accelerating force depends on the direction of the vectors of the forces generated by the fore- and hindlimbs and their relationship to the body's center of mass (CoM). In a standing dog, the CoM is found behind the sternum, level with the ninth intercostal space. If the vector points backwards, deceleration takes place, if it points forwards, acceleration takes place. The position of the paw relative to the dog's CoM during the stance phase plays a role here. The hindlimbs are behind the CoM, while the front limbs only reach a position which makes it possible to push the body directly at the end of the stance phase. The weak braking force exerted on the hindlimbs essentially results from the setting down of the paw counter to the direction of travel. The forces then quickly become positive and are subject to a continual increase during the stance phase. The dog is thus pushed forward by the hindlimb. The thrusting out of the neck during the trot shifts the CoM further forwards and, in relative terms, thus lends the forelimbs more thrust, too. The braking forces exerted in particular by the forelimbs are reduced by the fact that touchdown does not take place until after the backwards movement at the end of

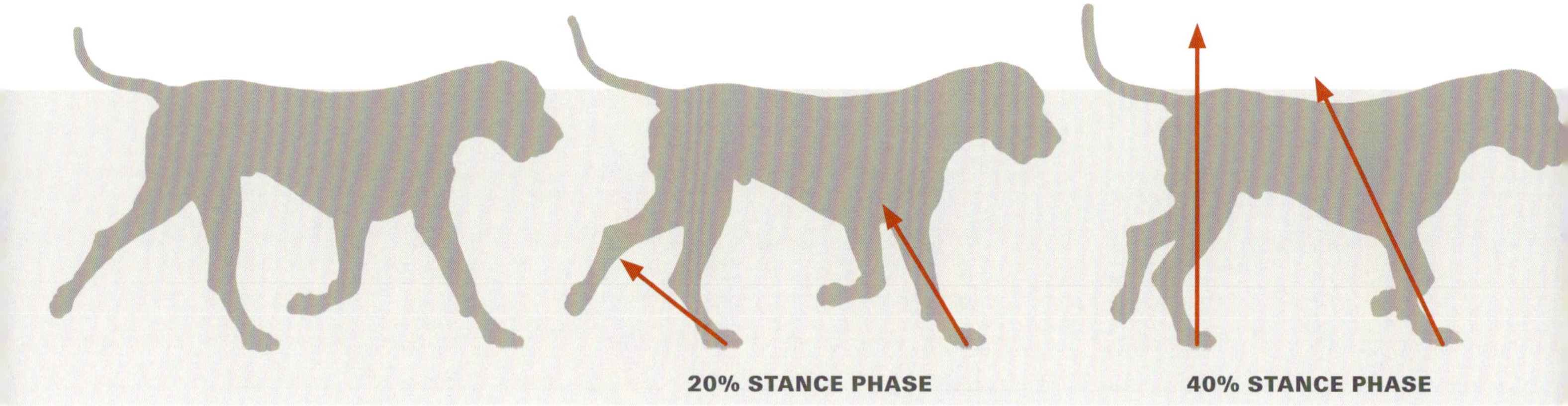

Direction and magnitude of the force vectors represented by their length at four moments in the stance phase.

the pre-swing phase has begun. The forward reach allows the limb to be accelerated backward to match the ground speed when the foot touched down. Otherwise the collisional loss of energy would be much greater. The exaggerated forward reach before touchdown in some breeds may look elegant, but it has no benefit whatsoever in terms of propulsion.

When a dog is running in a straight line, the mediolateral forces are, at under 3 percent of the normalized forces, extremely small. In a 31 kg Labrador Retriever they amounted to 8 percent in the forelimbs (Bertram et al. 1997). In Greyhounds they are even smaller at just 1.3 percent of body weight. This shows that the limbs are moved in one plane under their pivot point, parallel to the direction of travel. The mediolateral forces increase quickly, however, when the dog performs turning maneuvers or runs in a curved line (Usherwood & Wilson 2005). Unlike humans, who compensate for the centrifugal forces which occur when running in a curve with longer contact with the ground, the contact times in dogs are the same running in a curve as they are running in a straight line. This means that dogs do not slow down in a curve. At the same time, much higher mediolateral forces are generated at the front to compensate for the centrifugal forces, while the back limbs continue to generate non-braked thrust. To ensure that the paw maintains a stable contact point, the body is able to rotate around it. The forearm is capable of passive rotation outwards and of inward rotation.

GROUND REACTION FORCES DURING THE GALLOP

During the gallop, the ground reaction forces exerted by the limb which is placed first (the trailing limb) and that, which is placed second (the leading limb) differ. One previous study showed – in relation to the forelimbs only – that the leading limb developed greater braking forces and the trailing limb developed greater acceleration forces (Bryant et al. 1987). Walter & Carrier (2007) investigated the rotary gallop in six dogs (two Labrador Retrievers, one Weimaraner and three mongrels), which ran along a 40 m track at an average speed of 32 km/h. The authors confirmed that the trailing forelimb developed acceleration forces that were around 20 percent higher than those observed in the leading limb. They also found that in the leading limb, the braking and accelerating forces cancel each other out. The same study showed that though the trailing forelimb developed a greater vertical force (223 percent of body weight) than the leading limb (212 percent), because its stance phase is 5 percent shorter, which means that it is stiffer and that its position differs from that of the limb to be placed second, the vertical force impulse, i.e. the force over time, remains the same. In the hindlimbs, the situation is reversed. Here it is the leading limb which is subjected to the higher vertical forces (164 percent as opposed to 147 percent), but because the leading limb has a shorter period of contact with the ground, the vertical impulse are the same. In a comparison of the fore- and hindlimbs during the gallop, the vertical forces are much greater at the front than at the back – 39 percent greater in the trailing forelimb and 26 percent greater in the leading forelimb. The force impulse is 32 percent greater. The braking force impulse, too, is a whole 92 percent greater in the forelimbs than in the hindlimbs, despite the fact that the acceleration values are more or less the same. The trailing forelimb thus needs to be stiffer.

The horizontal forces observed in connection with the gallop also differ between the trailing and leading limb. The trailing forelimb develops a 12 percent greater acceleration force and an 11 percent smaller braking force than the leading forelimb, which is related to the fact that it lands further back and is more extended posteriorly. At the back it is the leading limb which develops an approximately 18 percent greater acceleration force and acceleration force impulse (Walter & Carrier 2007). While the hindlimbs provide the acceleration impulses necessary for loco-

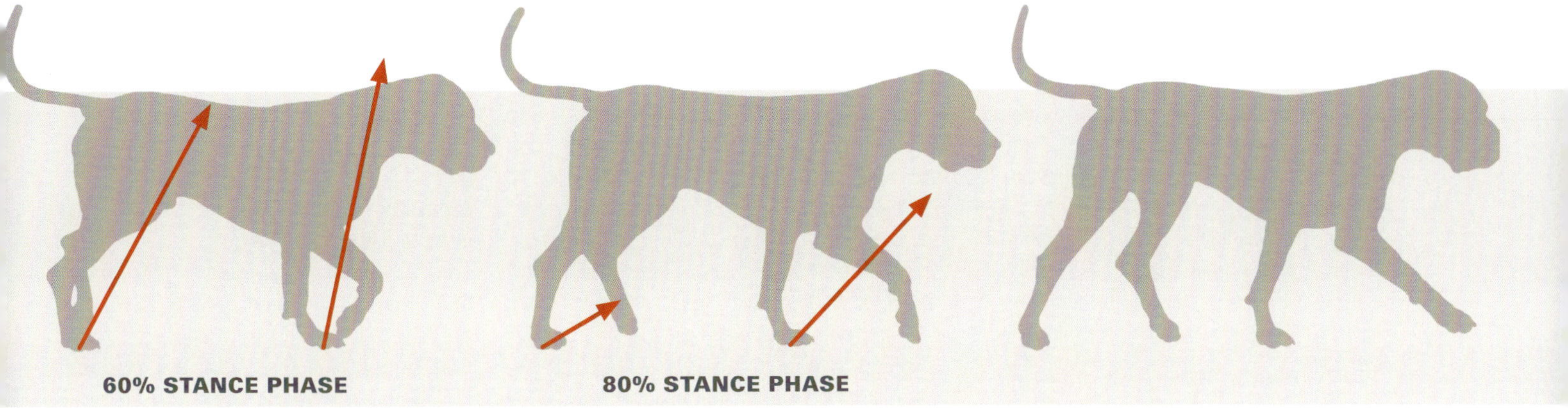

motion, the acceleration impulses at the front are almost zero.

The lateral forces are stronger in the gallop than in the symmetrical gaits, and at their maximum reach around 20 percent of the weight force in the forelimbs and 16 percent (trailing limb) or 12 percent (leading limb) in the hindlimbs.

During the first 20 percent of the stance phase, the front limbs push outwards, then briefly inwards, and then up until halfway through the stance phase outwards again. In the second half of the stance phase, the situation differs from dog to dog. The size of the forces depends, of course, on the placement relative to the center line of the limbs under the body. The hindlimb does not generate lateral forces until after the first third of the stance phase. These are mostly inwardly directed.

GROUND REACTION FORCES IN RAPID ACCELERATION

Although most research to date has concentrated on constant, cyclic locomotion in the various gaits and at various speeds, dog locomotion is actually characterized by sudden spurts, turns and braking maneuvers. It is not unusual for dogs to accelerate via a few leaps from standing to full blown galloping: they do not change gear slowly. Recently however, a study looked into what exactly happens when a dog displays rapid acceleration, and attention was dedicated to the initial push off and the first few steps. Dogs were motivated, either via the throwing of a tennis ball or by "an experimenter running with a hotdog" (Walter & Carrier 2009) to start moving at maximal speed. What kind of a dog wouldn't run full speed after a hotdog – despite the unfortunate name given to this delicacy!

The steps taken in a sprinting start do not evidence the typical footfall pattern of the gallop but resemble a "half bound", a gait typical of small animals (Fischer et al. 2002). The back limbs touch down simultaneously, but the front limbs move as they do in the gallop. Acceleration is not the only circumstance under which dogs push off with both legs simultaneously; they use the same technique in sustained periods of jumping such as running through water. Not only is pushing off synchronously a more effective means of acceleration, it also raises the body's center of mass. This is why long-backed breeds find quick starts and jumps more labor-intensive than short-backed dogs.

The seven dogs investigated by Walter & Carrier (three Labrador Retrievers, 3 mongrels and 1 Shih Tzu) reached a speed of 3.6 m/s (13 km/h) over the first two meters (Walter & Carrier 2009), accelerating from a position in which the hindlimbs were significantly bent and the front part of the body held high. The knee joints remained 20° more flexed than in the gallop, while the hip and ankle joints were more extended. Acceleration took place not just horizontally forwards but upwards and forwards. This had the effect of taking the strain off the forelimb, which during the first and second step cycle only bore 30 then 42 percent of the maximal weight force exerted during the gallop, which in that gait is, as we know, equal to double the body weight. Although the accelerating forces are not, in total, higher than in the gallop, the forelimbs contribute 43 percent to acceleration as the braking forces that would normally occur at the beginning of the stance phase are absent. The forward-directed forces generated by the hindlimbs during the initial acceleration are up to 50 percent greater and the vertical forces around 50 percent smaller than in the gallop.

In the Greyhound, the initial accelerations are, as one might expect, even higher, reaching up to 4.5 m/s (17 km/h). A study by Williams et al. (2009), which was unfortunately limited to the hindlimbs, showed that here too, this resulted from an increase of up to 30° in the degree of flexion of the joint angles, and particularly of the ankle joint. The shift in the vertical force as the result of the "crouch start" and the change in the way force impulse is distributed between the fore- and the hindlimbs also compensates for the strong angular force impulse around the body's center of mass which occurs during rapid acceleration. During acceleration the dog's nose is observed to move up and down and its head thrusts forward. During rapid braking, which in a Labrador Retriever trotting at 7 km/h can happen within two steps (Lee et al. 1999), similarly strong angular force impulses occur, but these are cushioned by a quick weight transfer. Nevertheless, this doesn't stop some dogs braking with their chins.

INVERSE DYNAMICS

The lengths to which it is necessary to go to measure the total force exerted by a limb are justifiable, but measuring the individual forces at work in each limb is a different matter. The experiments involved are so complex and invasive that it is no longer possible, afterwards, to draw conclusions about the normal joint forces at work during locomotion because measuring devices bias the result enormously.

One alternative is to turn to inverse dynamics, using an approximation to calculate the joint forces and net joint torques on the basis of what is known about the kinematics, ground reaction forces, mass distribution over the joint segments and mass moments of inertia. This method involves four steps:

1 measuring the limbs (morphometry)
2 measuring the joint angles (kinematics)
3 measuring, synchronously, the ground reaction forces and the kinematics (dynamics)
4 calculating from the results of steps 1-3 the net forces and net torques (inverse dynamics).

Of the three parameters used to calculate the inverse dynamics, the most important are the ground reaction forces (Nielsen et al. 2003). Predictably, the torques in the hip and knee joint increase in the first third of the stance phase and those in the ankle joint at the end of the stance phase, as speed increases (Colborne et al. 2006).

SELF-STABILIZATION

After a disruption in the motion sequence, humans and four-legged animals are rapidly able to self-correct, either via control mechanisms or through properties inherent in the mechanics known as dynamic self-stabilization mechanisms. Self-stabilization in this context means that a stable, repeating cycle is re-established simply via the feedback system incorporated into the mechanics of the legs. Thanks to "intelligent" mechanics, no further control is necessary (Ringrose 1997, Wagner & Blickhan 1999, Blickhan et al. 2007). Self-stabilization during locomotion explains why a dog is able to run on uneven ground more quickly than its reflex paths allow without faltering or stumbling. Amazingly, only two parameters have to be adjusted: the angle of attack of the the limb as a whole and the stiffness of the leg (Seyfarth et al. 2002). The angle of attack has been investigated in Greyhounds (Wentnik, 1979) and Airedale Terriers (Groß, 2008). According to Wentink (1979), this angle is 66° in the hindlimb in Greyhounds. Groß's investigations point to a very similar value of 70°. The touchdown angle of the forelimb is 68° - a result confirmed by the present study.

The stiffness of the leg can be calculated on the basis of functional limb length during the stance phase. The vertical displacement of the skin markers on the upper edge of the shoulder blade and the hip joint was measured at touchdown and takeoff and in the middle of the stance phase in 13 breeds. A basic pattern emerged, though breed-specific deviations were also observed. During the walk, the body in most breeds was raised in the middle of the stance phase and lowered again towards takeoff. The only breeds in which this did not occur were the Scottish and Bedlington Terriers and the Bearded Collie. During the trot, the body was lowered in the middle of the stance phase, though this was more clearly observable in the hip joint than in the shoulder blade. However, the vertical displacement during both the walk and the trot amounted to less than one percent of the withers or hip height, most probably because the marker on the upper edge of the shoulder blade did not sufficiently convey the low level of vertical displacement which occurs during the trot. Despite the greater vertical ground reaction forces which act upon them during the trot (60 percent), the forelimbs were, relatively speaking, stiffer than during the walk. The hindlimbs, on the other hand, were more compressed, like as spring. Limbs are transformed into "springy struts" (Wentink 1979).

ENERGY EXCHANGE DURING THE WALK AND THE TROT

The forces which generate joint torques have various causes and need to be compensated for at the joint by work performed by the muscles. When torques occur during acceleration, they actually need to be overcompensated for. The work involved is not particularly strenuous for the muscles as it is facilitated by a simple physical principle: when an egg is rolled along its long axis, there is a phase of "downhill rolling" which requires no extra energy, and a phase in which energy input is needed. Physically speaking, what happens during the first phase is that potential energy, which is generated by gravity, is transformed into kinetic energy. As this is the principle on which pendulums work too, the locomotor behavior of a limb has been compared to an inverse pendulum (Alexander 1976, Cavagna et al. 1977, Griffin et al. 2004, Usherwood et al. 2007). Cavagna et al. (1977) recognized early on that this pendulum-like energy exchange is attributable to the center of mass and not the individual limb, emphasizing that it is only made possible by the temporal coordination of the force action of all the limbs on the ground. Later authors oversimplified and attributed the pendulum action to the individual limb.

During the walk, the center of mass of the body is actively raised to a point above the limb, and as it moves back downwards the potential energy is transformed into kinetic energy. The inverse pendulum effect is highest in "stiff" limbs such as human legs. If at all, dogs only have "stiff" legs when performing a slow walk. As soon as they start walking quickly, the limbs impact and begin to behave elastically (Usherwood et al. 2007). The ability to transform potential energy into kinetic energy thus depends on the stiffness of the limbs, and stiffness decreases as speed increases. Interestingly, in the study run by Usherwood et al. (2007), one Bull Terrier was found to have a springy, elastic gait all the

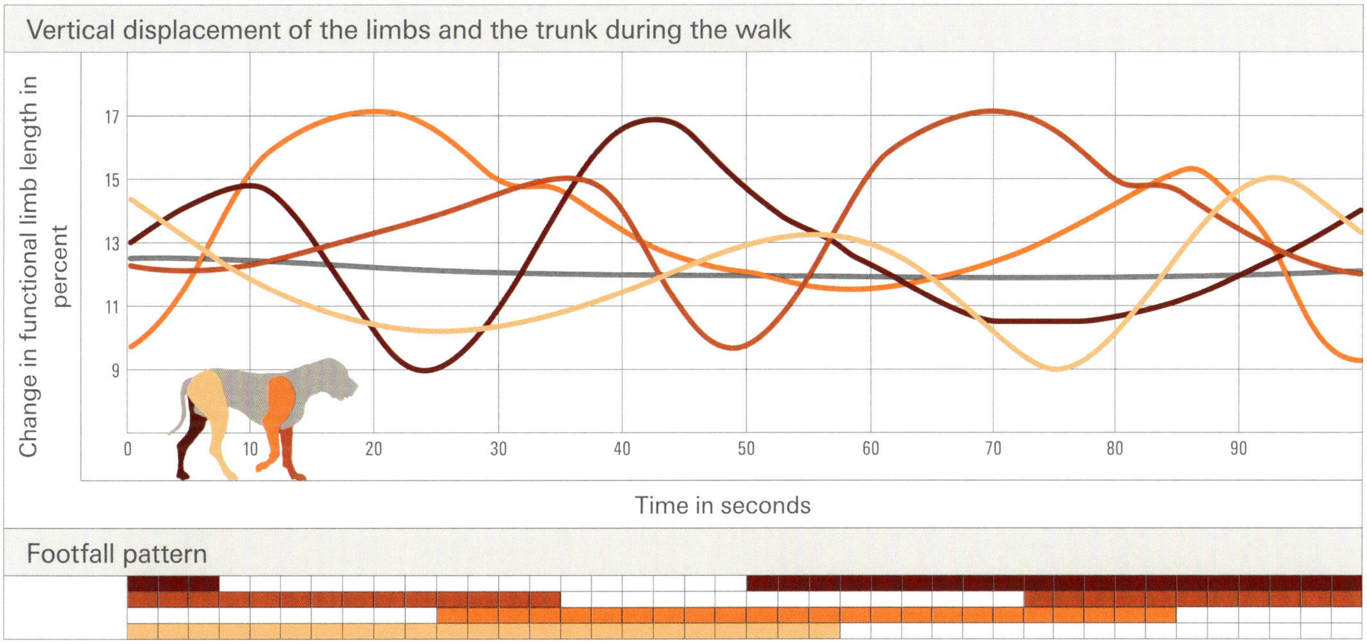

In the course of a step, each limb undergoes impaction (a shortening of the functional leg length) and then extension. The body's center of mass (gray line) does not exhibit significant vertical displacement as all four limbs together more or less cancel out its vertical displacement.

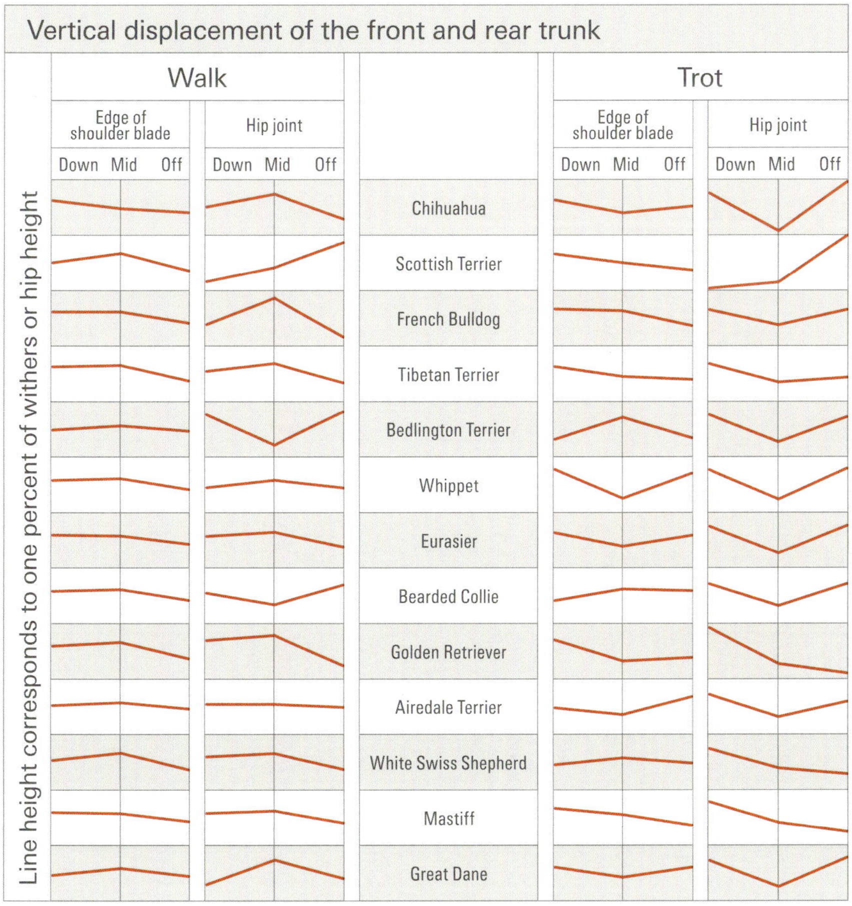

time, including during the slow walk. This study found the same elastic behavior in a number of breeds.

In the trot and the pace, the "running" gaits, the body's center of mass behaves differently. Rather than being raised in the first half of the stance phase, it is lowered. The vertical displacement corresponds in both the walk and the trot to around one percent of body height, and in the gallop this figure has been found to double (Jayes & Alexander 1978). The center of mass moves in a wave. In breeds that are sustained trotters, the course it follows is flatter than in breeds which are more adapted to the gallop. In the trot and the pace, instead of potential energy being transformed into kinetic energy, a different but equally effective "economy program" comes into play. Like in a rubber ball, elastic energy is stored during contact with the ground and then recovered later on in the stance phase. The elastic energy is generated by the gravity-induced impact of the joints and the passive stretching of the tendons that this causes. Like a suspension spring, the limb is compressed at touchdown and released at takeoff. The spring-mass model proposed by Blickhan

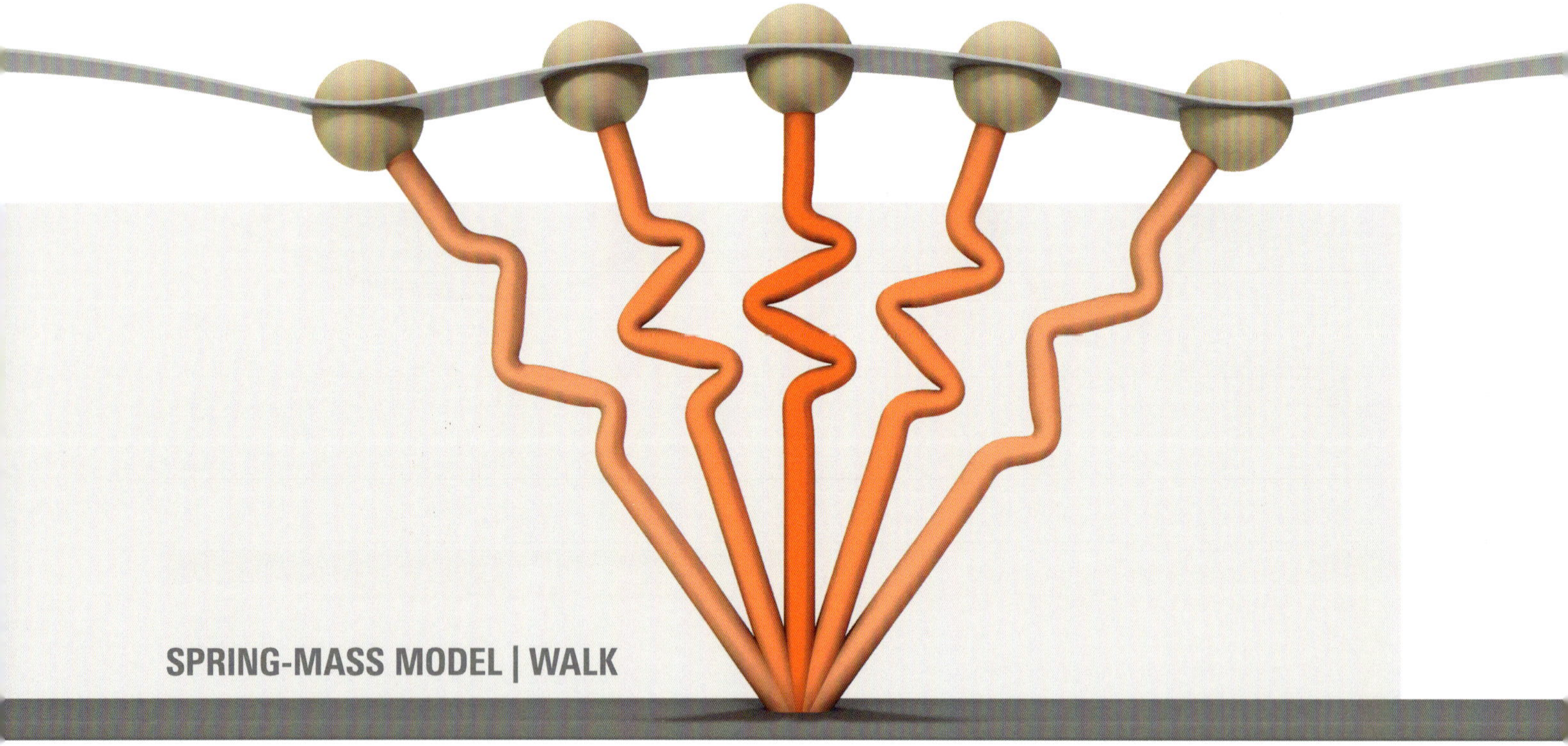

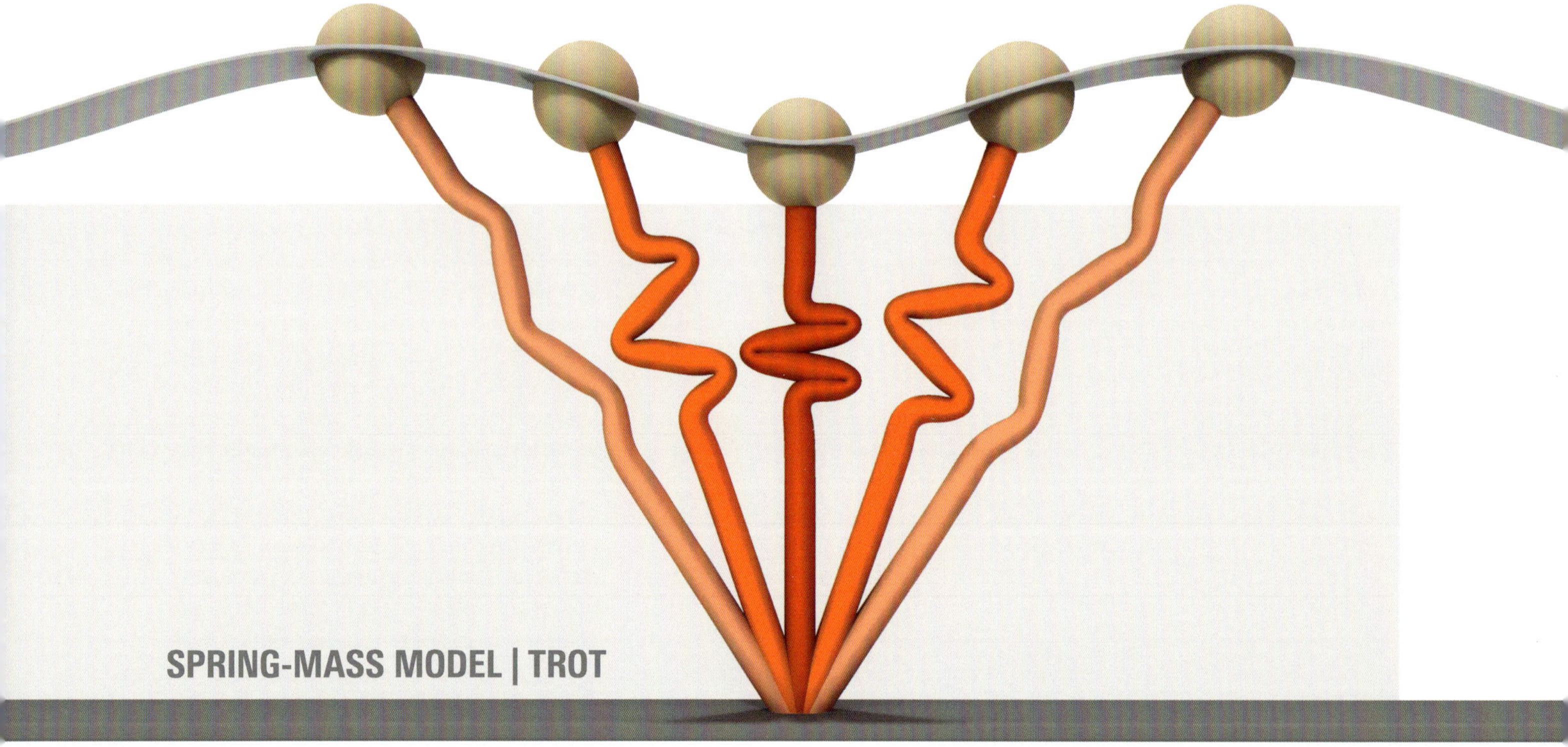

Blickhan's spring-mass model describes the behavior and motion of the body's center of mass (represented by the spheres) under the influence of gravity. The stiffness or compliance of the limbs differs between the walk and the trot, which means that in the walk, the body's center of mass is, ideally, raised, and in the trot it is lowered. The bobbing of the trunk is clearly visible in a trotting dog.

(1989) describes in an astoundingly simple way the behavior of limbs under the influence of gravity in terms of a point mass attached to a single spring. The model was met with some euphoria and led to the energetic aspects of running – or in the case of dogs, trotting – being explained rather generally in terms of the mechanism for the recovery of energy from elastic storage (Taylor 1994).

The gallop combines the two energy saving measures described above - the transformation of potential energy into kinetic energy, which "saves" 15-30 percent of the energy that would theoretically be expended (Cavagna et al. 1977), and the recovery of elastic energy. As should be clear from our analysis of the walk, however, the physical mechanisms in question cannot be tied neatly to specific gaits. Model calculations of energy consumption have shown that in a dog weighing 9.2 kg, the trot is the most energy-efficient gait for speeds up to around 13 km/h. Above this speed, however, the gallop becomes more efficient (Nanua & Waldron 1995). Relatively speaking, the higher the speed the dog is travelling at, the less energy the gallop costs and the more the trot costs.

In another, very complex study, Gregersen et al. (1998) investigated how much work is performed by every joint of the fore- and hindlimbs and how much of this work "costs nothing" thanks to the recovery of elastic energy. The authors had three dogs perform the trot (3 m/s) and a fast gallop (8 m/s) on a 50 m long trackway fitted with a ground-level platform for measuring the ground reaction forces. In addition, they recorded the activity of two selected muscles (M. semimembranosus, M. triceps brachii caput longum, see figure on page 102-103)) in four other dogs moving on a treadmill. The work of the joints was divided into positive and negative work following the conventional classification of flexion as negative and extension as positive. The experiment showed clearly that the two types of work were distributed very unevenly across the joints. As a basic principle it can be said that the distal joints perform a large proportion of the negative work. In the gallop, for example, the wrist and ankle were responsible for 81 percent of the total negative work, and the ankle alone was responsible for 98 percent of the negative work of the hindlimbs (Gregersen et al. 1998). On the other hand, it is precisely these two joints which, thanks to the passive stretching of powerful sinews such as the Achilles Tendon, recover the most elastic energy – 97 percent to be precise. The negative work therefore costs almost nothing. The heavy muscle masses, which are situated close to the centre of the body, only perform positive work (e.g. the hip extensors during the trot and the gallop). They are, in the main, fleshy muscles with long fascicles (see p. 118), which means that energy recovery is lower. In the elbow, up to 96 percent of the work can come from the recovery of elastic energy in the trot and less than 60 percent in the gallop; in the shoulder this changes to 38 percent in the trot and 49 percent in the gallop. In the knee joint, the differences between the gaits are particularly striking: 60 percent recovery in the trot versus less than 5 percent in the gallop (Alexander 1984, Gregersen et al. 1998).

The investigations into the dynamics of dog locomotion presented here change our understanding fundamentally. Locomotion is no longer the result of an action generated by muscle power, but each gait sees gravity-induced energy, be it potential energy or stored elastic energy, recovered in a variety of ways from the movement itself. Gravity is a dog's best friend! In the trot in particular, the dog's favorite gait, three quarters of the work performed is recovered and harnessed – it is no wonder, with such a low energy input, that dogs rarely tire. It must also be emphasized, however, that any kind of maneuver involving acceleration and deceleration does require extensive muscle work. The energy-saving mechanisms only come into play during regular, constant locomotion.

Regular locomotion of this kind does not require a complex control system or extensive brainwork: simple compliant mechanisms, the mechanical intelligence of the limbs, can, without requiring much additional energy, compensate for disturbances caused by factors such as uneven terrain and guarantee self-stabilizing movement.

Dogs have a four-wheel drive. The propulsive forces supplied by the forelimbs are close to those of the hindlimb in both the trot and gallop. The horizontal forces from both the fore- and hindlimbs contribute to propulsion of the body. During the trot and the gallop, the braking and accelerating forces in the hindlimbs are , in net terms, balanced. The forelimbs contribute more to propulsion during spontaneous movements such as rapid accelerations. Dogs do not lose speed when moving in a curve. The high centrifugal forces which occur are absorbed by the passive rotation of the radius against the ulna. The maximum vertical forces are 50 percent higher in the forelimbs than in the hindlimbs. During the trot, one forelimb has to bear more than the weight of the body, and during the gallop more than double the body weight. In terms of body weight, the maximum ground reaction force is higher in small breeds than in large breeds.

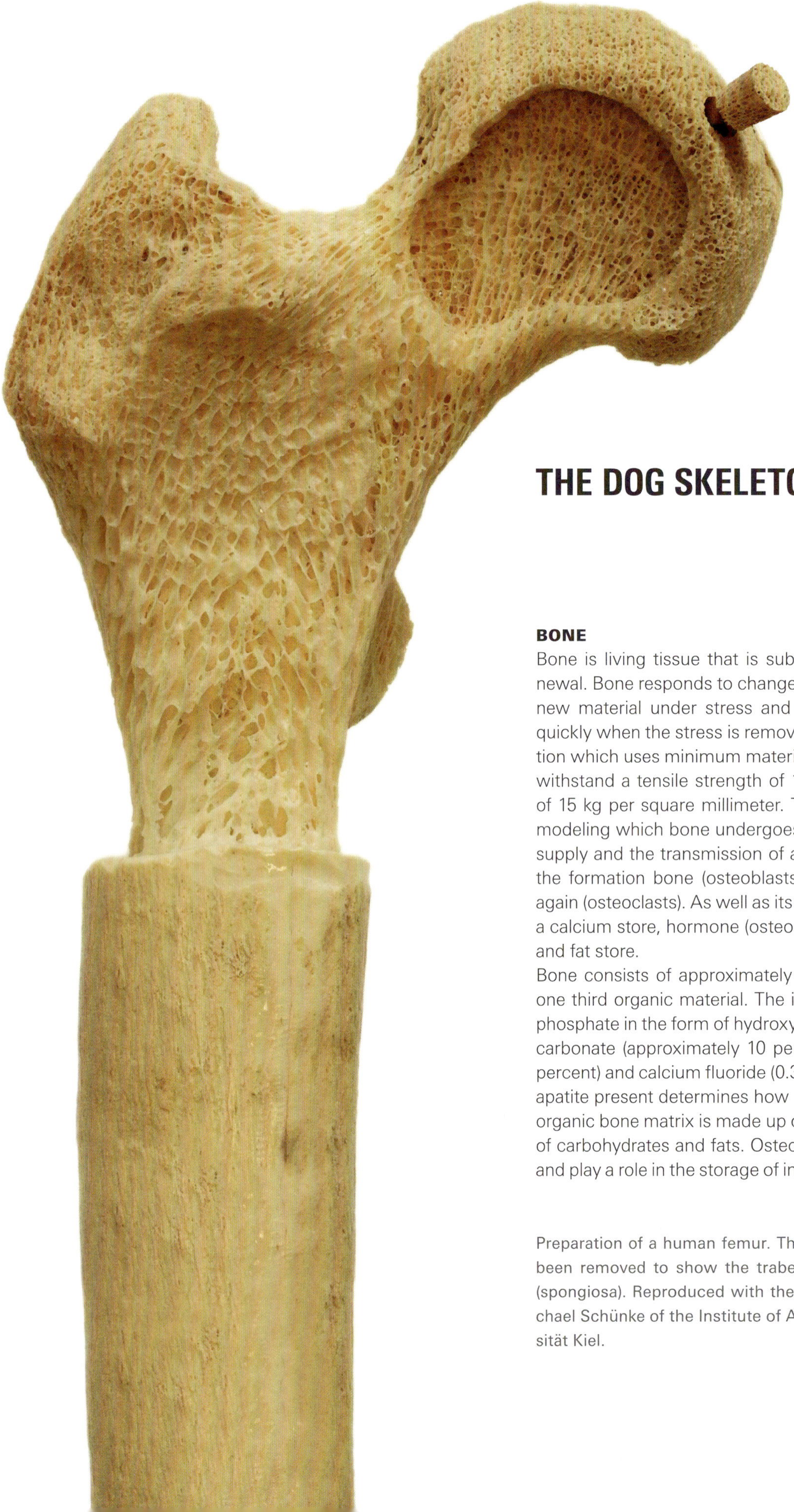

THE DOG SKELETON

BONE

Bone is living tissue that is subject to constant depletion and renewal. Bone responds to changes in mechanical demands, forming new material under stress and breaking it down again relatively quickly when the stress is removed. Bone is a lightweight construction which uses minimum material for maximum stability. Bone can withstand a tensile strength of 10 kg and a compression strength of 15 kg per square millimeter. The constant stress-dependent remodeling which bone undergoes requires sensitivity, a good blood supply and the transmission of a signal to the cells responsible for the formation bone (osteoblasts) and those which break it down again (osteoclasts). As well as its skeletal function, bone also acts as a calcium store, hormone (osteocalcin) producer, blood synthesizer and fat store.

Bone consists of approximately two thirds inorganic material and one third organic material. The inorganic components are calcium phosphate in the form of hydroxylapatite (up to 90 percent), calcium carbonate (approximately 10 percent), magnesium phosphate (1.5 percent) and calcium fluoride (0.3 percent). The amount of hydroxylapatite present determines how hard the bone is. 90 percent of the organic bone matrix is made up of collagen fibers. The rest consists of carbohydrates and fats. Osteoblasts produce the collagen fibers and play a role in the storage of inorganic minerals, especially hydro-

Preparation of a human femur. The bony cortex of the upper part has been removed to show the trabecular structure of the spongy layer (spongiosa). Reproduced with the kind permission of Prof. Dr. Dr. Michael Schünke of the Institute of Anatomy, Christian-Albrechts-Universität Kiel.

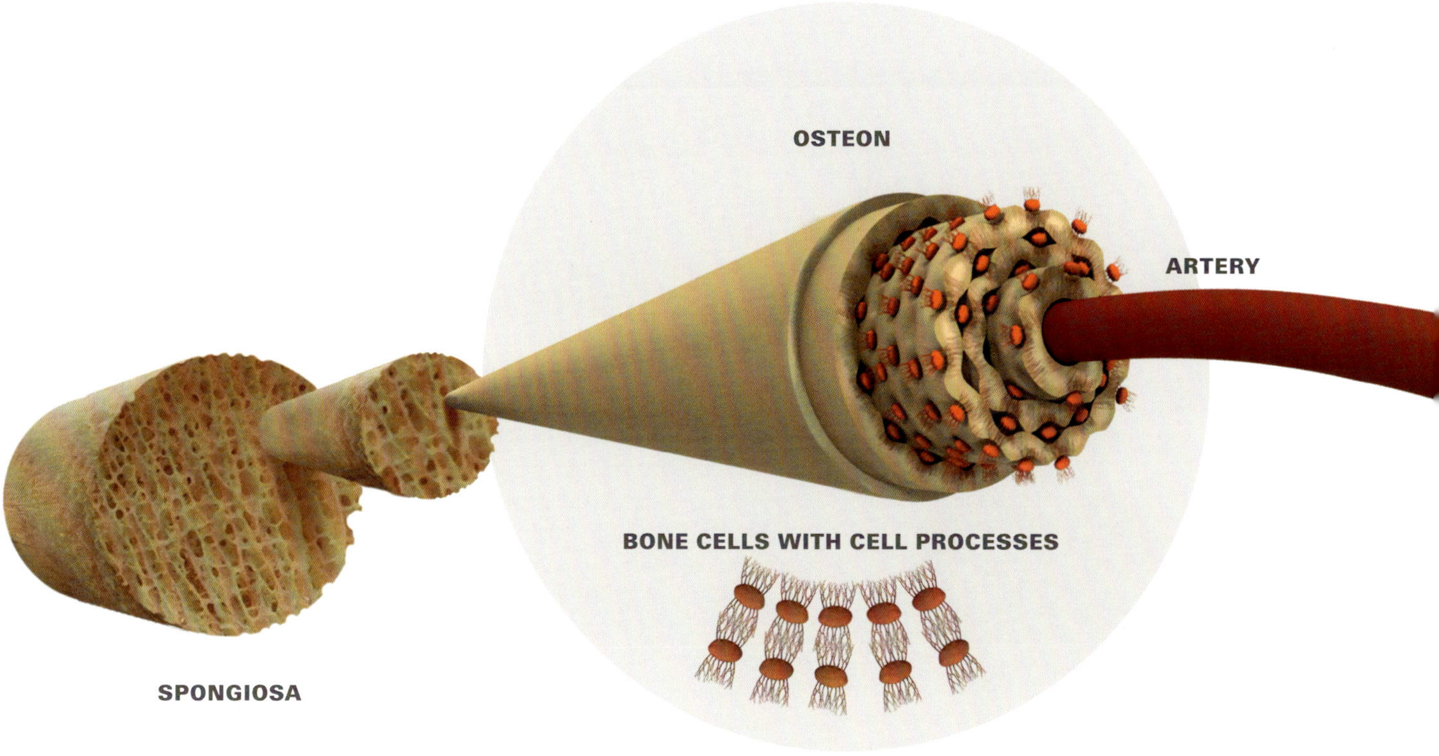

xylapatite. The ending "blast" indicates a cell which is still maturing. After around three days, an osteoblast becomes a mature bone cell (osteocyte). Osteoclasts are giant cells with multiple nuclei which break down the mineralized matrix of the bone with the help of enzymes. Young bones build up bone mass more quickly than old bones. Calcium is deposited in bone cells in a process known as calcification. Poor blood flow and thus a reduction in calcium supply causes bone density to decrease with age. The degree to which this happens varies from bone to bone. As tests on astronauts who have spent long periods at zero gravity have shown, lifelong exposure to mechanical/physical strain is essential in maintaining the equilibrium between bone formation and resorption. The frequency of exposure to strain is more important than its intensity. Low but regular strain stimulates bone growth more than high but infrequent strain. Bone growth and the bone formation/resorption equilibrium depend on dynamic, not static strain. Lying on the sofa leads to the resorption of bone.

The elements which sense what kind of strain a bone is under are the bone cells themselves. Bone cells are sensitive to pressure and are connected to each other via long processes. These processes, which are filled with liquid, are able to detect local changes in strain. Bone under strain is "squeezed" like a sponge and the flow of liquid inside the processes changes. This releases messenger substances such as nitric oxide (NO), an important signal molecule which also plays a role as a signal transmitter in the blood vessels and other areas of the body. In 1998, findings about the function of NO were awarded the Nobel Prize.

As a rule, bone consists of a hard cortex with a sponge-like interior (spongiosa). The spongiosa appears to be unevenly distributed throughout the bone, but in actual fact the trabeculae, which are tiny lattice-shaped beams, rearrange themselves according to the direction of tensile strain and compressive stress, as Badoux & Hoegeveen (1976) proved in connection with the femoral head in dogs.

The trabeculae of the spongiosa are interspersed with red or yellow bone marrow. In simplified terms, a distinction is drawn between short and flat bones such as the scapula (shoulder blade) or the carpal bones, and long bones such as the femur (thigh bone) or tibia and fibula (lower leg bones). Long bones are filled with yellow, fatty marrow. The interior of short bones and flat bones is completely taken up by trabecular spongiosa, which in turn is filled with red, hemopoietic (blood-forming) bone marrow. Bone tissue is well supplied with blood, as evidenced by the amount of blood loss which occurs when a bone is broken. Blood circulates via arteries which branch inside the bone marrow cavity, whereas venous drainage takes place from the inside out. The arteries in the bone run through special cavities known as osteons which are approximately 10 cm long but only 200 micrometers wide.

Another type of bone is known as sesamoid bone. Sesamoid bones are ossifications embedded in certain tendons. The best-known bone of this type is the knee cap (patella). Sesamoid bones are usually found close to joints. They prevent tendons from being loaded in shear, act as spacers, improve the angle of attack of the tendons and increase their moment arms.

BONE GROWTH

During the fetal period, cartilaginous precursors to the long bones are laid down which in other respects are very similar to the bones as they will later be. Ossification begins in what is approximately the middle of the cartilaginous shaft in the form of an outer layer which encloses the cartilage cylindrically and spreads slowly from the middle to the end of the bones. This rigid bony cortex has two important consequences for the cartilage inside. Firstly, the cells of the cartilage (chondrocytes), which at this stage in development are still dividing rapidly, are forced to arrange themselves on top of each other to form columnar cartilage as the bony cortex does not permit outward expansion. The growing cartilage is thus forced into a particular direction and makes the forming bone grow lengthways. As the cortex spreads and becomes thicker, the chondrocytes within it stop being supplied with nourishment. The reason for this is that cartilage is generally free of blood vessels and must thus be nourished via diffusion. As the bony cortex forms a barrier which limits the diffusion of nutrients, however, the chondrocytes at the center of the shaft suffer the most. At this point, blood vessels sprout through gaps in the bony cortex, bringing with them not just an excellent supply of nutrients but numerous new cells, allowing the process of enchondral ossification to begin. The new arrivals include cartilage-eating cells and bone-forming cells. The former start to break down the cartilage, the latter to build up bone. This process is centered initially around any remaining longitudinal bars of cartilage which calcified when the nutrient supply was cut off as described above. The whole process then spreads to the ends of the bones. Eventually, blood vessels sprout at the top and bottom end of the future bones, allowing enchondral ossification to take place here too. These

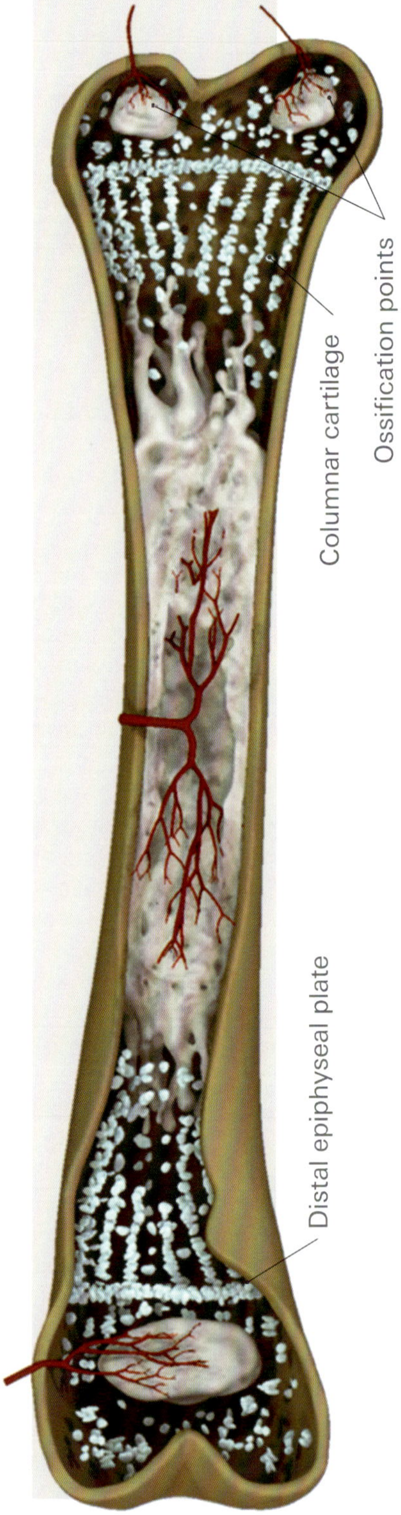

ossification centers will later form the articular ends of the bone and are known as epiphyses. Between the epiphyses and the growing bone shaft remains a zone of the original cartilage which is known as epiphyseal cartilage. As long as the chondrocytes here continue to divide and form new cartilage, the bone can grow lengthways.

The number of ossification centers and epiphyses is variable in long bones, and differs from bone to bone (Hare 1959, 1960, 1961, Schröder 1978). The humerus, for instance, has four to six such centers, while the radius generally just has two.

The ossification centers appear at very different times. In the forelimbs of German Shepherds, for instance, the proximal growth plates of the upper arm appear within 14–16 days of birth, and the distal growth plates within 49–63 days (Schroeder 1978). As a result, bones do not grow equally at both ends. Approximately 80 percent of longitudinal growth in the humerus takes place at the top end and only around 20 percent at the distal end (Henschel, 1973a). In the radius, 40 percent of growth takes place proximally and 60 percent distally, while the ulna, which only has one growth plate, only grows distally. To ensure that the ulna grows at the same rate as the radius despite only having one epiphyseal plate, the epiphysis of the ulna is conical and not flat, which makes the growth surface one and a half times bigger.

About 90 percent of longitudinal growth takes place in the first six to seven months of life (Riser und Shirer 1965). Hare (1960) noticed that although the different epiphyses close in the same order in different breeds, the actual timing of closure can vary greatly within litters, breeds and from breed to breed. Epiphyseal closure marks the end of longitudinal bone growth, while the lateral growth caused by the setting down of new layers and remodeling within

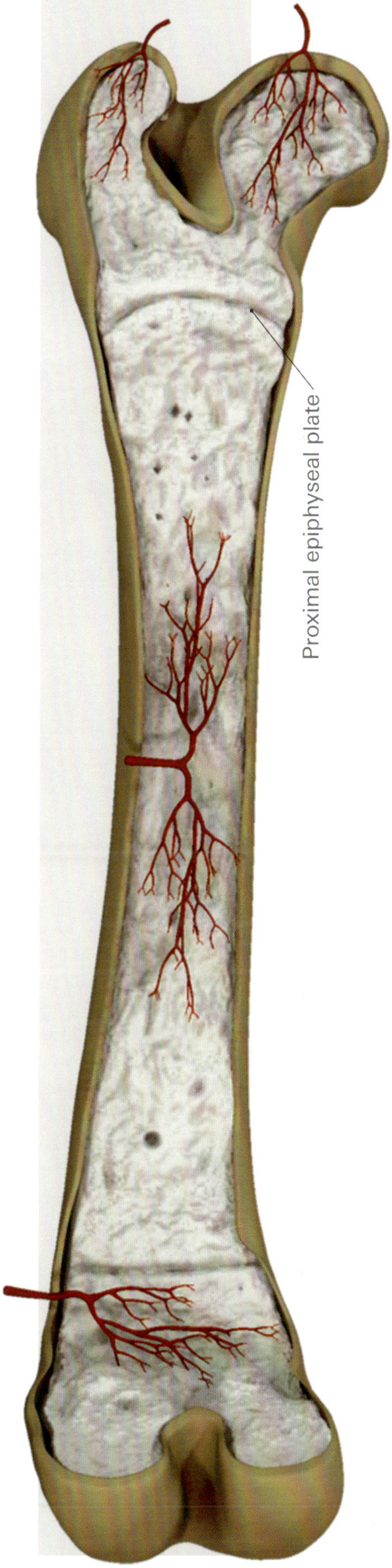

the bone continues throughout life.

The scapula starts off as a flat bone with various ossification points; the scapular spine does not develop until later. Previous studies into the development and growth of the forelimbs have tended to neglect the scapula. In evolutionary terms, the scapula belongs to the shoulder girdle and thus, strictly speaking, to the posterior skull region. This means that though it has evolved to be functionally part of the forelimb, the way in which it develops has not yet caught up. The mechanism by which the growth of the scapula is coupled to the growth of the arm has never yet been investigated.

ACHONDROPLASIA

If the epiphyses close too early in development, prematurely putting a stop to endochondral ossification, the result is dwarfism. This abnormality is also known as achondroplasia, chondrodystrophia and chondrodysplasia. The short limbs of Dachshund, Pembroke Welsh Corgiesand Basset Hound are the result of this growth disorder. In the Basset Hound, all the epiphyseal plates close much earlier than in breeds such as the German Shepherd, for example. The epiphyseal plates in the bones of the forelimbs of Basset Hounds close just two thirds of the way through the normal developmental period (Hitz 1973). These dogs are chondrodystrophic dwarves.

According to Parker et al. (2009), achondroplasia is caused in all breeds by mutations in just one gene. The authors argue on this basis that short-leggedness probably only evolved once. If this is the case, all of today's short-legged breeds descend from just one short-legged ancestor. The gene in question is responsible for a growth factor which in turn controls the regulation of growth and differentiation processes.

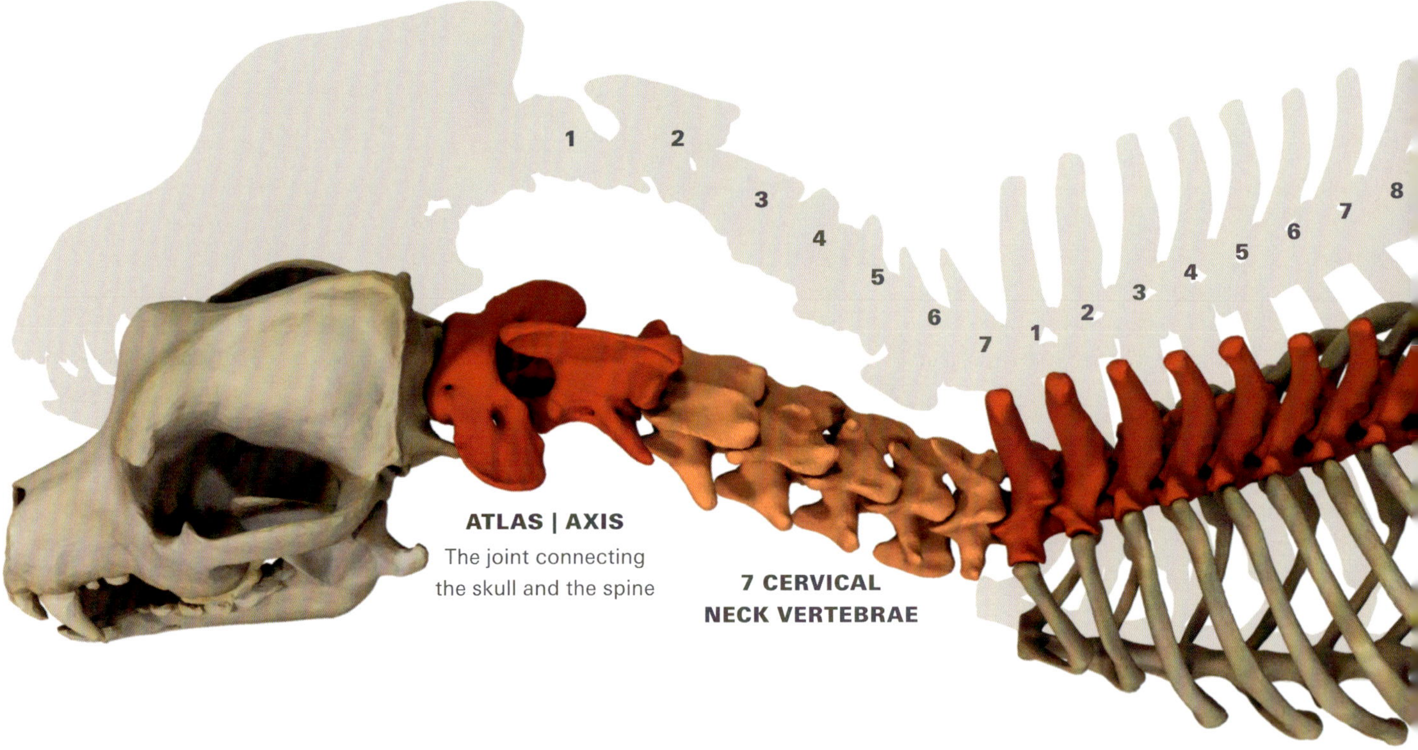

SPINE
As in all mammals, the spine in dogs is divided into structurally and functionally distinct regions. The first two cervical (neck) vertebrae make up the "head joint", or atlanto-occipital and atlanto-axial joints. This joint complex enables the head to move independently of the neck and permits turning, nodding and all the possible combinations of these movements. The joint complex only evolved somewhere in the early history of mammals more than 200 millions years ago. When birds turn their heads it is mainly the result of tiny, accumulative movements in the neck joints.

CERVICAL SPINE
The cervical spine describes a steep curve. Illustrations often show the neck as a straight continuation of the back, but in reality it curves right the way down to the shoulder joint, at which point it curves back up to the withers. Generally, this is the way the head is carried. Only when the dog is trotting or galloping the neck thrusts forward so that the head hangs from it like an outstretched cantilever.

THORACIC SPINE
The thoracic spine is a relatively inflexible section of the spine whose main job is locomotion, providing a solid base for the thoracic extrinsic appendicular muscles. Ventilation is a function that was added to the basal locomotor function.. The thorax, which is fairly stiff, is expanded and contracted by muscles, and the diaphragm, also a muscle, closes the thoracic cavity off from the abdomen to form a dome-like space. Dogs have 13 thoracic vertebrae and, accordingly, 13 pairs of ribs. The first seven to nine ribs are directly connected with the breastbone (sternum) via a section of cartilage. A distinction is made between these "true" ribs and the "false" ribs that follow which are only connected to the sternum via a cartilaginous costal arch. The thirteenth rib is "free", ending in muscle without being connected to the breastbone at all. The diaphragm is always attached to the last pair of "false" ribs. The breastbone is made up several segments of bone which correspond to the vertebrae in the way they are arranged.
Numerous observations in a range of different mammals, including dogs (Bramble Carrier 1983) have shown that there is a connection between the step cycle and the rhythm of breathing. Breathing in and out, the ventilation of the lungs, is coupled to step frequency in a strict 2:1 ratio in symmetrical gaits and a 1:1 ratio in the gallop, and has been observed to replace diaphragmatic respiration thanks to a pump piston mechanism created by the internal organs of the abdomen being pushed passively forwards because of their inertia when a forelimb makes contact with the ground and being pulled back during acceleration.
Experiments by Ainsworth et al. (1996, 1997) seem to find partly contradicting evidence, but the coupling of ventilation to stride is after all evident. The issue that has generated the controversy is the extent to which the locomotor movements drive the ventilation. Uncoupling is rare during trotting, but it is next to impossible to get dogs to breathe in an unsycncronized pattern during galloping. In the real world they most probably never uncouple for more than a step or two.

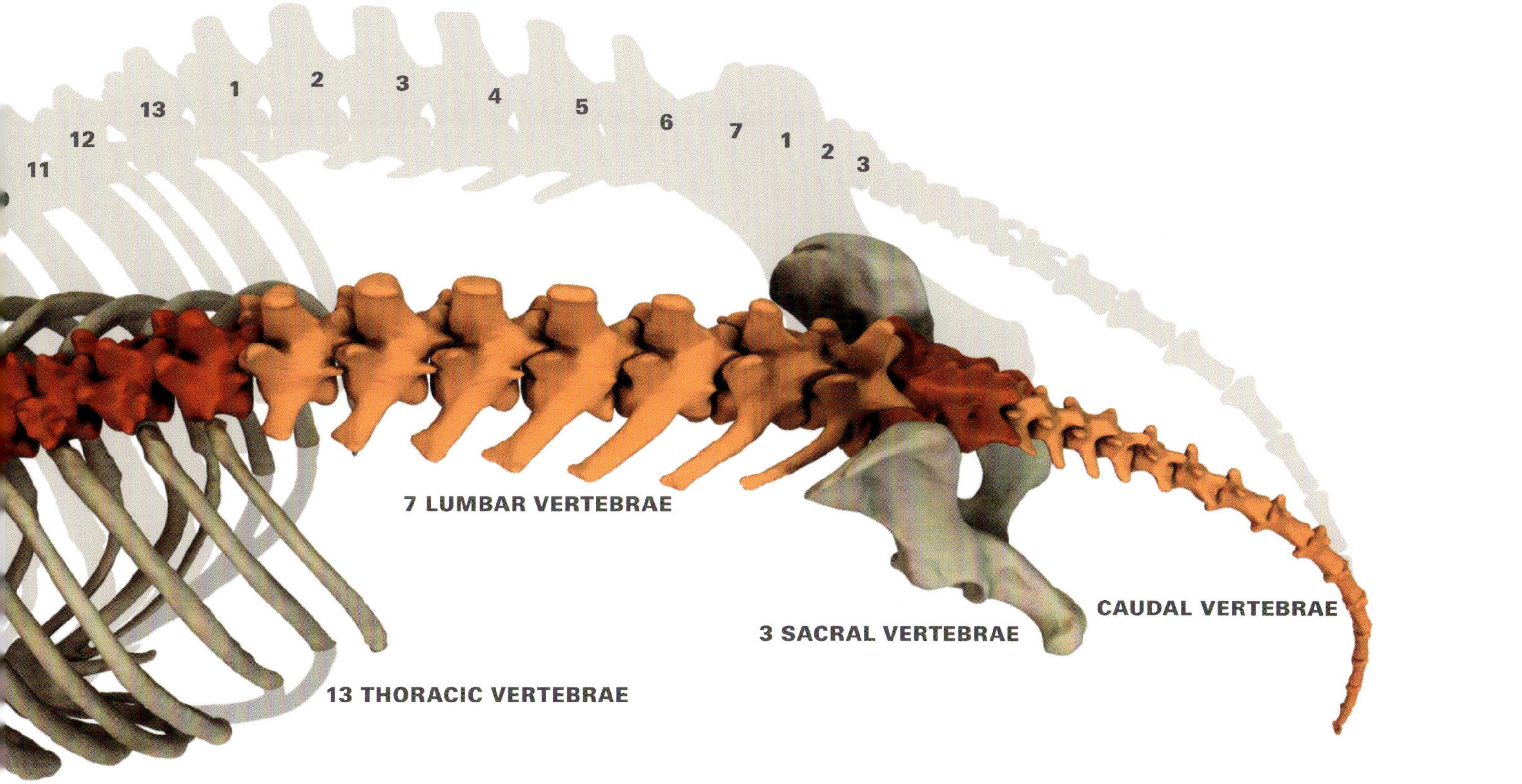

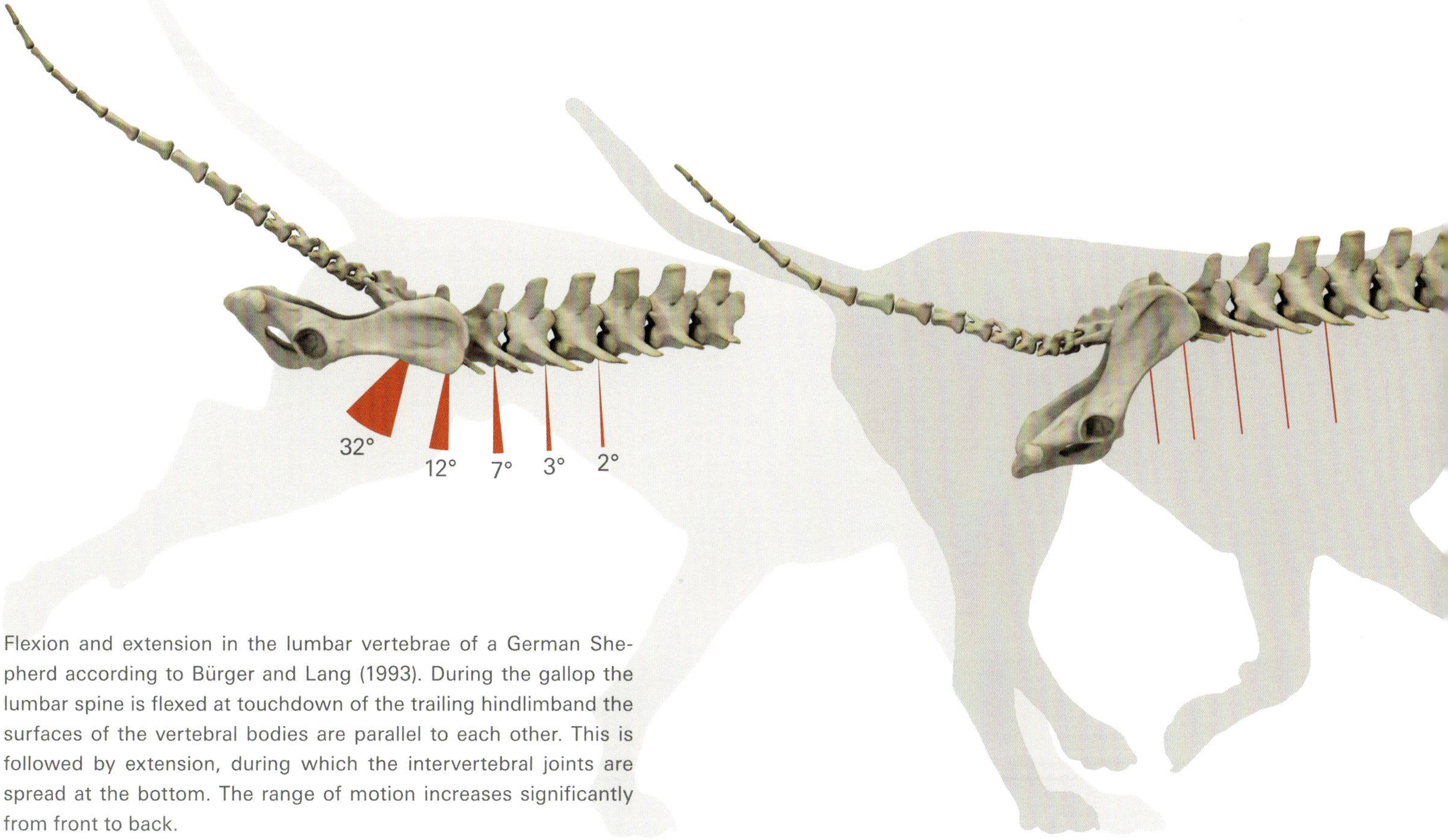

Flexion and extension in the lumbar vertebrae of a German Shepherd according to Bürger and Lang (1993). During the gallop the lumbar spine is flexed at touchdown of the trailing hindlimband the surfaces of the vertebral bodies are parallel to each other. This is followed by extension, during which the intervertebral joints are spread at the bottom. The range of motion increases significantly from front to back.

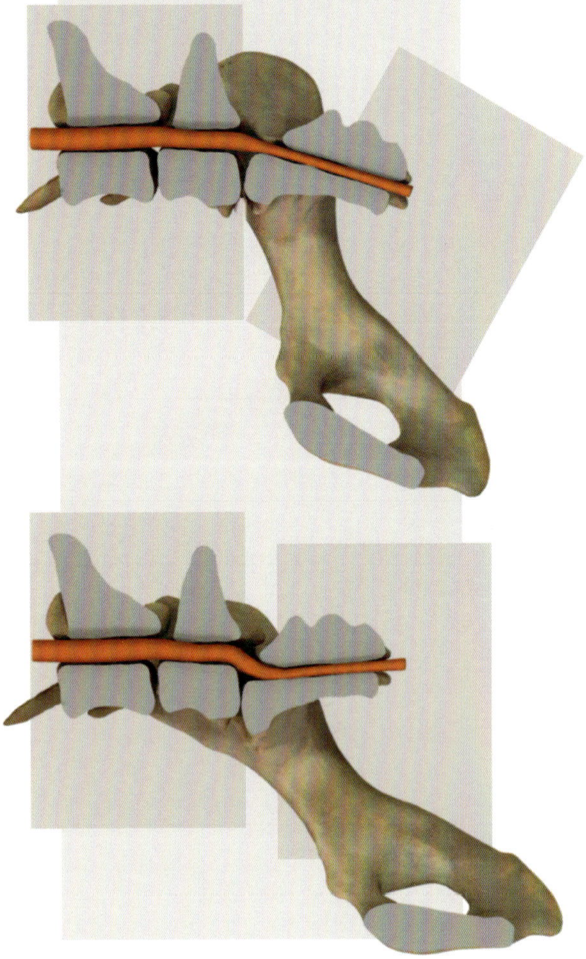

Movement in the sacroiliac joint takes the form of tiltimg and translation.

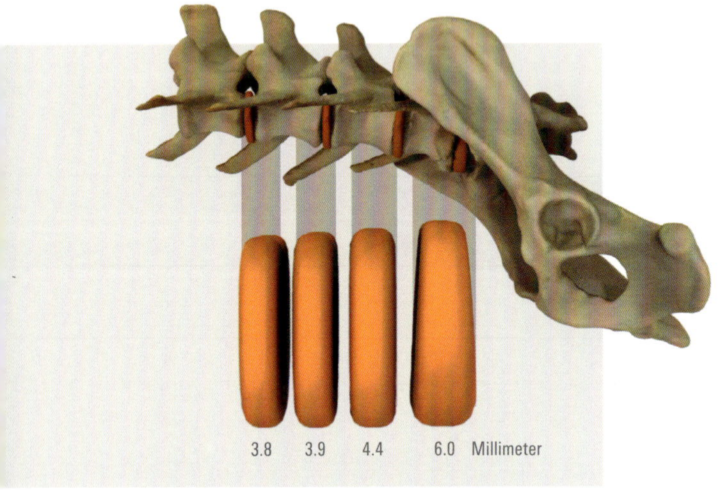

Increase in the thickness of the intervertebral discs from the 4th lumbar vertebra to the sacroiliac joint according to Benninger et al. (2006).

LUMBAR SPINE

The lumbar spine is extremely interesting in terms of locomotion. As the result of the steep curvature caused by the summation of movements in the intervertebral joints, the lumbar spine can flex and extend by over 50° in the gallop, thus contributing over 50 percent to stride length. In dogs and all other mammals except horses, the gallop is mainly the product of the lumbar spine. Only one study has calculated the precise motion of the lumbar spine during a stride, and that concentrated on the joint between the second and third lumbar vertebrae (Schendel 1995). As the data were collected after an invasive operation, their reliability is disputed, however. In the study in question, lateral movement in the joint investigated amounted during the walk to 4-6°, rotations around the longitudinal axis and flexions and extensions were minimal. In a different study, the flexibility at maximal extension and maximal flexion of the section of the spine from the third lumbar vertebra (L3) to the 1st sacral vertebra (S1) was investigated using the preserved spines of 13 German Shepherds (Bürger Lang 1993). The average range of movement was 60°, although one specimen only managed 42° and another an impressive 74°. This extraordinary flexibility is not distributed uniformly across the five intervertebral joints, but increases dramatically towards the pelvis. Flexibility in the joints between L3 and L4 is just 3°, increasing to almost 12° between L6 and L7 and reaching a peak of 32° between the 7th lumbar vertebra (L7) and the sacrum (lumbosacral joint). In one of the specimens, this value was over 40°. A significant difference was also observed in the specimens between male and female dogs, with the flexibility of the lumbosacral joint in bitches generally being around 9° greater.

Similar ranges of excursion have also been observed in the lower lumbar vertebrae of other breeds. Benninger et al. (2006) found in a comparison between nine German Shepherds and other breeds of comparable weight that other breeds exhibit even greater flexibility than German Shepherds.

Usually, especially in the highly flexible lumbosacral joint, the vertebrae are rotated against each other around a non-fixed rotational axis, which results in them being tilted in opposite directions. In addition, they can be shunted against each other. The anatomical basis of this astounding flexibility is the shape of the intervertebral discs and the way in which the articular surface of the vertebrae is constructed. The critical feature of the intervertebral discs is the ratio of cross-sectional area to thickness. The thickness increases between L4-L5 and L7-S1 by 58 percent, but the cross-sectional area only increases by 20 percent, which means that in total the ratio of cross-sectional area to thickness decreases by 25 percent (Benninger et al. 2006). This points to a significant increase in flexibility, as the thicker an intervertebral disc is in proportion to its cross-sectional area, the greater the range of possible flexion and extension. While the intervertebral discs of the lumbar spine are uniform in shape, that of the lumbosacral joint is cuneiform, with the widest part formed by the lower edge.

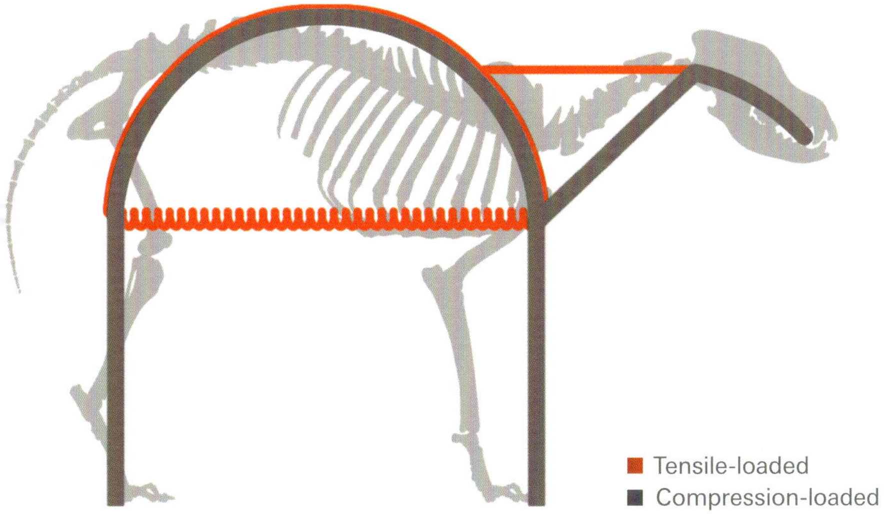

BRIDGE ARCH SKELETON

In terms of statics, the skeleton of the trunk of mammals is a bowstring construction, with the spine forming a compression-loaded "bridge arch". The apex of the pressure-resistant arch, or bow, is level with the distal thoracic vertebrae. The arch is made up of the vertebral bodies and intervertebral discs, upon which tension is exerted by the dorsal processes and their ligaments, the back muscles and the dorsal ligament. The "string" of the bow is made up at the front by the relatively stiff thorax and at the back by the abdominal muscles. The neck and head jut out like a cantilever beyond the front column (the forelimb) and form a reverse bow in relation to the trunk. The neck muscles and the nuchal ligament provide the tension on the dorsal side and the cervical vertebrae absorb the pressure on the ventral side.

SACRUM

In the sacrum, the transverse processes and the costal processes of three vertebrae grow together completely, leaving only the exit points of the spinal nerves free. The ossification process generally takes place during the first eighteen months of life. The sacrum is connected to the pelvis via the sacroiliac joint, which consists of two parts, a crescent-shaped true synovial joint filled with joint fluid whose surfaces are covered with cartilage, and a taut ligamentous apparatus (syndesmosis). The sacroiliac joint is extremely inflexible, with a maximum range of movement of 7° (4° extension and 3° flexion (Gregory et al. 1986)).

In the sacroiliac joint, the propulsion generated by the hind legs is transmitted to the spine and thus to the trunk. The joint is an important shock absorber and bears the whole mass of the rear body when the dog is standing. This explains the differences in the sacroiliac joints of dogs of different weights that are observed not just between dogs of different breeds but also between adults and juveniles of the same breed – the joint adapts to the load it has to bear. In a study involving 145 fully grown dogs and 45 juveniles, ranging in size from Yorkshire Terriers to Rottweilers, Breit Künzel (2001) noted that although body weight increases by a factor of seven, the area of the joint increases by a factor of just six. This means that the load borne by the sacroiliac joint in larger breeds is disproportionately high, reaching up to 2.7 times the dog's body weight. The shape of the joint also changes, though interestingly the authors only observed these changes in Rottweilers, Bernese Mountain Dogs and German Shepherds, in which the joint becomes concave. The angle of inclination of the sacrum and the point during growth at which ossification takes place are age and size-dependent. The load-dependent modeling of the joint takes place during the first year of life from about 12 weeks of age (Breit Künzel 2001). After the first year, no more age or sex-specific changes are observed.

TAIL

As well as being a control mechanism, the tail plays an additional role in the expressive behavior of dogs. There is also anecdotal evidence of it having a further possibly very important function. In a VDH Academy workshop, one participant reported that one of her litters had contained short and long-tailed puppies. The puppies with the longer tails got to their feet 1-2 days earlier than those with the short tails, an important observation that has yet to be scientifically investigated. Even in breeds where tails are docked, the puppies born with long tails have been observed to stand up two to three days earlier than the others.

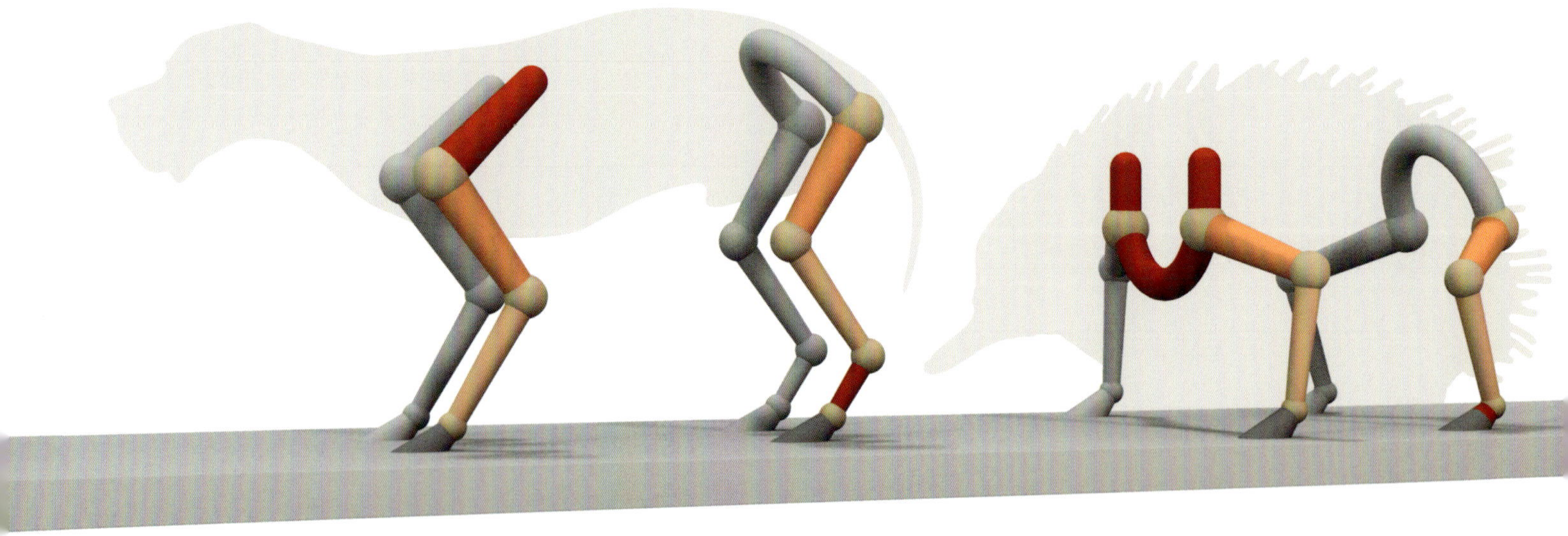

PLACENTAL MAMMAL: DOG **MONOTREME: ECHIDNA**

DURING MAMMALIAN EVOLUTION, THE FORELIMBS (INCLUDING THE SHOULDER BLADE) BECAME SEPARATED FROM THE SKELETON OF THE TRUNK AND ARE NOW ONLY HELD IN PLACE BY MUSCULAR SLINGS.

As in modern reptiles, the limbs of the ancestors of both placental mammals and recent egg-laying mammals (monotremes) exhibited a sprawled posture. The double-flexed mammalian limb evolved through inward rotation from this sprawled posture into the body axis. However, the "new" mammalian limb combines different elements in the fore- and the hindlimbs. In the hindlimb, an element (the tarsus) that was already present was transformed via the formation of a new joint, the upper ankle, between the tibia and the anklebone. In addition, the metatarsus underwent a process of elongation. The forelimb, on the other hand, was redesigned completely. In the course of the evolution of mammals, the shoulder blade, or scapula, became a functional component of the forelimb. The scapula is now located in the same plane as the upper and forearm and is used in propulsion. The pivot points of the fore- and hindlimbs are more or less on a level with each other. This caused a complete shakeup in the function of the fore- and hindlimb segments: the femur now corresponds functionally to the scapula, the lower leg to the upper arm and the hind foot to the forearm and hand. The uppermost pivot point of the forelimb is located towards the top of the scapula, and that of the hindlimb in the hip joint – at least in the walk and the trot. In these gaits the movements of the scapula and femur contribute over 60 percent to step length.

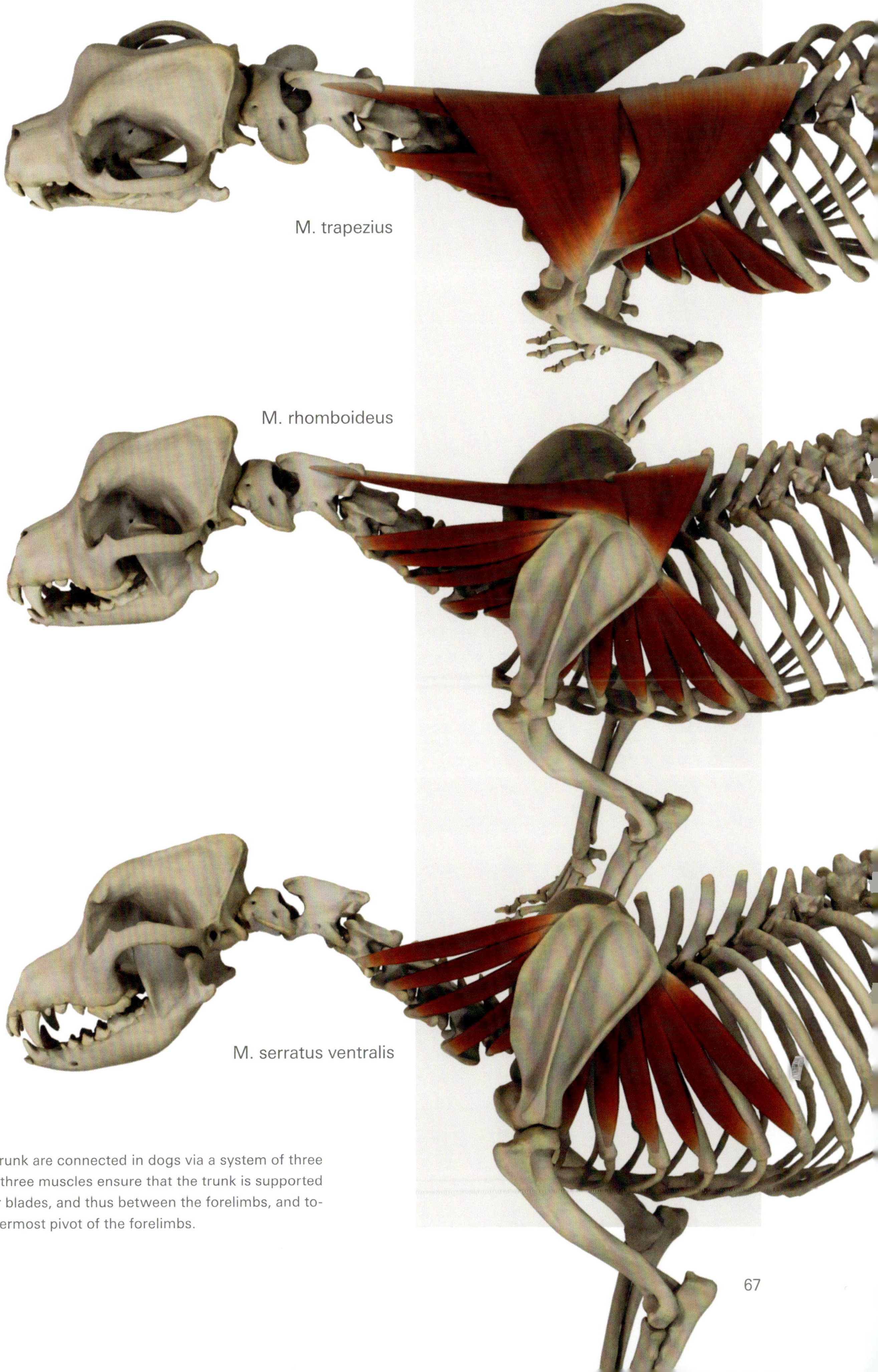

The scapula and the trunk are connected in dogs via a system of three large muscles. These three muscles ensure that the trunk is supported between the shoulder blades, and thus between the forelimbs, and together create the uppermost pivot of the forelimbs.

SHOULDER BLADE
The shoulder blade (scapula) is an almost flat bone. Its outer surface is divided into two by the scapular spine (spina scapulae), a longitudinal reinforcement that counteracts the bending stress of the shoulder blade.

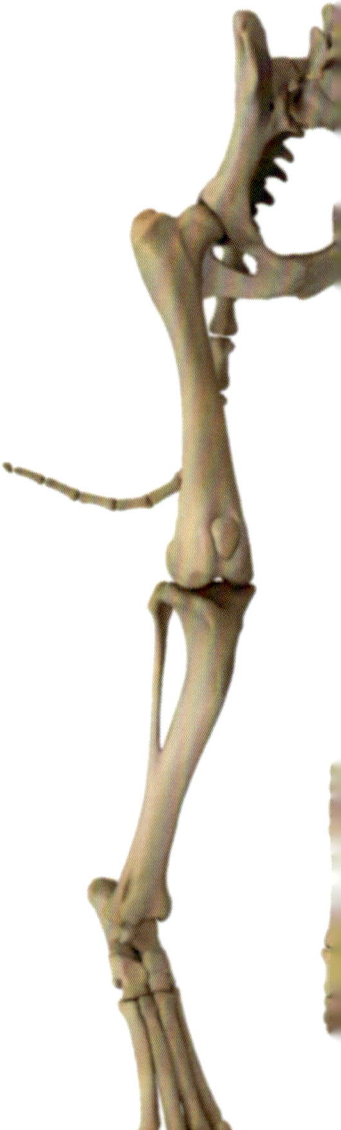

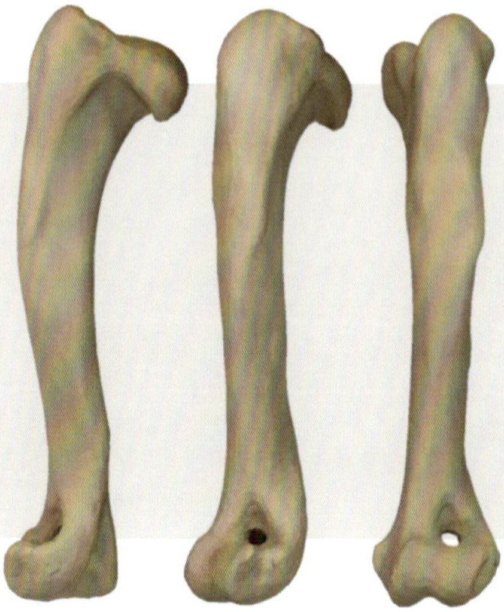

UPPER ARM
The upper arm (humerus) is a long bone. Work to failure and maximum force before failure have shown that the humerus is the strongest bone (Kemp et al 2005). It is divided into a head (caput humeri), a shaft (corpus humeri) and an articular roller.

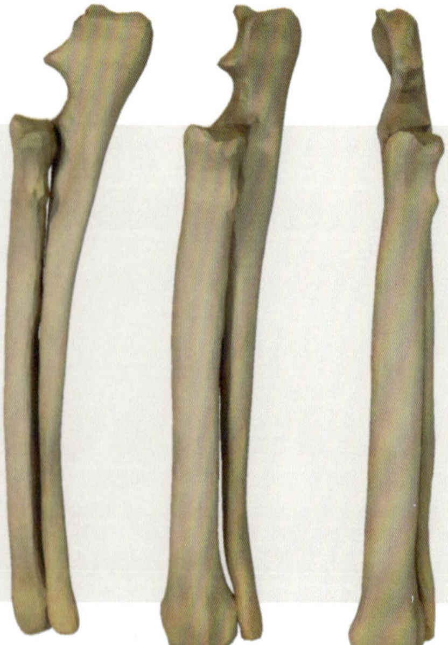

FOREARM
The forearm is made up of two bones called the radius and the ulna, which cross each other. This distinctive arrangement is known as the pronation position, and in it, both elements are largely fixed. They can only be moved against each other passively.

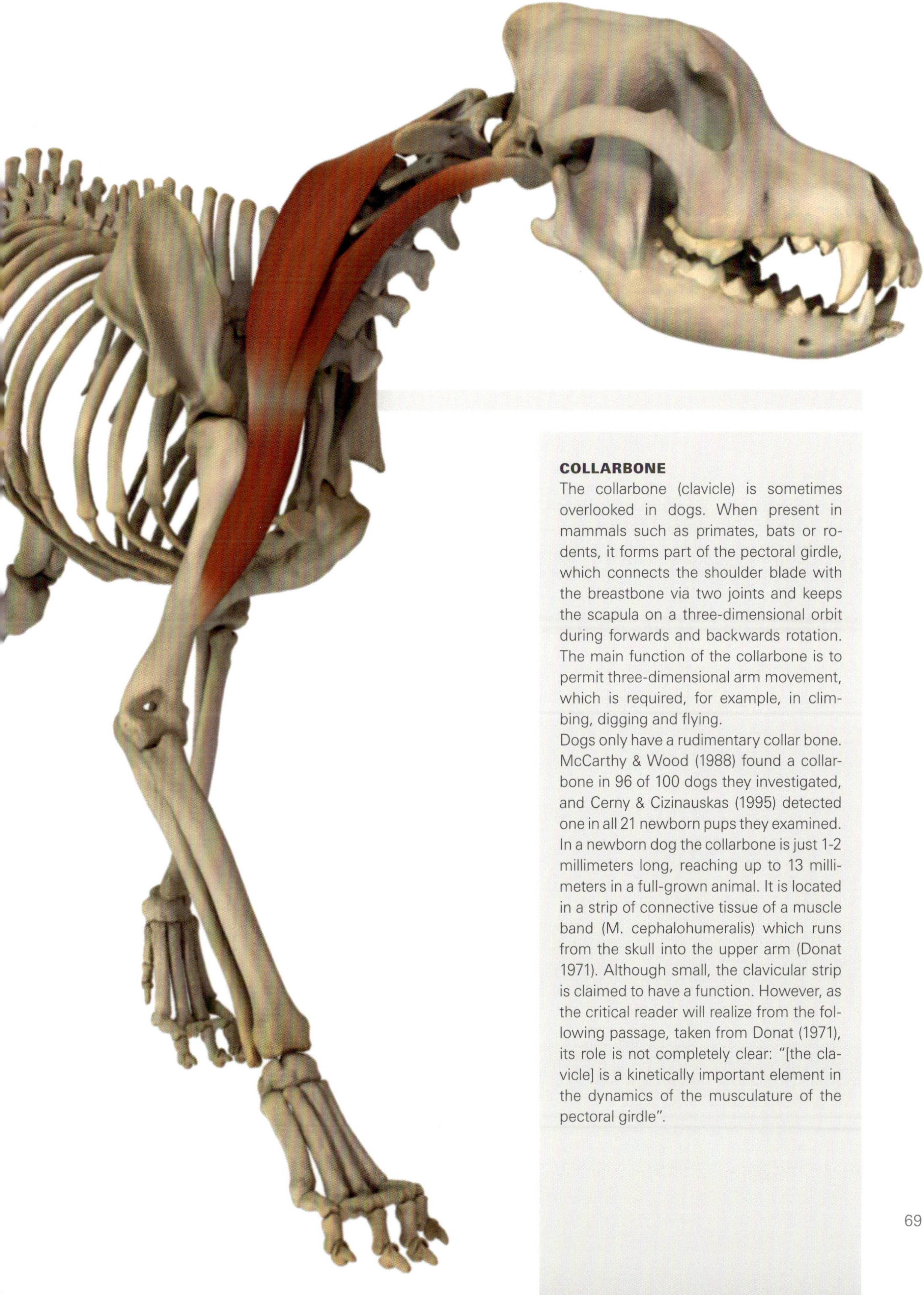

COLLARBONE

The collarbone (clavicle) is sometimes overlooked in dogs. When present in mammals such as primates, bats or rodents, it forms part of the pectoral girdle, which connects the shoulder blade with the breastbone via two joints and keeps the scapula on a three-dimensional orbit during forwards and backwards rotation. The main function of the collarbone is to permit three-dimensional arm movement, which is required, for example, in climbing, digging and flying.

Dogs only have a rudimentary collar bone. McCarthy & Wood (1988) found a collarbone in 96 of 100 dogs they investigated, and Cerny & Cizinauskas (1995) detected one in all 21 newborn pups they examined. In a newborn dog the collarbone is just 1-2 millimeters long, reaching up to 13 millimeters in a full-grown animal. It is located in a strip of connective tissue of a muscle band (M. cephalohumeralis) which runs from the skull into the upper arm (Donat 1971). Although small, the clavicular strip is claimed to have a function. However, as the critical reader will realize from the following passage, taken from Donat (1971), its role is not completely clear: "[the clavicle] is a kinetically important element in the dynamics of the musculature of the pectoral girdle".

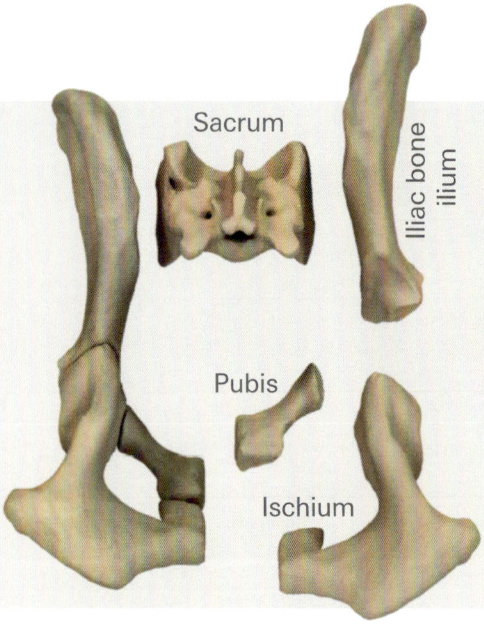

PELVIS

The pelvis is fixed to the spine via the sacroiliac joint. It is made up of three bones (the iliac bone, the ischium and the pubic bone) which meet in the acetabulum. Richter & Löffler (1976) observed some breed-specific differences in the pelvis, with that of the Dachshund in particular deviating from the other breeds investigated (German Shepherd, Boxer, Spaniel, Poodle) in its almost square shape and the breadth of the wings of the ilium.

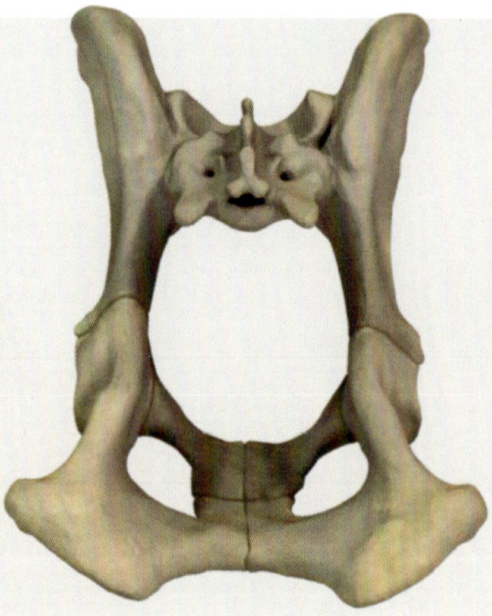

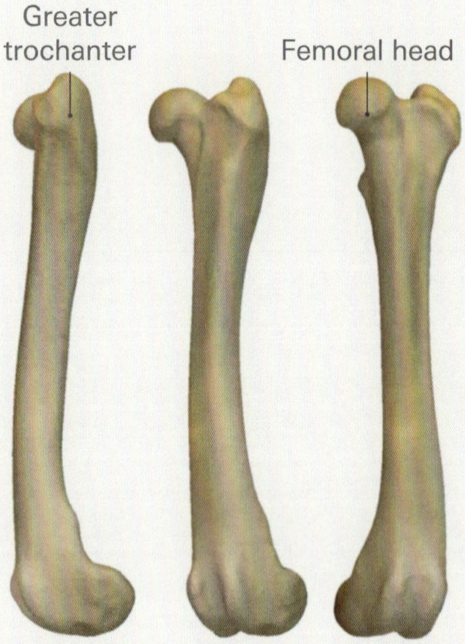

FEMUR

The femur is the second strongest bone in the body. The femoral head (caput ossis femoris) is clearly offset from the femoral shaft by a short "neck". In most dogs, the greater trochanter (trochanter major) extends until it is level with the femoral head, though this is not the case in German Shepherds, Boxers and Poodles. In Dachshunds, the greater trochanter extends beyond the femoral head (Richter und Loeffler, 1976). Distally the femur has two condyli which form the posterior knee joint together with the tibia. The anterior knee joint is formed by the femoral groove and the knee cap (patella).

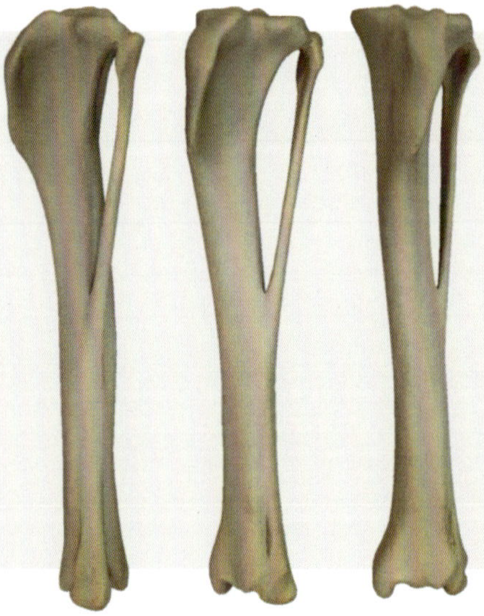

LOWER LEG

The lower leg is made up of two long bones, the tibia (shin bone) and the fibula (calf bone). The fibula is reduced in dogs and coadunate with the tibia.

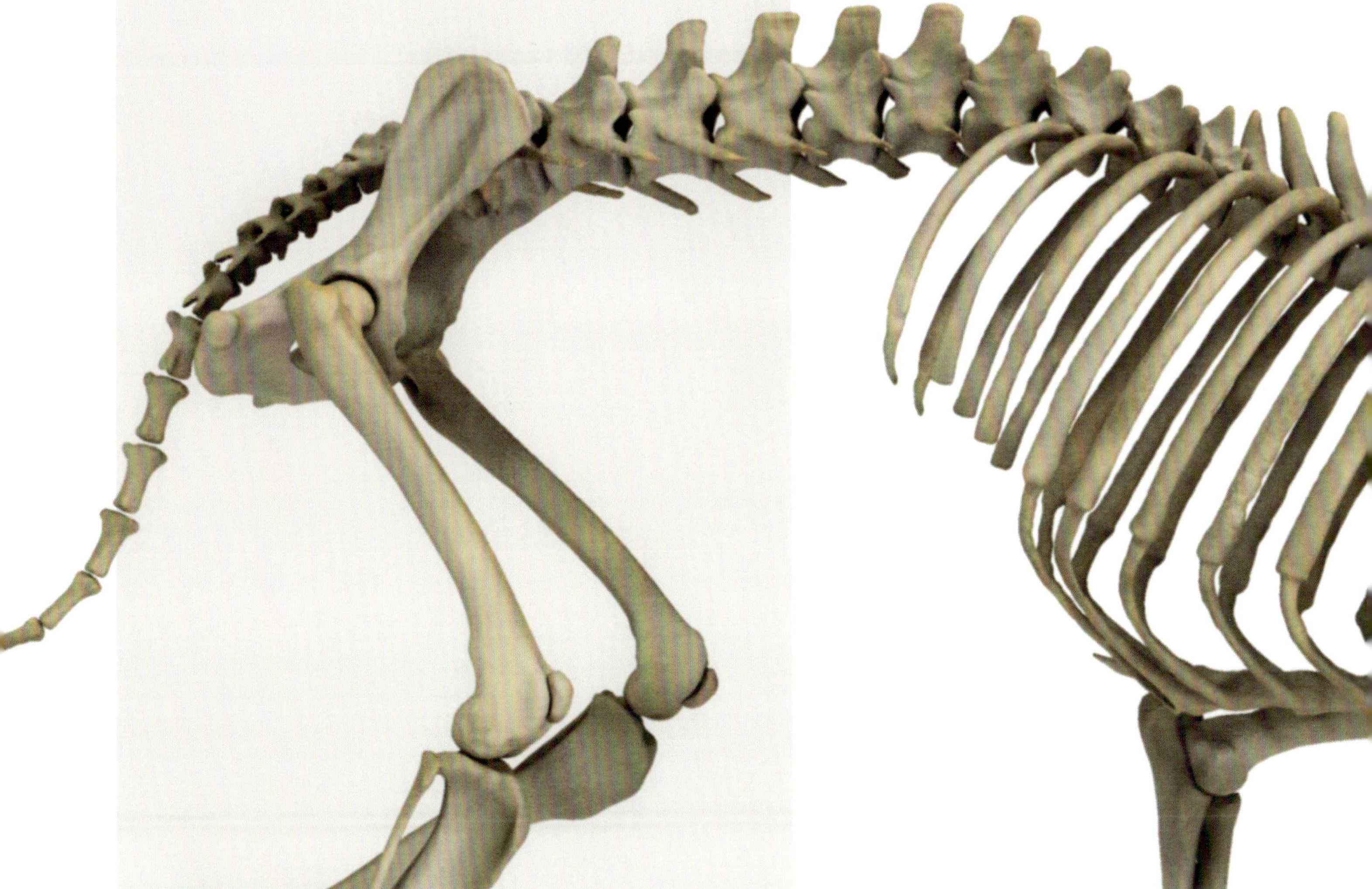

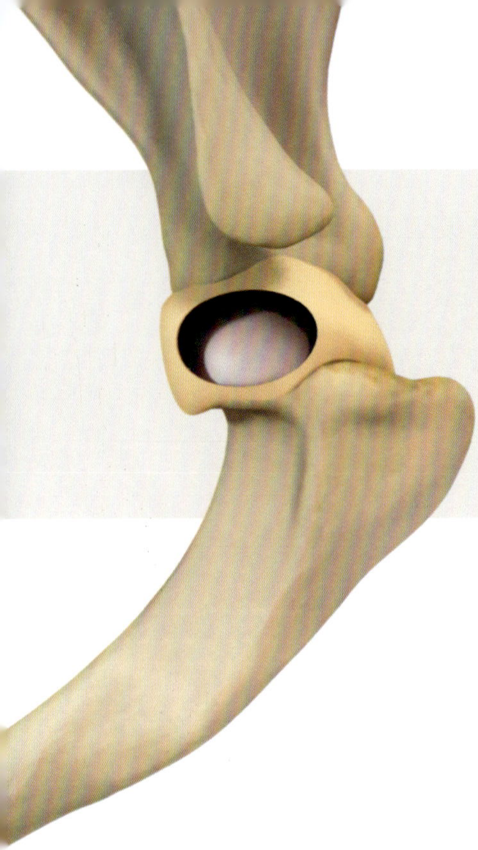

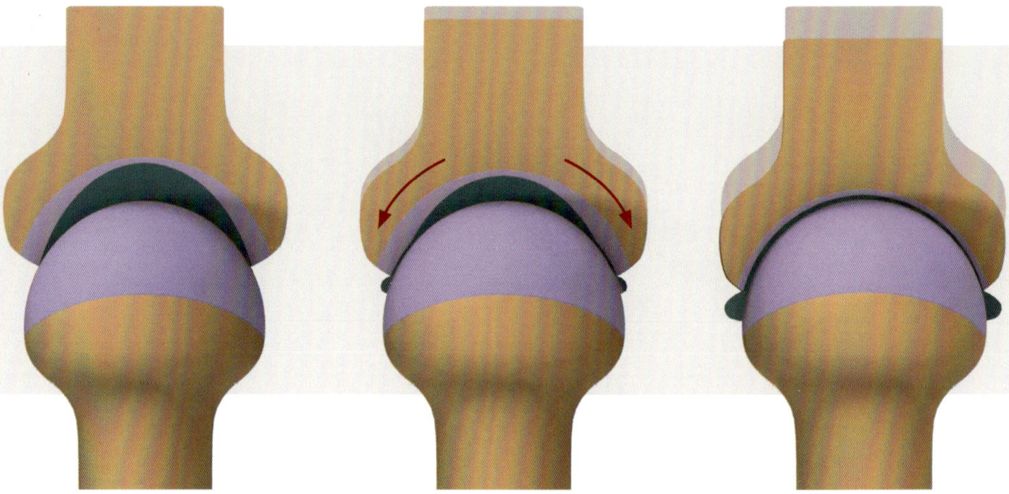

The behavior of an incongruent joint under strain. As the load increases, the synovial fluid is pushed out to the edge of the joint, where the strain is greater than in the center and the cartilage thus thicker. In the bone beneath the layer of articular cartilage (subchondral bone), tensile strain develops (arrow).

JOINTS AND ARTICULAR CARTILAGE

A joint is defined as the flexible connection between two or more skeletal elements. "True" joints, or diarthroses, are all those joints which possess an intra-articular space, or joint cavity, between two cartilage-covered articular surfaces. Synarthroses are fusions of bone, such as that found in the sacrum, of cartilage - as seen in the breastbone, or connections of taut connective tissue such as that observed in the sacroiliac joint. While forces are transmitted in true joints as compressive strain and tensile strain, they are transmitted in synarthroses through transverse, or shear strain (Kubein-Meesenburg et al. 1990). Furthermore, the rotational axes are relatively fixed in true joints, while in synarthroses the rotational axis moves according to the forces at play.

Articular cartilage is a soft, visco-elastic material with numerous pores. Only 10 percent of the total volume of cartilage is actually made up of cartilage cells. 70 – 80 percent is water and only 20 – 30 percent is made up of solid matter, mostly collagen. This ratio of solid to liquid components enables articular cartilage to withstand high levels of strain, and in particular the strain exerted by large compressive forces. Cartilage responds to compressive strain with a visco-elastic deformation: the incompressible watery matrix is pushed into the firm matrix of collagen fibers. The greater the strain, the stiffer it thus becomes. The thickness of articular cartilage varies from joint to joint and within the different regions of a joint. It also decreases with age. Articular cartilage needs to be subjected to continuous, differentiated strain as it is nourished by joint movement and otherwise turns into fibrous cartilage or bone, which have different biochemical and mechanical properties. While the animal is still growing, articular cartilage is supplied with blood via blood vessels which originate directly in the bone. As the layer of bone underneath the cartilage increases in thickness, however, this form of nutrient supply is severely compromised. From now on, the articular cartilage obtains most of its nourishment via diffusion from the joint fluid. Damaged cartilage cannot be replaced, as practically no cell division takes place in adult tissue.

Although cartilage is ten times more malleable than trabecular, or spongy bone, which in turn is ten times more malleable than the compact outer layer of bone, because it is only present in a thin layer cartilage only plays a minor role in shock absorption. The layer of bone underneath the cartilage (subchondral bone bed) and the spongy bone beneath that are the main shock absorbers.

True joints are hermetically enclosed by an articular capsule made of connective tissue. The innermost layer of this capsule excretes joint fluid (synovia), a viscous liquid less than a milliliter deep which nourishes the articular cartilage and, under normal strain, prevents the two articular surfaces from coming into direct contact.

Many diarthroses are physiologically incongruent, which means that they do not fit together exactly when not under strain. The incongruence optimizes stress distribution by spreading the pressure across the joint while the concave part is simultaneously subjected to tensile strain. The outlying regions of the joint play a particularly important role in pressure distribution (Eckstein et al. 1993). While the area of contact between the two elements of the joint is only small when the strain is low, as soon as the joint is subjected to greater strain the concave articular surface is stretched and its convex counterpart pushed deeper into it. The surface area available for pressure transmission thus increases, and the isolated peak pressure points which would occur in a congruent ball and socket joint, in which pressure is centered, are avoided. This elastic increase in the surface of the joint is made possible by the sponge-like ability of cartilage to expel fluid under pressure and absorb it when the pressure is removed. One side effect of this is that articular cartilage is particularly efficient in taking up the glucose that is present in synovial fluid. This in turn highlights the importance of regular, dynamic strain for the health of the cartilage and thus of the joint.

A few years ago, a method was developed in Munich (initially in the department of human anatomy but which soon took hold in the department of veterinary anatomy) to identify patterns of load transmission in the joint by analyzing the articular cartilage and the subchondral bone plate beneath it (e.g. Müller-Gerbl 1991, Eckstein et al. 1993, Maierl & Böttcher 1999, Dickomeit 2002, Eller 2003, Lieser 2003, Maierl 2003, Kaiser 2006, Winhard 2007). These structures reflect long-term patterns of strain and the way they are dealt with by the joint. By measuring, among other things, variations in the thickness of cartilage across the joint (a reflection of compressive strain) and investigating the arrangement of bones and split lines of cartilage, which in turn shed light on the preferred orientation of the collagen fibers and thus the direction of principal strain, these studies provided an important insight into the behavior under strain of individual joints and explained the way in which physiologically incongruent joints work.

SHOULDER JOINT

The shoulder joint is an anatomically incongruent joint in which the humeral head articulates with the much smaller socket of the shoulder blade. The concave articular surface of the shoulder blade and the articular surface of the humeral head, which is twice as big, make for a significant range of motion which is limited to some extent by passive mechanisms (ligaments) and active mechanisms (muscles) (Bardet 1998). As always, upon closer inspection the articular surfaces are revealed to vary – within certain parameters - even within the same breed (in this case Beagles (Ruchay 2008)). Overall, the bone is denser in the glenoid cavity than in the humeral head. Typically of an incongruent joint, bone density and articular cartilage thickness are lower in the region of the cavity roof than at the border (Eller 2003). The maximum possible range of motion has been calculated for various breeds (Rauscher 1986: German Shepherd, Wirehaired Dachshund, Longhaired Dachshund; Mai 1995: German Shepherd, Dachshund, Great Dane and Doberman). The results vary. In the German Shepherd a maximum extension of 138° (Rauscher 1986) or 145° (Mai 1995) occurs, in the Doberman and Great Dane it is 149°, in the Wirehaired Dachshund 128° (Rauscher 1986) or 146° (Mai 1995) and in the Longhaired Dachshund 115° (Rauscher 1986) or 142° (Mai 1995). The median range of motion was 91° in the German Shepherd and 76° or 66° in the Dachshund according to Rauscher (1986), and 90° in the Great Dane, 98° in the Doberman, 100° in the German Shepherd, 102° in the Longhaired Dachshund and 108° in the Wirehaired Dachshund according to Mai (1995). Newton (1985) reports median flexion and extension values of up to 145° in ten mongrels. Abduction and adduction and rotation around the longitudinal axis of the upper arm differ by up to 40° between breeds (Mai 1995).

The standing angle was measured by Lott et al. (1990) to be 129° in the German Shepherd and 135° in the Boxer. According to

Bone density distribution on the articular surface of the glenoid cavity (shoulder socket) and the humeral head (bulbous head of the upper arm) according to Eller (2003). The density of subchondral bone reflects long term stress patterns. The colors of the density map reflect differences in density distribution and are thus relative values. Density increases from blue to red.

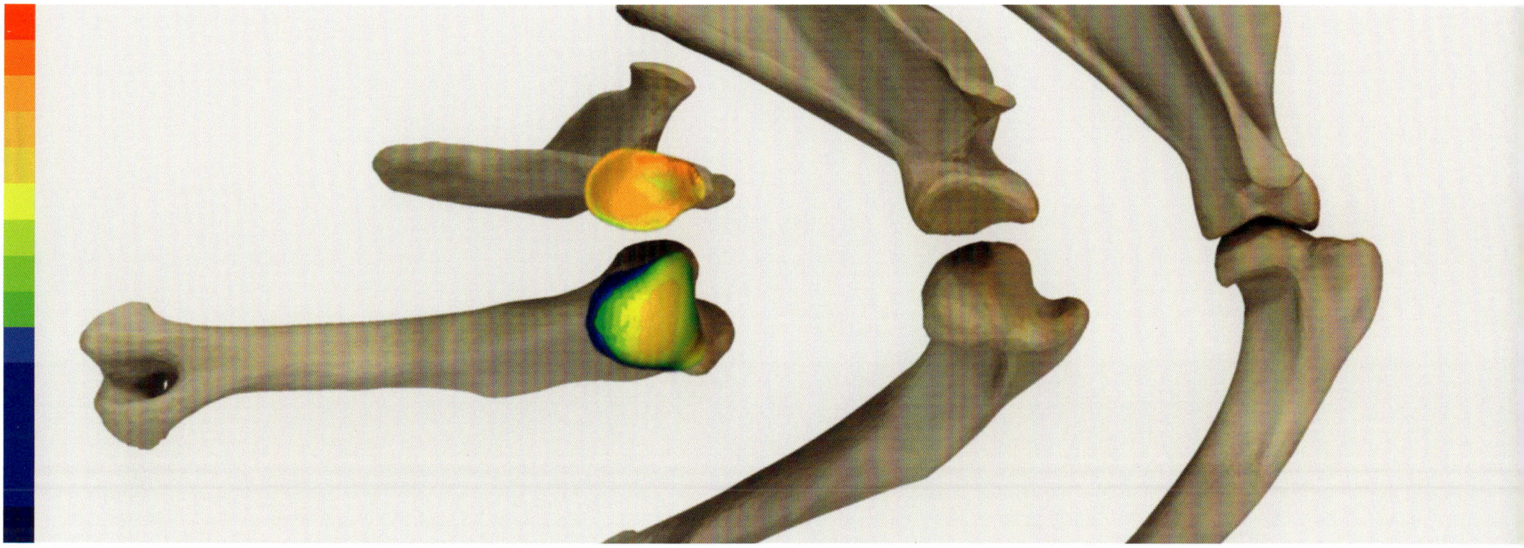

Size of contact surface under increasing loads after Eller (2003).

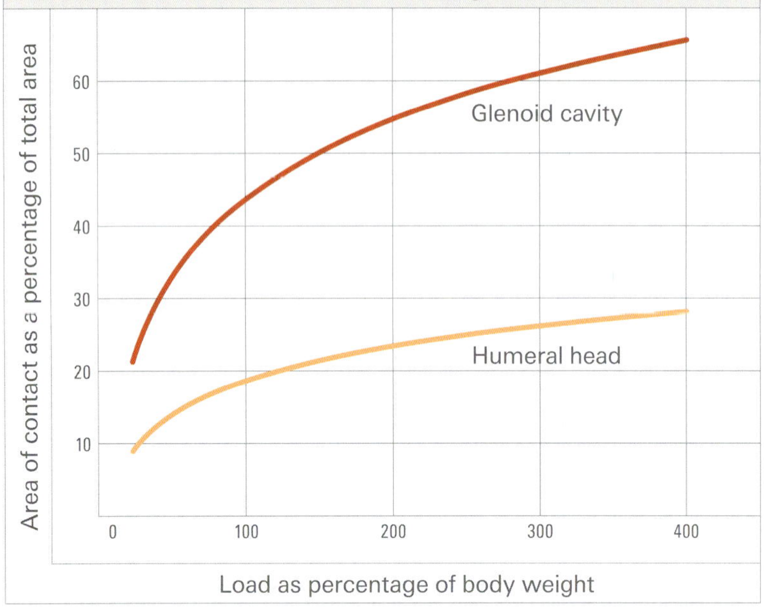

Rauscher (1986), it is 104° in the German Shepherd, 84° in the Wirehaired Dachshund and 82° in the Longhaired Dachshund. Mai (1995) came up with angles of 131° (Great Dane), 129° (Dobermann), 112° (German Shepherd), 105° (Wirehaired Dachshund) and 92° (Longhaired Dachshund). The angle of 90° prescribed in many breed standards including that for German Shepherds was not evidenced by a single German Shepherd. Roos et al. (1993), who published the standing angles obtained by Rauscher, correctly pointed out that the 90° angle specified in breed standards only applies to the "outer appearance of the muscular shoulder – upper arm region" and is therefore only a rough estimate.

Kinzel et al. (1976) and Scherrer et al. (1979) explored the extent to which the maximum possible range of motion corresponds to the actual range of motion during locomotion using complex and invasive measuring techniques.

When dogs are walking, the articular surfaces of the shoulder blade and upper arm slide backwards. During the stance phase of the walk, maximum amplitude amounts to less than 30°, while abduction, adduction and rotation amount to less than 10°.

The range of motion measured by means of our high speed X-ray videography technique is also lower than 30° (see page 94). Moreover, the strain is not distributed equally over the articular surfaces during these small joint movements – some regions are subjected to much greater stress than others. As the strain increases, so too does the area across which pressure is distributed.

Interestingly, the risk zone for osteoarthritic lesions lies outside the range of motion during locomotion of the two articular surfaces and is found in the area bordering the regions of the joint which are not subjected to strain (Scherrer et al. 1979). This underlines the importance of motion for healthy joints.

In a unique experiment conducted in the department of veterinary anatomy at Munich University, the size of the areas of the joint that came into contact was measured under increasing loads. The results showed that at stress levels corresponding to 50 percent of body weight, as would occur when the dog was standing, for instance, a good 30 percent of the articular surface of the glenoid cavity is involved in load-bearing, compared to just 15 percent of the humeral head. Even when loads reach four times body weight, as is the case in the gallop, just 65% and almost 30% of the respective articular surfaces are involved in load-bearing (Eller 2003). Moreover, the strain placed on these areas of the joints during locomotion is of a much shorter duration than is the case in experiments involving a material testing machine.

Intermittent
Constant

The load bearing area in the shoulder joint as measured invasively by Scherrer et al. (1979). The risk zone for osteoarthritic lesions is shown in black on the right hand side.

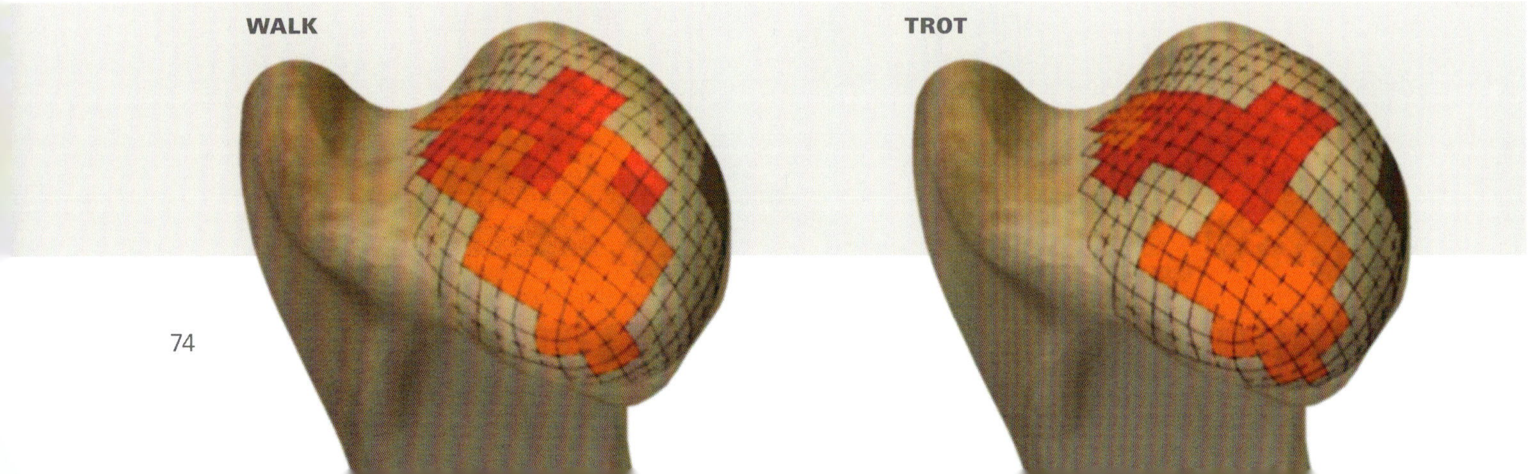

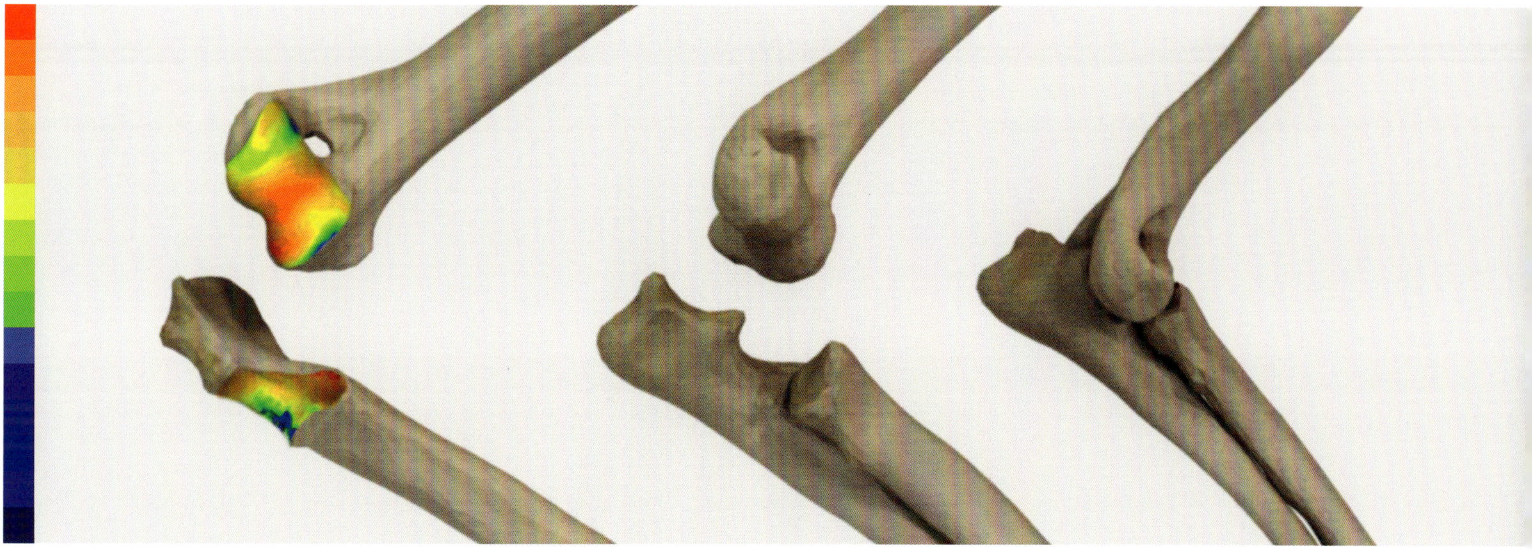

Bone density distribution across the articular surfaces of the elbow joint according to Maierl (2003).

ELBOW JOINT

The elbow joint is made up of three sub-joints, two of which connect the upper arm with the two lower arm bones, ulna and radius, and one of which connects these two bones with each other. The first two sub-joints together form a physiologically incongruent hinge joint; the third one is a swivel joint. The elbow joint does not permit sideways movement.

Even the movement which takes place in a seemingly simple hinge joint is still complex. According to Maierl (2003), who dedicated an entire habilitation thesis to the elbow, the forearm rotates during locomotion around a central pivot point in the elbow joint. As well as vertical forces, significant crosswise forces also occur which place particular strain on the medial coronoid process, which plays an important role in the disorder elbow dysplasia.

The standing angle in the elbow joint varies between 152° in Boxers, 150° in Great Danes, 140°-150° in Poodles, 138° in Dobermans, 137°-150° in German Shepherds, 127°-124° in Dachshunds and 120° in Basset Hounds (Rauscher 1986, Lott et al. 1990, Mai 1995). The maximum angle of excursion is 140° in the Dachshund and around 165° in other breeds; the maximum angle of flexion 20° in Dobermanns, 22°-28° in German Shepherds, 27°-29° in Dachshunds and up to 34° in Great Danes.

The widely accepted view that the humero-radial joint bears 75 – 80 percent of the forces exerted on the elbow joint is disputed by Maierl (2003). Although the large contact area on the radial head is subject to high levels of compressive stress, Maierl found out over the course of various studies, some funded by the GKF, that the main load is borne by the ulna. The same result was obtained, independently, by Mason et al. (2005). The greatest stress is borne by the outlying regions of the concave articular surfaces. Accordingly, this is where the highest density bone is found, and the articular cartilage is also thicker around the edges of the articular surfaces than it is in their centers (Winhard 2007). In addition, the ulna has the proportionally largest articular contact area in relation to the total area of the joint.

Under stress levels of less than 100 newtons, only a few areas, particularly in the caudal region of the joint, come into contact with the articular roller of the upper arm, but the number increases as load increases (Maierl 2003). The maximum vertical forces which occur in the joint correspond to 57 percent of body weight (Nielsen et al. 2003), though Maierl (2003) reports significantly higher joint reaction forces.

FOREARM OR RADIOULNAR JOINT

In dogs, the ulna and the radius are fixed together in the pronation position and rotate only against each other passively during locomotion. When dogs run in a curve, the paws are anchored firmly to the ground, they themselves do not slide "round the bend". Rotation takes place around them and around the radius to which they connect, and so the ulna rotates with the body around the radius. The paw on the outside of the curve needs to be rotated inwards and that on the inside rotates outwards. Rauscher (1986) and Roos et al. (1992) investigated passive forearm rotation in dogs and found that in the German Shepherd, inward rotation exceeds the anatomically normal position by 18° and outward rotation exceeds it by 50°; in the Wirehaired Dachshund the values are 28° and 48°, and in the Longhaired Dachshund 32° and 46°. Dogs are therefore well equipped for pronation beyond the anatomically normal position, which can be seen as a prerequisite for the rapid changes of direction required when chasing prey, for example.

PASTERN

Dogs have seven carpal bones, like all carnivores, in which two of the carpals are fused to form one element called the scapholunatum.

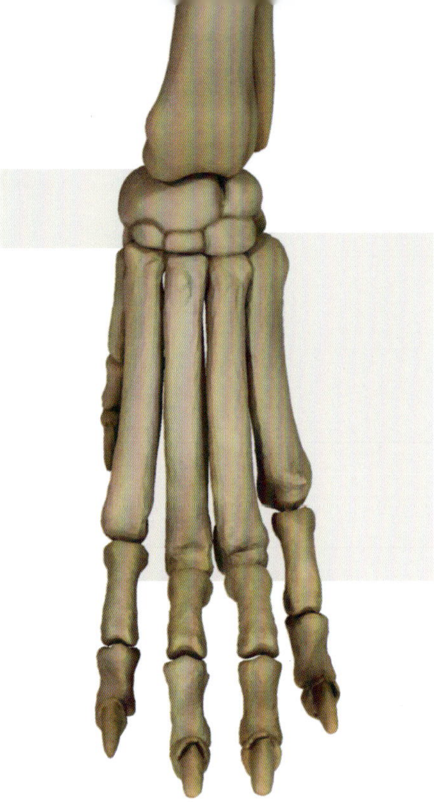

METACARPAL BONES

The first metacarpal bone is much shorter than the other four. The third and fourth are approximately the same length.

PASTERN AND CARPAL JOINTS

All the joints between the forearm and the front pastern (metacarpus) are collectively termed the carpal joint. The carpus is made up of seven separate bones, which is a useful feature for shock absorption, among other things, as the bones can move slightly to avoid colliding with each other when subjected to strain. All the bones in one row usually behave as a uniform articular surface. The most important joints in a longitudinal direction are the radiocarpal or wrist joint, a hinge joint with limited lateral flexibility whose range of flexion and extension amounts to 95° (Roos et al. 1992); the midcarpal joint, that between the first and second rows of carpal bones, another hinge joint with limited lateral movement whose range of flexion and extension amounts to 45°; and the carpometacarpal joint, a taut joint between the second row of carpal bones and the metacarpals whose range of motion amounts to just 10°. Rotational movements such as pronation and supination are not possible in the carpal joint but take place in the distal radioulnar joint, except in Dachshunds, where up to 8° of rotation can occur in the carpal joint. Mai (1995) reports a range of motion of between 165° and 206° for the carpal joint as a whole in Great Danes, German Shepherds, Dobermans, Wirehaired Dachshunds and Longhaired Dachshunds, and Rauscher (1986) gives values of 176-190° for the German Shepherd, Wirehaired Dachshund and Longhaired Dachshund. The standing angle constitutes a hyperextension of the hand which is least pronounced in Great Danes at 202° but more pronounced in the other breeds investigated at between 217° and 219° (Mai 1995). The degree of hyperextension increases with age (Kaiser 2006).

Wrist flexion is associated in most dogs with slight abduction, an "automatism" which turns the front paw outwards shortly after takeoff. Before touchdown the paw rotates inwards in slight supination and usually touches down pointing inwards with the outer edge first. These movements are the result of the interconnections between the various joints involved: wrist flexion is inevitably accompanied by carpal rotation.

As in the other joints, the average size of the contact surface increases with the amount of body weight to be borne (Kaiser 2006), in the wrist joint from 30 to 50 percent, in the midcarpal joint from 28 to 45 percent and in the carpometacarpal joint from 29 to 44 percent. Even under great strain, however, a maximum of only half the articular surface is involved in weight-bearing. The maximum vertical forces reach 60 percent of body weight in the wrist and 62 percent of bodyweight in the carpometacarpal joint (Nielsen et al. 2003).

HIP JOINT

In terms of shape, the hip joint is a ball and socket joint. The semispherical femoral head sits in the acetabulum as it would in a hinge joint, although during locomotion in particular, the range of possible mobility is never exhausted. The acetabulum (from the Latin for vinegar cup) has a Ω-shaped surface, but not all of the cavity is covered with articular cartilage. The almost semispherical femoral head (caput ossis femoris) is nearly completely covered by a layer of articular cartilage several millimeter thick and is significantly larger than the acetabulum. Differences in the hip joint are observed from breed to breed (Richter & Löffler 1976).

The recess of the "Ω" is home to the femoral head ligament, which is more a leftover from the blood supply system than a structure with any real mechanical function. During the femoral growth phase, the femoral head is supplied via blood

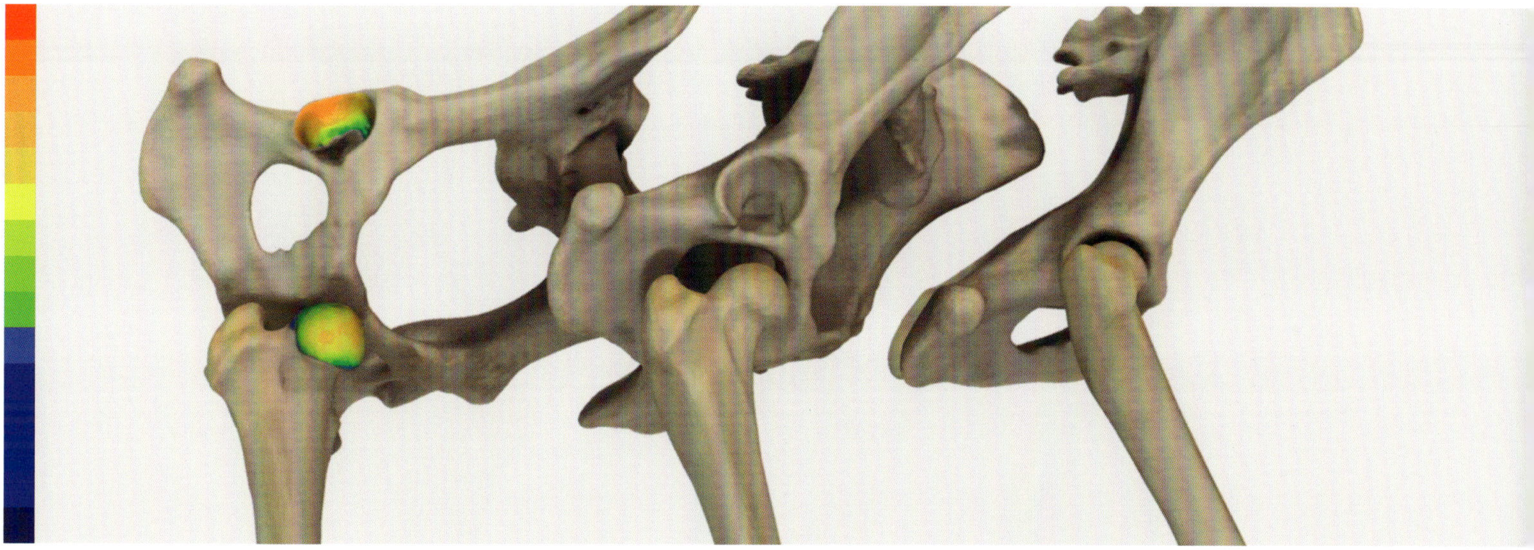

Bone density distribution across the articular surfaces of the hip joint according to Lieser (2003). It is particularly striking in the hip joint how strain is greater on the outer edges of the socket (acetabulum). This is typical of incongruent joints.

vessels which run through this ligament rather than via vessels in the leg.

The hip joint has three axes of motion, some of which, however, are extremely limited in dogs. During locomotion in particular, no significant movement except flexion and extension occurs. The stress-bearing area during locomotion is not the entire surface of the sphere but is limited to a belt around it. As the load increases, the contact area can more than double, as Lieser (2003) found out by investigating the hip joints of 43 mainly large dogs (body mass > 20kg) whose average age was 6. Even under loads of more than four times body weight, just 55 percent of the articular surface of the femoral head is called into play (Lieser 2003).

According to calculations by Shahar & Banks-Sills (2002), the load borne by the hip joint corresponds maximally to body weight during the three-legged support phase at walk. Gait analyses carried out using instrumented hip endoprostheses produced values for four dogs which ranged between 0.8 and 1.65 times body weight (Page et al. 1993). During rapid gaits and when jumping, load levels can reach several times body weight. Prieur (1980) measured in a 30 kg dog moving at the moderate speed of 7 km/h that the

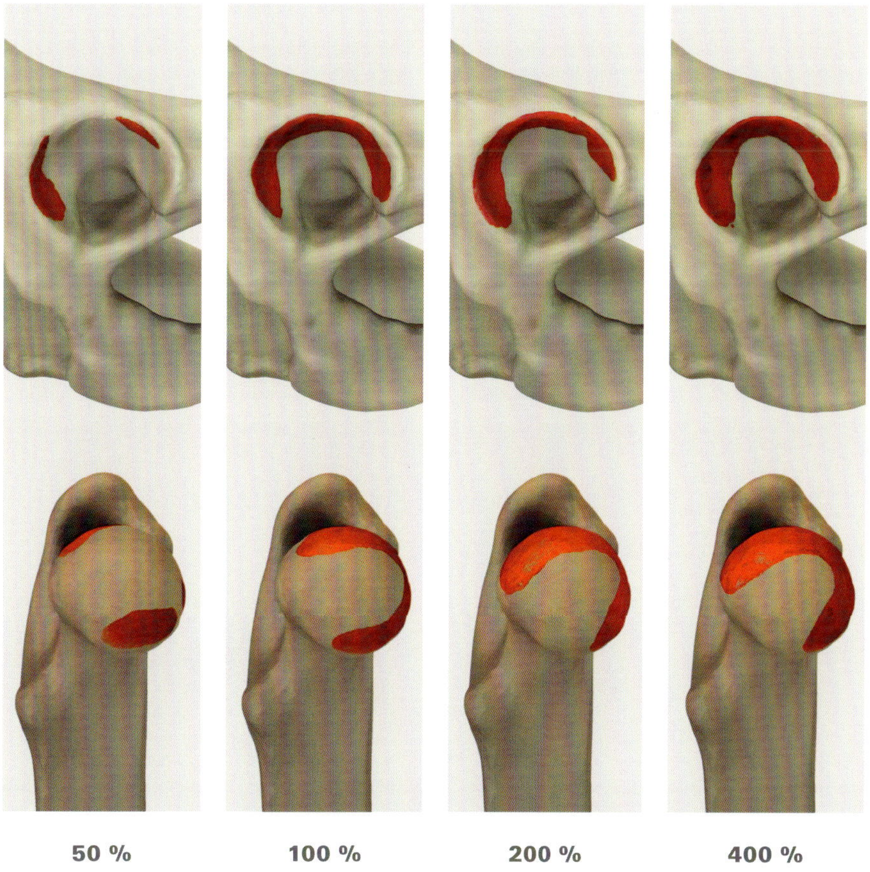

Load-dependent size of the articular contact surfaces in relation to body weight according to Lieser (2003).

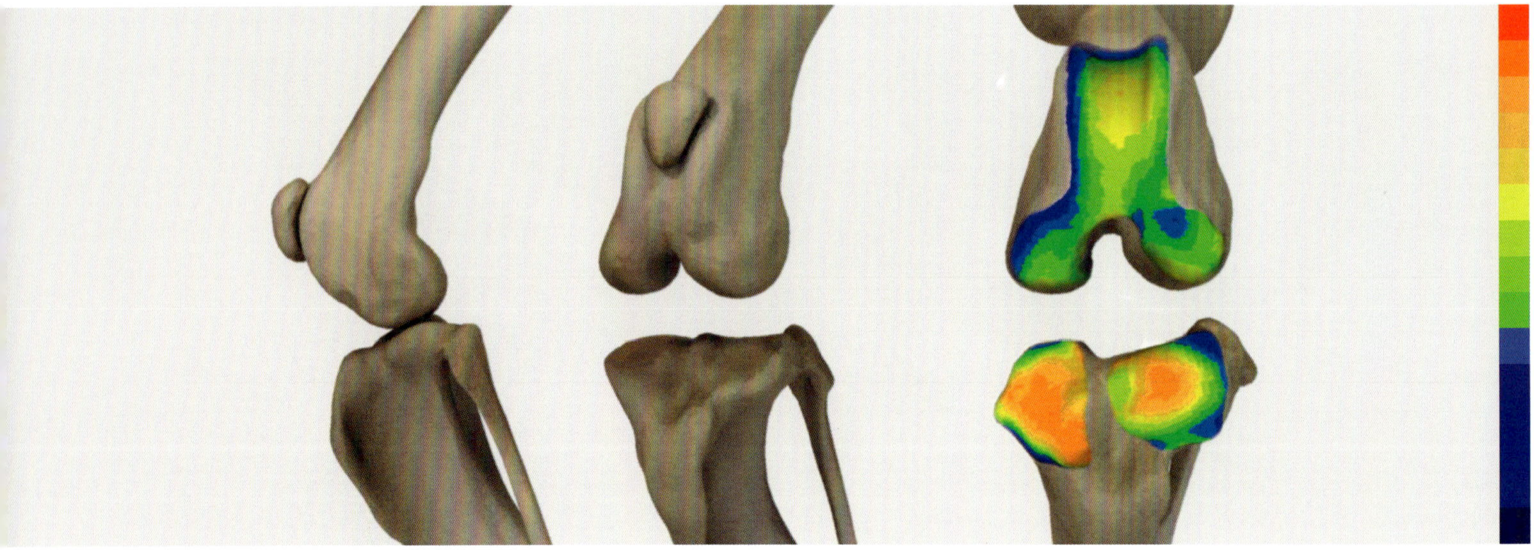

Bone density distribution across the articular surfaces of the knee joint according to Riegert (2004).

load exerted on the hip joint corresponded to six times body weight. This load is, of course, distributed unevenly across the articular surfaces of the femoral head and the acetabulum (Lieser 2003). Load transmission in the acetabulum takes place mainly via the outer edges of the joint. As is always the case in incongruent joints, the greater the load, the more the femoral head stretches and widens the acetabulum. The inner edge of the articular surface only comes into play under very great levels of stress (Lieser 2003).

Strain affects the inner and outer sides of the femur in different ways, compressing the medial side and subjecting the outer side to tensile elongation (Page et al. 1993). This is reflected in the alignment of the trabeculae (Prieur 1980). It is rare in the literature to find information about the standing angles of the hindlimbs, and those values which are provided vary greatly. According to Adrian et al. (1966), the standing angles in various investigated breeds, including Beagles and Collies, are 110° (hip joint), 140° (knee joint) and 150° (ankle joint). According to Lott et al. (1970), the standing angles for German Shepherds and Boxers are 126° and 124° (hip joint), 123° and 133° (knee joint) and 124° and 146° (ankle joint). Newton (1985) claims that the maximum range of motion in the hip is 70°- 80° flexion, 80°- 90° extension, 70°- 80° abduction, 30°- 40° adduction, 50°- 60 inward rotation and 80°- 90° outward rotation. During locomotion, breed-specific patterns of abduction and adduction occur.

Left tibia viewed from above with inner and outer menisci.

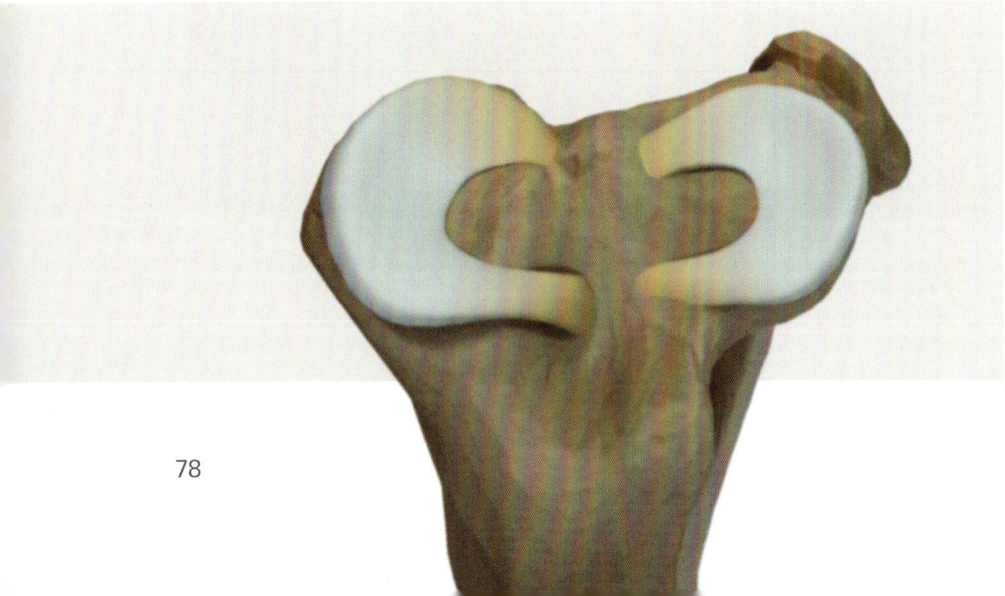

KNEE JOINT

The knee joint is the most complex leg joint, both in terms of its movement and in terms of the way in which it is constructed. It is a chambered joint and consists of three sub-joints, that between the femur and the tibia (femorotibial joint), that between the femur and the patella, or kneecap (femoropatellar joint), and the proximal connection between the tibia and the fibula. Between the articular surfaces of the femur and the tibia, two discs

TARSUS

The tarsus is made up of seven bones (tarsal bones) which are arranged in two staggered rows.

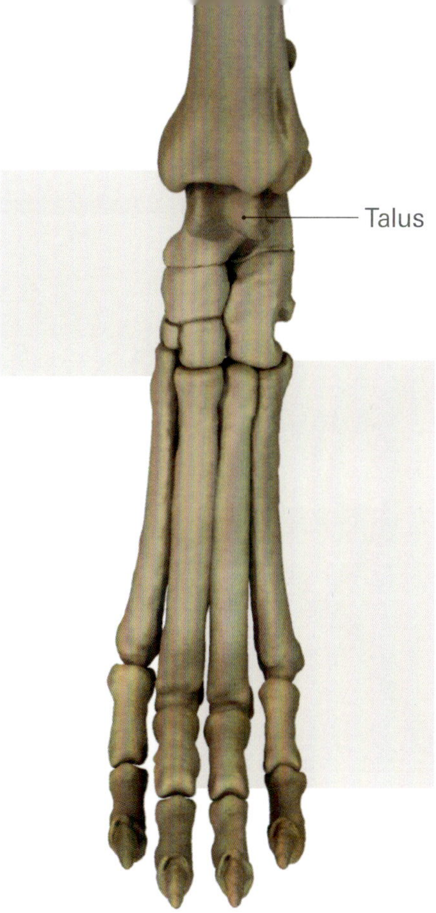

METATARSALS

Dogs only possess four metatarsal bones. The two outer bones and the two inner bones are the same length.

of fibrocartilage known as the menisci (singular meniscus) are located. These go some way to compensating for the incongruence between the extremely circumflex femoral condyles and the almost level articular surface of the tibia.

The menisci have a number of useful functions, "firstly creating consistently congruent, movable and flexible sockets (instantaneous glenoid), secondly having a shock-absorbing effect and acting as elastic cushions during the load-bearing phase, thirdly acting as a wedge to prevent the over-extension and over-flexion of the joint, fourthly functioning as a pressing spring to generate just the right level of tension in the ligaments connecting the femur and the tibia, fifthly enlarging the compressed articular surfaces to improve the distribution of pressure and sixthly ensuring that the joint capsule cannot become trapped in the joint space." This wonderful description by Schreiber (1947, p. 737) says it all.

The motion of the knee joint is a complex combination of gliding, rotation and flexion/extension. The femoropatellar joint is a gliding joint in which the kneecap glides like a sled over the articular surface of the femur. The kneecap is actually a sesamoid bone, an ossification in the end ligament of the femoral extensors. As well as flexion and extension, the knee also permits, depending on its position, a small degree of rotation (Riegert 2004, Vasseur & Arnoczky 1981, Tashman et al. 2003). Extension and flexion occur, necessarily, by means of "drawer movement", for which reason the knee joint is also described as a roll-glide joint. The two menisci slide a whole centimeter forwards when the knee is extended, and move back again when the knee is flexed. The maximum amplitude/range of motion is 130°, and the knee joint is capable of a maximum 150° extension.

Rotational movement in the outstretched knee joint is minimal as it is counteracted by the lateral ligament. When the knee is bent this ligament is completely relaxed and inward rotation is possible. As soon as it is stretched, any inward rotation is "automatically" turned back outwards by the increasing tension in this ligament. The rotation is limited by the medial and lateral ligaments and the cruciate ligaments, and too extreme a rotation can lead to trauma and the tearing of one or several of the ligaments.

ANKLE JOINT

The ankle joint is a composite joint which connects the tibia and fibula with the tarsal bones. Of the various joints which comprise the ankle, the articulation between the tibia and the cannon bone (talus) is the most important for locomotion, forming a cochlear joint which is capable of an approximately 25° deviation from a straight axis.

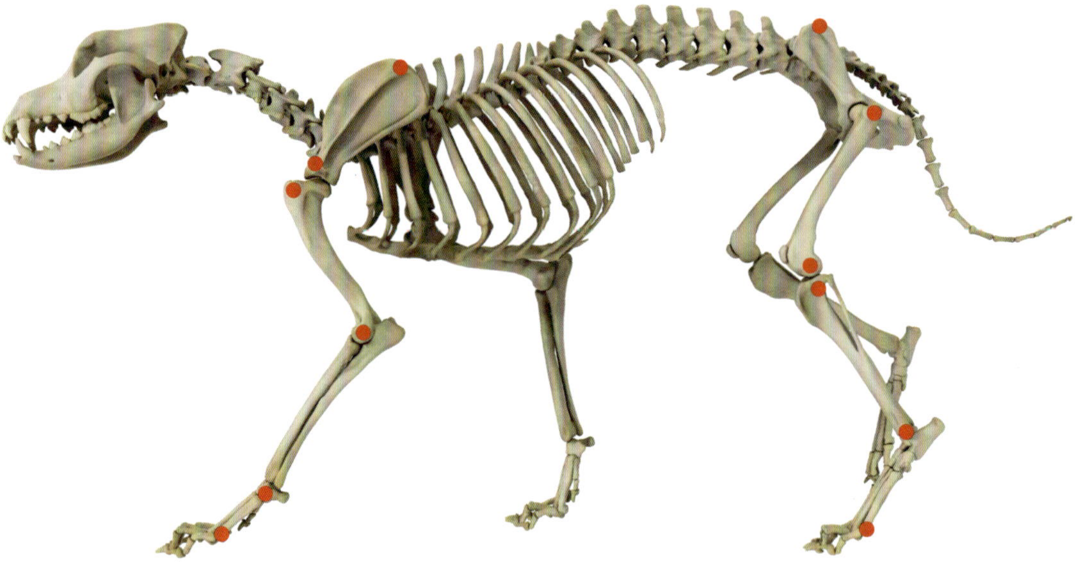

PROPORTIONS OF THE 30 BREEDS INVOLVED IN THE JENA STUDY

Before they were filmed, all the dogs involved in the study were measured. The values presented here for each breed are median values. In addition, we also present the standard deviation values. These provide information about the degree to which the measurements varied between individuals. Where the standard deviation value is high, the individual measurements deviate more significantly from the calculated median value than where the standard deviation value is low.

It is obvious that the upper arm of a Standard Schnauzer will be shorter in absolute terms than that of a Great Dane. In order to calculate whether its relative length is greater or less than that of a Great Dane, however, we need to relate this measurement to the proportions of the entire forelimb. All the limb segments were added together to find the total anatomical length of the limb. The proportions are calculated on the basis of the percentage which each limb segment contributes to the total length. Unlike the anatomical limb length, the functional limb length is the minimum distance between the paw in ground contact and the upper edge of the shoulder blade (scapula) or hip joint. There is a difference, then, between anatomical and functional limb length.

COMPARISON BETWEEN FORE- AND HINDLIMBS

A cross-breed comparison of the ratio between anatomical forelimb length and anatomical hindlimb length revealed surprising uniformity. In all the breeds investigated, the forelimb is somewhat longer than the hindlimb (indicated by quotient greater than one). In the German Shepherd, a breed in which a distinction is made between individuals with a straight back and individuals with a sloping back, only very minor differences are observed between the two lines.

The functional limb length was measured from standing, with the front paws positioned perpendicular to the shoulder and the back paws directly under the hip joint. In a comparison between the functional lengths of the fore- and hindlimbs, the functional length of the hindlimbs was also revealed to be shorter than that of the forelimbs in all the breeds investigated. The functional length of the hindlimbs is the shortest in relation to that of the forelimbs in the two German Shepherd lines, though the differences between other breeds are very minor. Swiss Mountain Dogs have strikingly long hindlimbs, however.

As the anatomical limb length expresses the greatest distance which can theoretically be spanned by a limb, the relationship between anatomical and functional limb length is a measure of the degree of angulation of a limb. If the anatomical limb length is significantly greater than the functional limb length, the dog's limbs display a high degree of angulation. If the two values are similar, the limbs are particularly straight. The most angulated limbs are displayed by Dachshunds and other small breeds. Airedale Terriers and Bernese Mountain Dogs have the straightest legs. Mastiffs have very angulated forelimbs and straight hindlimbs, while the opposite is true of German Shepherds.

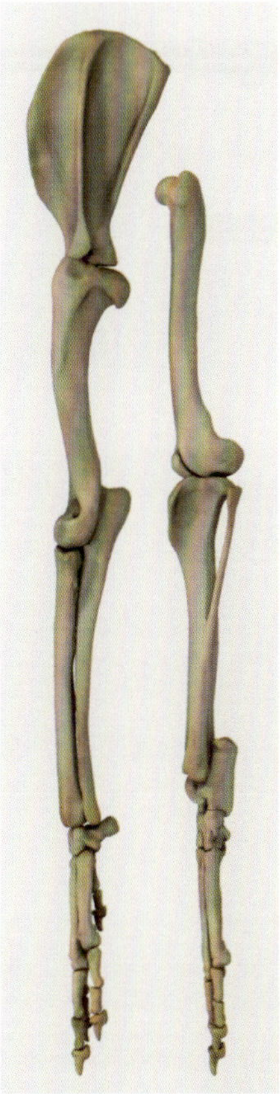

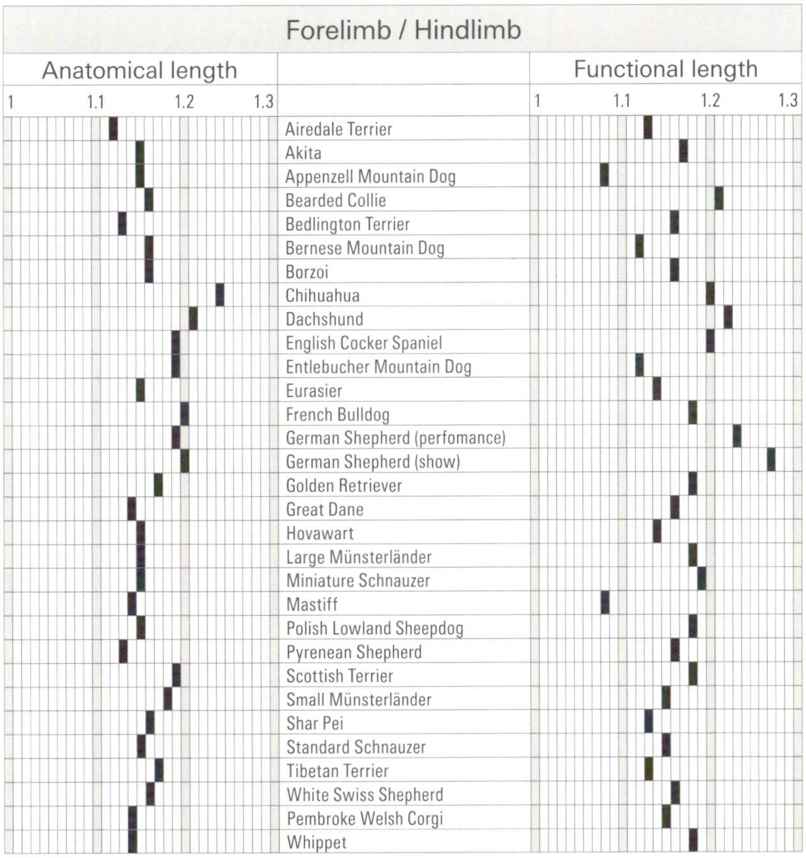
Forelimb / Hindlimb
Anatomical length
Functional length
1
1.1
1.2
1.3
Airedale Terrier
Akita
Appenzell Mountain Dog
Bearded Collie
Bedlington Terrier
Bernese Mountain Dog
Borzoi
Chihuahua
Dachshund
English Cocker Spaniel
Entlebucher Mountain Dog
Eurasier
French Bulldog
German Shepherd (perfomance)
German Shepherd (show)
Golden Retriever
Great Dane
Hovawart
Large Münsterländer
Miniature Schnauzer
Mastiff
Polish Lowland Sheepdog
Pyrenean Shepherd
Scottish Terrier
Small Münsterländer
Shar Pei
Standard Schnauzer
Tibetan Terrier
White Swiss Shepherd
Pembroke Welsh Corgi
Whippet

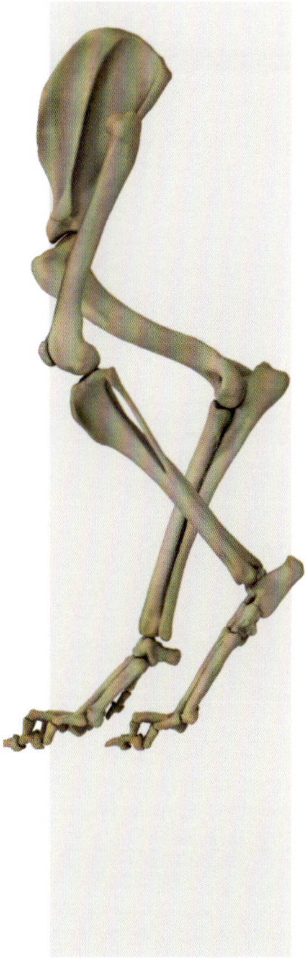

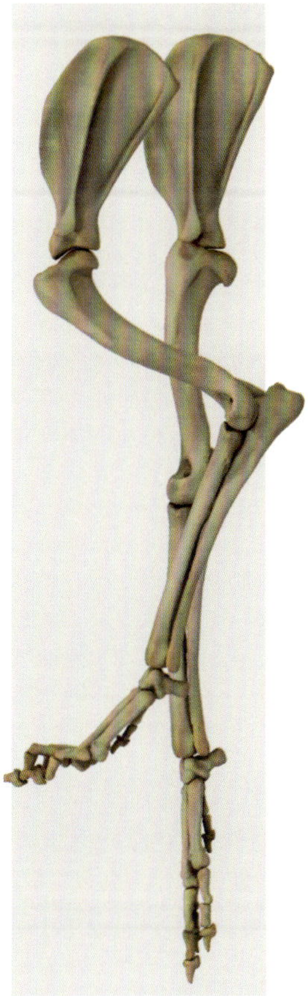

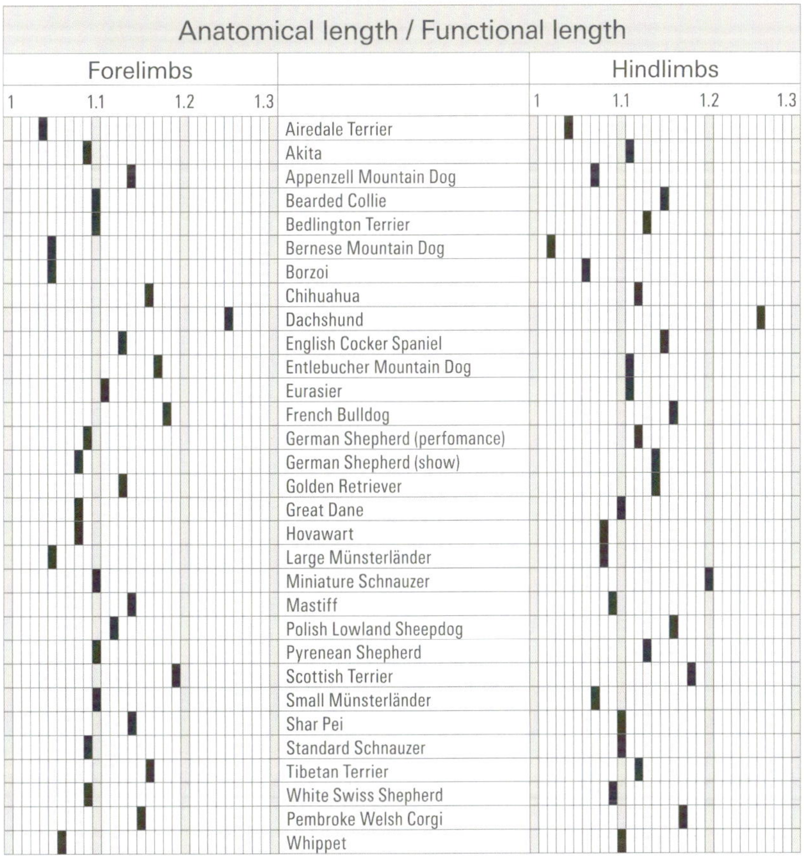
Anatomical length / Functional length
Forelimbs
Hindlimbs
1
1.1
1.2
1.3
Airedale Terrier
Akita
Appenzell Mountain Dog
Bearded Collie
Bedlington Terrier
Bernese Mountain Dog
Borzoi
Chihuahua
Dachshund
English Cocker Spaniel
Entlebucher Mountain Dog
Eurasier
French Bulldog
German Shepherd (perfomance)
German Shepherd (show)
Golden Retriever
Great Dane
Hovawart
Large Münsterländer
Miniature Schnauzer
Mastiff
Polish Lowland Sheepdog
Pyrenean Shepherd
Scottish Terrier
Small Münsterländer
Shar Pei
Standard Schnauzer
Tibetan Terrier
White Swiss Shepherd
Pembroke Welsh Corgi
Whippet

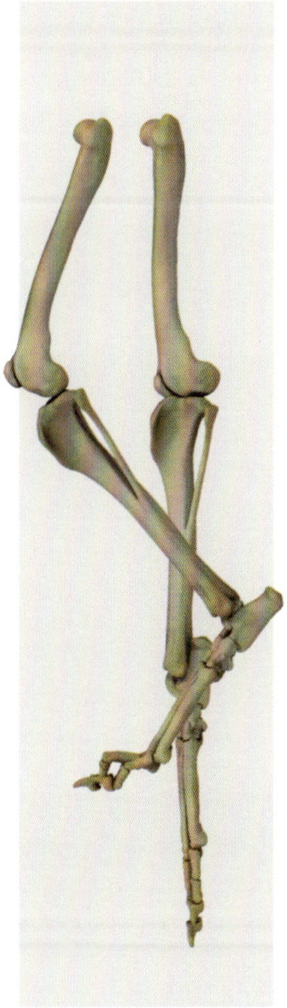

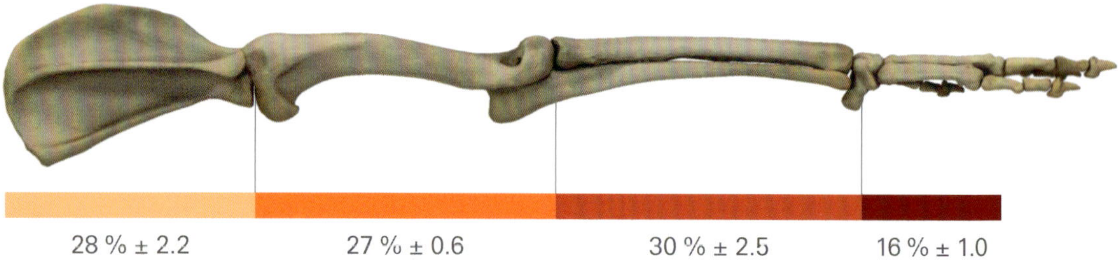

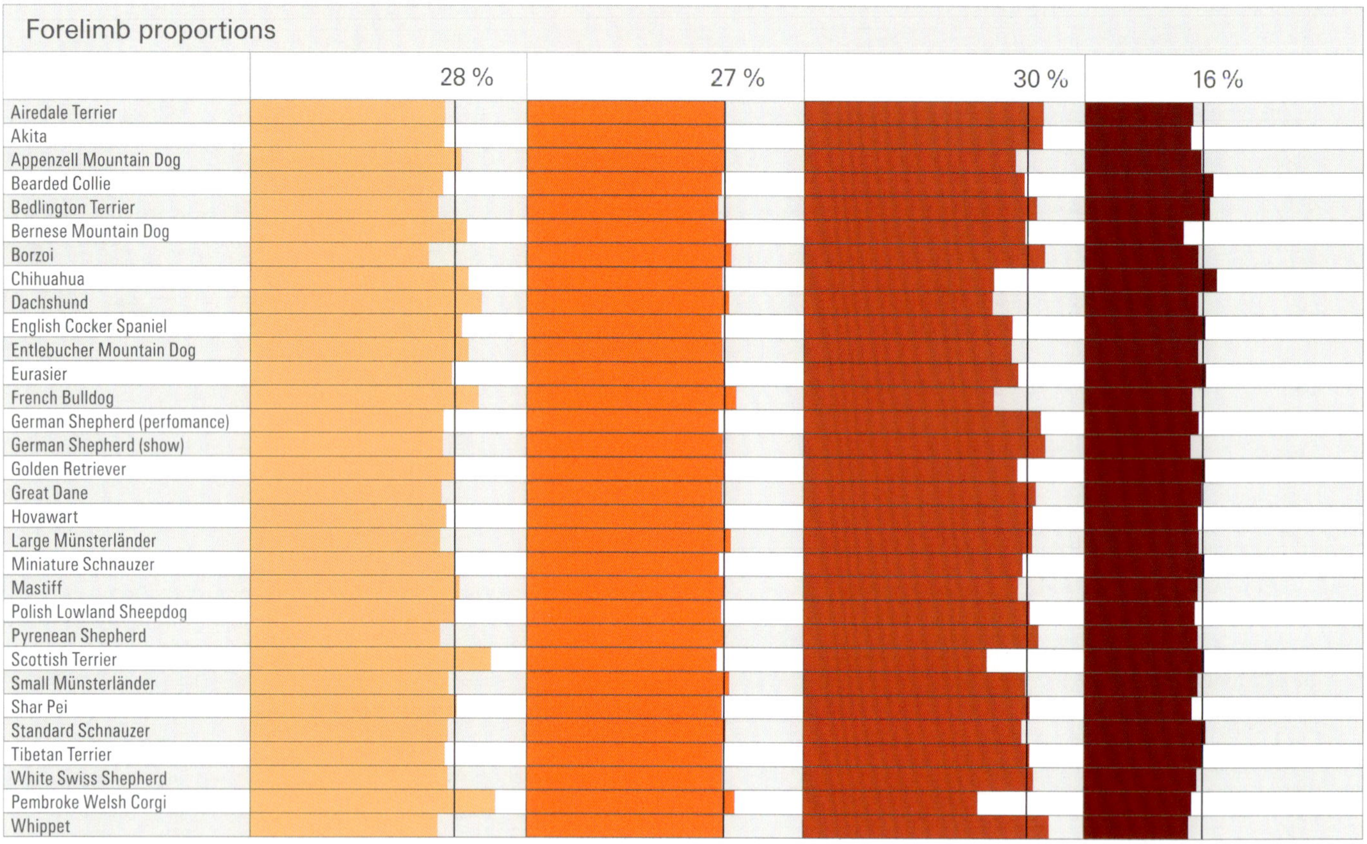

FORELIMB PROPORTIONS

On average, the shoulder blade contributes 28 percent to the anatomical limb length. In small breeds, the shoulder blade is proportionally longer. In the Pembroke Welsh Corgi it contributes 34 percent, in the Scottish Terrier and the Dachshund 33 percent and in the French Bulldog 31 percent to total anatomical limb length. Borzois have the smallest shoulder blade, making up 24 percent of anatomical limb length. The contribution of the shoulder blade in all the other 25 breeds deviates by just 2 percent from the median value. The median contribution of the upper arm is 27 percent of the anatomical limb length. Standard deviation is just 0.7, which indicates that the values are fairly uniform across all the breeds. There is no obvious link between the proportional length of the upper arm and the size of the dog. The longest upper arm, relatively speaking, is that of the French Bulldog, which contributes 29 percent to anatomical limb length. The shortest, that of the Scottish Terrier, contributes 26 percent.

The median value for the forearm is 30 percent of anatomical limb length, though this is the forelimb segment which varies the most in terms of proportional length, with standard deviation at 2.5 percent. The breeds which have a proportionally long shoulder blade have a proportionally short forearm and vice versa. Both German Shepherd lines, the Akita and the Airedale Terrier belong to the breeds with proportionally long forearms, while Swiss Mountain Dogs (with the exception of the Bernese Mountain Dog) are among those with proportionally short forearms.

The pastern (metacarpus) makes up 16 percent, on average, of the anatomical forelimb length. The standard deviation is 1.0, which means that none of the individual values differs greatly from the median. The Bernese Mountain Dog has the proportionally shortest pastern, while the Chihuahua has the longest at 18 percent of anatomical forelimb length.

These measurements show that the forelimbs of all breeds consist of two relatively uniform segments, the upper arm and the pastern and two variable segments, the shoulder blade and the forearm.

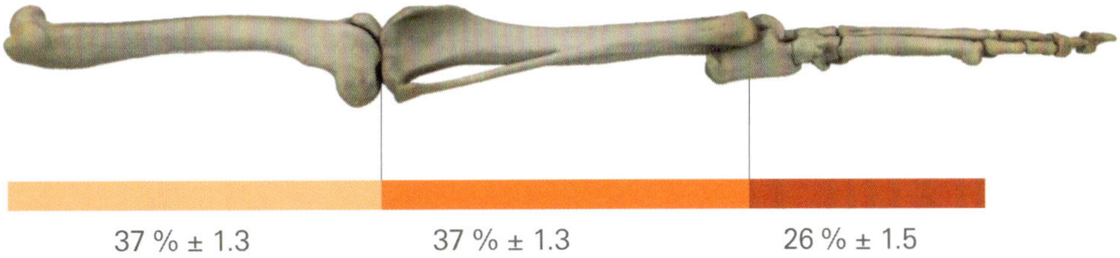

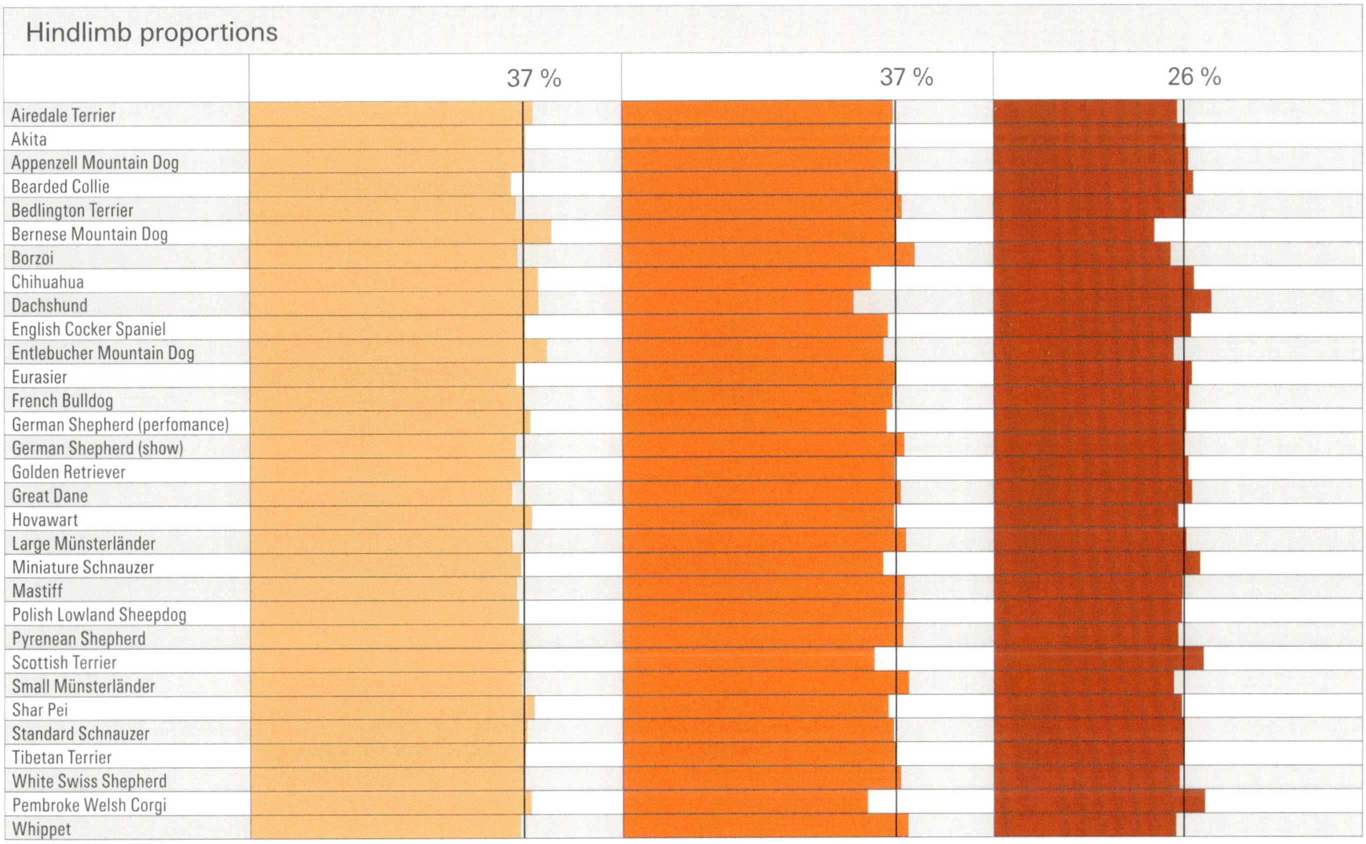

HINDLIMB PROPORTIONS

The femur contributes 37 percent to anatomical hindlimb length on average, a value which is relatively uniform across the breeds, with standard deviation at 1.3. The medium-sized breeds Entlebucher Mountain Dog and Bernese Mountain Dog have the proportionally longest femur at 41 percent, and the Chihuahua and the Dachshund have the second longest at 39 percent of anatomical limb length. This shows that the proportional length of the femur is not related to the size of the dog. Large Muensterlaenders, Great Danes and Bearded Collies have the proportionally shortest femurs at 35 percent of total anatomical limb length.

The median value for the lower leg is, like that for the femur, 37 percent of total anatomical length, though the standard deviation is higher which means that there is greater variation in the individual values observed. Dachshunds, Pembroke Welsh Corgis, Chihuahuas and Scottish Terriers have the proportionally shortest lower leg; Borzois, Whippets, and Large and Small Münsterländers the proportionally longest.

Although the median values for the proportional length of the femur and lower leg are the same, the two elements were only actually approximately the same length in just over a third (12 of 30) of the breeds investigated. In the two German Shepherd lines, the proportional length of the femur and that of the lower leg differed by two percent. In the show line the proportional length of the femur is two percent greater than that of the lower leg, and vice versa in the performance line. The contribution of the hind pastern (metatarsus) to anatomical limb length is 26 percent. The short-legged breeds Dachshund, Pembroke Welsh Corgi, Scottish Terrier and Chihuahua have the proportionally longest metatarsus, though the proportions are similar in the much larger Bearded Collie, Golden Retriever and Great Dane. The proportionally shortest metatarsus is that of the Bernese Mountain Dog.

The three segments of the hindlimb all display a similar level of variability, as illustrated by the uniformity of the standard deviation values.

MEDIAN VALUES

It is important to remember that median values sometimes fail to reflect the real range of variation. Usually, the variability within a breed (the range of actual values obtained for each dog) is greater than the differences between the median values given here. The proportional length of the upper arm, for instance, varied within the French Bulldogs we measured between 26 and 32 percent, and within the Scottish Terriers between 21 and 30 percent. The median values, on the other hand, imply that French Bulldogs have the longest upper arm, relatively speaking, and Scottish Terriers the shortest. In reality, however, there are many Scottish Terriers with a proportionally longer upper arm than many French Bulldogs. The same applies to the proportional lengths of the femur and lower leg in the two German Shepherd lines. The median values imply that the femur of the show line is proportionally longer and its lower leg proportionally shorter than those of the performance line. The actual values, on the other hand, overlap. In the show line, the proportional length of the femur ranges between 33 and 39 percent, and that of the performance line between 36 and 41 percent. The relative length of the lower leg varies in the show line between 36 and 41 percent and in the performance line between 32 and 39 percent.

A consideration of the proportional length of the limb segments allows the following conclusions to be drawn: the proportional length of the forearm is always greater or equal to the proportional length of the shoulder blade. The only exception to this is seen in the small, short-legged breeds in which the shoulder blade is proportionally longer. This comes down to changes in bone growth. Because the bone stops growing so early in these breeds, juvenile proportions are maintained.
The proportional length of the femur and lower leg is almost the same. Unlike in the case of the forelimb, the breeds which deviate from this pattern cannot be allocated to any clear group. In the small breeds Chihuahua, Dachshund, Scottish Terrier and Pembroke Welsh Corgi the femur is proportionally longer than the lower leg, but this is also the case in the Entlebucher Mountain Dog, the Bernese Mountain Dog and the Shar Pei. In Borzois and Large Münsterländers, on the other hand, the lower leg is proportionally longer than the femur.

PROPORTIONS OF OTHER CANIDS

The dog's descent from the wolf is still evident in its body proportions. The results obtained by Wayne (1986b), who carried out a study into the proportions of 118 dogs of 65 different breeds and who investigated the skeletons of 27 wild canid species, show that most of the proportions found in domestic dogs correspond to those found in wild species of the same body size.
The long bones of the limbs behave isometrically in all canids. If a jackal is X percent smaller than a wolf, its upper arm and forearm and femur and lower leg will be smaller than the limb segments of the wolf by exactly the same percent. The length of the long bones thus decreases and increases with geometric similitude (see p. 30).
Lumer (1940) came to recognize the size-independent relationship between the long bones in a study in which he measured 146 dogs of 18 different breeds. He was the first to show that the radius and the upper arm and the tibia and the femur scale at 1:1 (1.098 to 1.090). In smaller dogs with a femur length of less than 12.5 cm, however, the metacarpal bones become longer than average and the metatarsal bones shorter.
Casinos (1986), who took measurements from 108 dogs of 63 different breeds and from 12 wolves, confirmed the geometric similitude of the long bones in canids. However, his results indicated that the femur deviates slightly from this relationship of similarity. The study carried out by Müller (1955) into 63 dogs of 21 different breeds and that performed by Unkel-Mohrmann (1999) into the German Shepherd showed the same thing. An investigation into Dachshunds by Detels (1980) revealed that the length of the upper arm accounted for 89 percent of the length of the shoulder blade, while in the Dachshunds we measured, the average was 87 percent. It is not possible from a comparison of just two lengths to say which of the two has been reduced or extended – this requires knowledge of the relationship to anatomical limb length. When this is taken into account, it becomes clear that in Dachshunds the upper arm has not become shorter in relation to the shoulder blade, but that the shoulder blade has become relatively longer. Detels (1980) showed that in Shorthaired and Wirehaired Dachshunds the femur is 16 percent longer than the lower leg, in the Long-haired

Dachshund it is 14 percent longer, and in the Dachshunds in our study it was 20 percent longer.
The shoulder blade, which is not a long bone, deviates from the geometric relationship described above. In large breeds, the length of the shoulder blade corresponds to that in wolves, but smaller breeds have a disproportionately long shoulder blade – longer than that in wild canids of a similar size (Wayne 1986b). This applies to all small and medium-sized dog breeds. The proportional length of the shoulder blade in Poodles, for instance, is 10 percent longer than in wolves, and in the Boxer it is almost 17 percent longer (Brückner 1971). In general it can be said that statements about whether the upper arm is proportionally long or proportionally short have no validity when they are based on a simple comparison of the upper and forearm and do not take into account the anatomical length of the entire forelimb including the shoulder blade. Only then does it become clear that it is the length of the upper arm which is the most constant. Scapula and lower arm length can vary according to body size.
The literature contains a number of other misapprehensions which, once written down, become widely accepted and lead to long-lasting misunderstandings. Brown (1986, p. 27), for example, writes that the shoulder blade of a dog with limbs of normal length is always 10 - 20 percent shorter than its upper arm. Aldington Stockmann (no date, p. 138) claim that "the humerus is *longer than the shoulder blade* in all breeds (with a small number of exceptions)" (italics original). The measurements we obtained for our 327 dogs show, however, that this is only the case in five of the 30 breeds investigated. In ten breeds the shoulder blade and upper arm are the same length and in 15 breeds the shoulder blade is longer than the upper arm. Aldington Stockmann (no date, p. 146) also claim that "on average, the hind legs are always around 6 – 10 percent longer than the forelimbs". This is only true when the shoulder blade is not taken into account in the calculation of limb length, yet the authors claim on p. 79 following that the shoulder blade is included in their calculations. Other discrepancies with our data are found in the statement that "*in actual fact there are no differences between the breeds in the relationship between the length of the femur and that of the hindlimb*" (italics original), in the claim that "the femur contributes 42 percent on average to the total length of the hindlimb", and in the assertion that "the lower leg is slightly shorter than the femur in most breeds" (all citations p. 146).
Even the common division of dogs into gallopers and trotters on the basis of differences in their proportions must be viewed critically in the light of the present study, which reveals that the show line of the German Shepherd ranks between Whippets and Borzois in terms of forearm length.
According to Aldington Stockmann (no date), the terrier front is created by a proportionally short upper arm, which may hold true for the Bedlington Terrier but does not apply to the Airedale Terrier, which in actual fact has the most elongated front limbs of any dog so far investigated.

DIFFERENCES IN PROPORTIONS ARE DIFFERENCES IN GROWTH

Differences in limb bone proportions in dogs are the result either of differences in the duration of the growth phase or differences in the intensity of growth during a defined growth phase. Schlegel discovered in 1913 that the proportional lengths of the long bones (humerus, ulna and radius, femur and tibia) in newborn puppies of different breeds are more or less the same – even in the Dachshund. The femur and upper arm are the same length, the radius is approximately 20 percent shorter and the tibia 10 – 15 percent shorter. This shows that the femur and the upper arm grow more quickly or more intensively in utero than the radius and tibia. The same values are observed in wolf and jackal cubs (Coppinger Coppinger 2001). In adult wolves the upper arm and the radius are, like the femur and the tibia, exactly the same length. The radius and the tibia must therefore grow more quickly or for a longer period after birth than the upper arm and femur, until they have caught up and reached the same length (Schlegel 1913, Weise 1965).
The differences in the proportions of the long bones in different breeds do not come about until the postnatal phase of bone growth. Selective breeding indirectly extends the growth process or stops it prematurely at different points in different breeds. In dogs whose forearm is shorter than their upper arm when fully grown, the growth process is interrupted early. These animals display puppy-like proportions. In breeds with longer forearms or lower legs, however, the growth process is extended.
Only a few studies into the postnatal growth of the long bones in dogs are based on measurements which really begin on the day of birth. Results are available for the Basset Hound, Beagle, Dachshund, Greyhound and Miniature Poodle (Hitz 1973, Wayne 1986a, b, Weise 1965). A comparison between Greyhounds and Beagles reveals that at birth, the total length of the femur of the two breeds differs by just 20 percent. After 50 days, the length difference has reached 203 percent, with the Greyhound femur now twice the length of that of the Beagle. However, by the time the dogs are fully grown, the difference has decreased to 183 percent. These results show that the rate of growth of the femur differs in these two breeds. Wayne (1986a, b) argues that growth begins later in the Beagle, rather than generally being slower overall.
In small and medium-sized breeds, growth temporarily ceases directly after birth, while in larger breeds, the limb segments continue to grow uninterrupted. This means that larger breeds always have a head start on smaller ones, although their growth curves do eventually follow an identical course (according to Wayne 1986b, this is the case from 40 days after birth).
The most thorough study into the postnatal growth of dogs was carried out by Weise (1965), who investigated a litter of Miniature Poodles and various Longhaired Dachshunds. The results indicate that growth is subject to a high level of variability. The growth curves of the Poodle littermates were not identical but revealed differences in the speed of growth – periods of slow growth and growth spurts

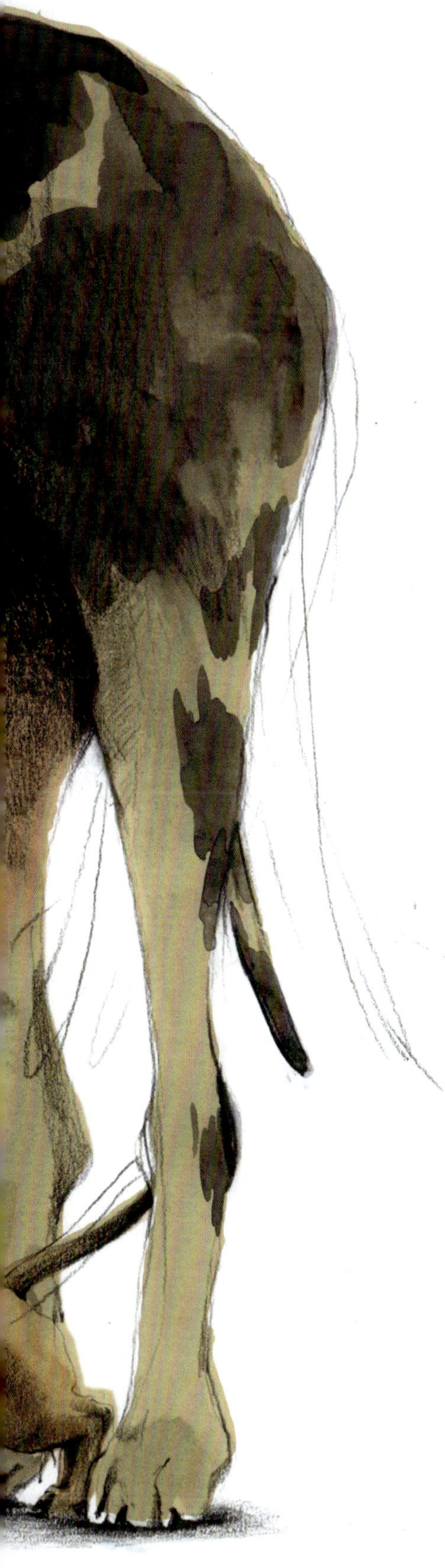

which occurred at different times in different individuals. After one month, the difference between the smallest and the largest pup of the litter was five centimeters, and after eight and a half months this gap had increased to 16 centimeters. Furthermore, the point at which growth finally stopped also differed from individual to individual. The first pup of the litter was fully grown after 166 days, the last after 220 days. However, the dog with the longest growth phase was not the largest in the litter: the largest animal was fully grown after 177 days. The duration of the growth phase is, therefore, not the decisive factor in the size of a dog.

The duration of the growth phase also differs from limb segment to limb segment. The upper arm of one of the littermates was fully grown at 14 centimeters after four and a half months, while another of the pups needed six months for its upper arm to reach the same length, this particular segment having experienced a one and a half month break in growth. These differences in growth rates mean that the upper arm and forearm can reach their adult size at the same time, though generally the forearm grows for a longer period than the upper. The pastern stops growing between three and a half and six months after birth. Weise confirms that the observations made in relation to the forelimbs also apply to the hindlimbs.

The results of an investigation into different litters of Miniature Poodles from the same kennel show that the upper arm stops growing between 140 and 243 days after birth, the forearm between 177 and 243 days after birth, the metacarpal and fingers between 105 and 176 days after birth, the femur between 124 and 224 days after birth, the lower leg between 117 and 243 days after birth and the metatarsal and toes between 116 and 176 days after birth. These results tied in with those obtained for the Longhaired Dachshunds, in which Weise (1965) observed uncoordinated series of growth spurts and growth stops in the various limb segments. Despite the size difference between Dachshunds and Miniature Poodles, the limb segments of the Dachshund are not uniformly smaller than those of the Miniature Poodle. The pastern and toes are only slightly shorter, while the upper and forearm and lower leg and, in particular, the femur, are much shorter. As mentioned above, the point at which growth stops does not seem to be connected with the size of the animal. "Sudden spurts in lengthways growth can temporarily or permanently change the size rankings of littermates. 'Giants' and 'dwarves' can usually only be recognized relatively late as such." (Weise 1965 p. 265).

According to a study carried out by Schroeder (1978) into two German Shepherd litters, significant differences between littermates arise in this breed, too. These differences stem from the variability in the point at which the growth plates (epiphyses) close. There is great variability even in the point of closure of the proximal and distal growth plates of the same bone. The proximal growth plate of the radius closes between six and a half and nine and a half months after birth while the distal growth plate of the radius and the single (distal) growth plate of the ulna close between nine and eleven months after birth.

The limb proportions of dogs, to conclude this chapter, correspond to those of wolves and other canids of the same size. Canids as a whole display unusual uniformity in the proportional lengths of their limb segments. The ratio of the various segments of a limb (shoulder blade, upper arm, forearm, hand; femur, lower leg, foot) to its total anatomical length remains constant not just among individuals of the same breed, but hardly differs between breeds either. The functional middle segment in particular (upper arm, lower leg), and here the upper arm especially, is practically the same length, proportionally speaking, in all dogs. Regardless of whether a dog has long forelimbs or short ones, the proportional length of the upper arm is always the same. Changes in proportions, as seen for example in some short-legged breeds, are always the result of postnatal differences in growth. At birth, dogs hardly differ in the relative lengths of their limb segments. The post-natal lengthways growth of the limb bones can differ significantly however, in terms of duration and intensity, even among pups from the same litter.

MOTION SEQUENCE

Locomotion is a continual, cyclical sequence of repeated movements. The movement performed by the forelimb is essentially the same in all gaits, while the movement of the hindlimb differs between the symmetrical gaits walk and trot on the one hand and the gallop on the other. A motion cycle consists in dogs of the retraction (backwards movement) of the limbs during the stance phase and their protraction (forwards movement) in the forward swing phase. In terms of locomotion, limbs consist of three potentially propulsive elements. In the forelimbs these are, in decreasing order of importance the ground distance traveled by the limb, the shoulder blade, upper arm and forearm. The forefoot can be neglected. In the hindlimb these elements are the femur, lower leg and hindfoot.

PIVOT POINT OF THE LIMBS

For the fore- and hindlimbs to be able to work uniformly, their pivot points need to be located on a level with each other. The pivot point of the hindlimb is located in a form-fitting joint, the hip joint. As the forelimb is only connected to the trunk via muscles, it does not move round a fixed point but round what is known as an instantaneous center of rotation, which in

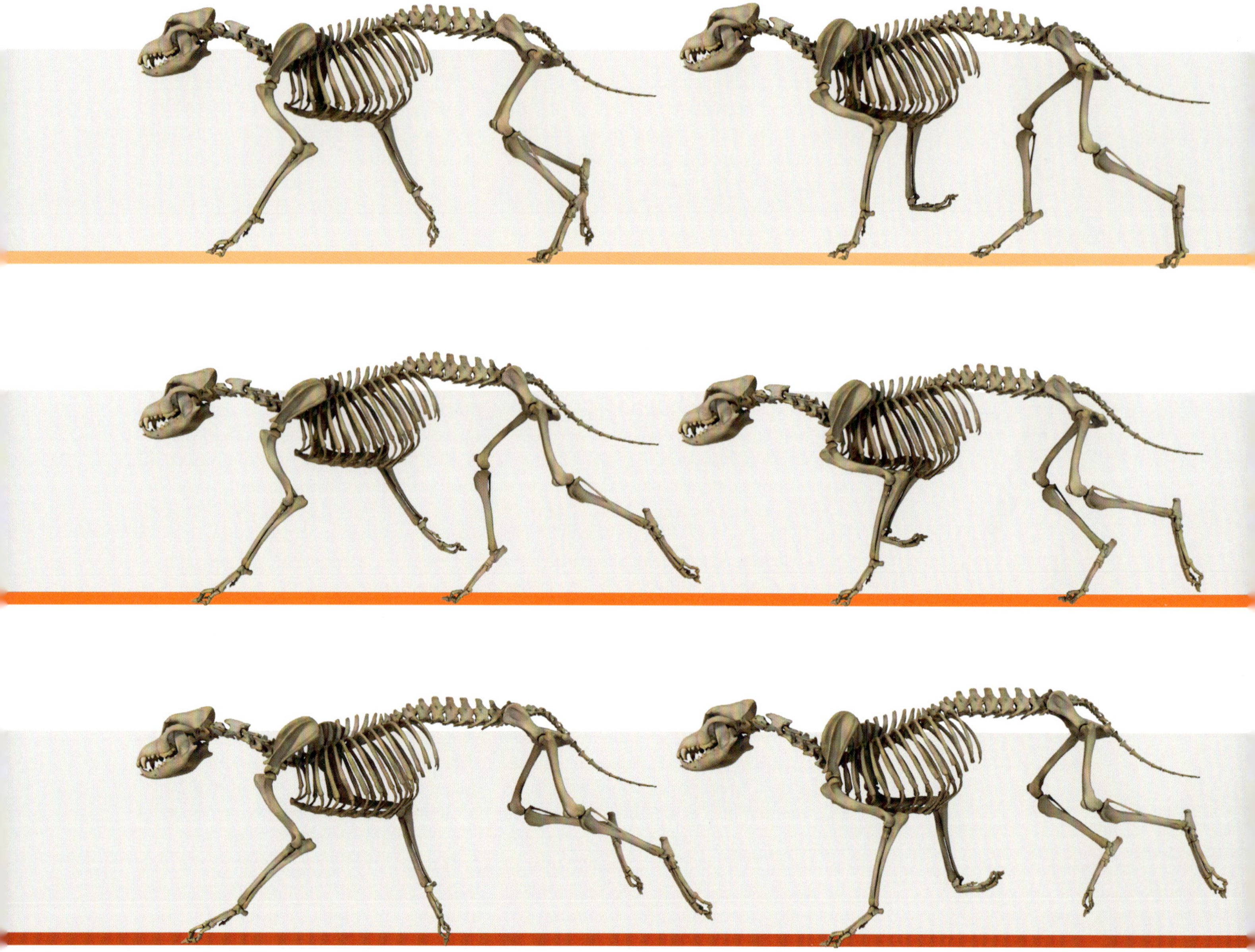

the X-ray film can be identified with a fair amount of precision on the upper edge of the shoulder blade. The instantaneous center of rotation is the equivalent of a force transmission joint, in which the position of the center of rotation is a function of the level of activity of the various extrinsic shoulder muscles and the force impulse on the limb. In kinematics, an instantaneous center of rotation is said to exist when the motion of a body consists of a combination of translation and rotation. A translation of the shoulder blade along the thoracic cage can be observed at the end of the stance phase in particular. The scapular pivot point moves up and down during a step cycle. The pivot points of the limbs on the same side of the body change height at the same time. When the hip is low, the front pivot point is also low.

STANCE PHASE

The motion that takes place during the stance phase consists of the largely uniform retraction of the limbs and their three elements at almost constant angular velocities (Owen et al. 2004). The first and third elements move in parallel. Fischer & Witte (1998) described this principle of movement as a "pantograph leg". As in this once-popular drawing aid, which was

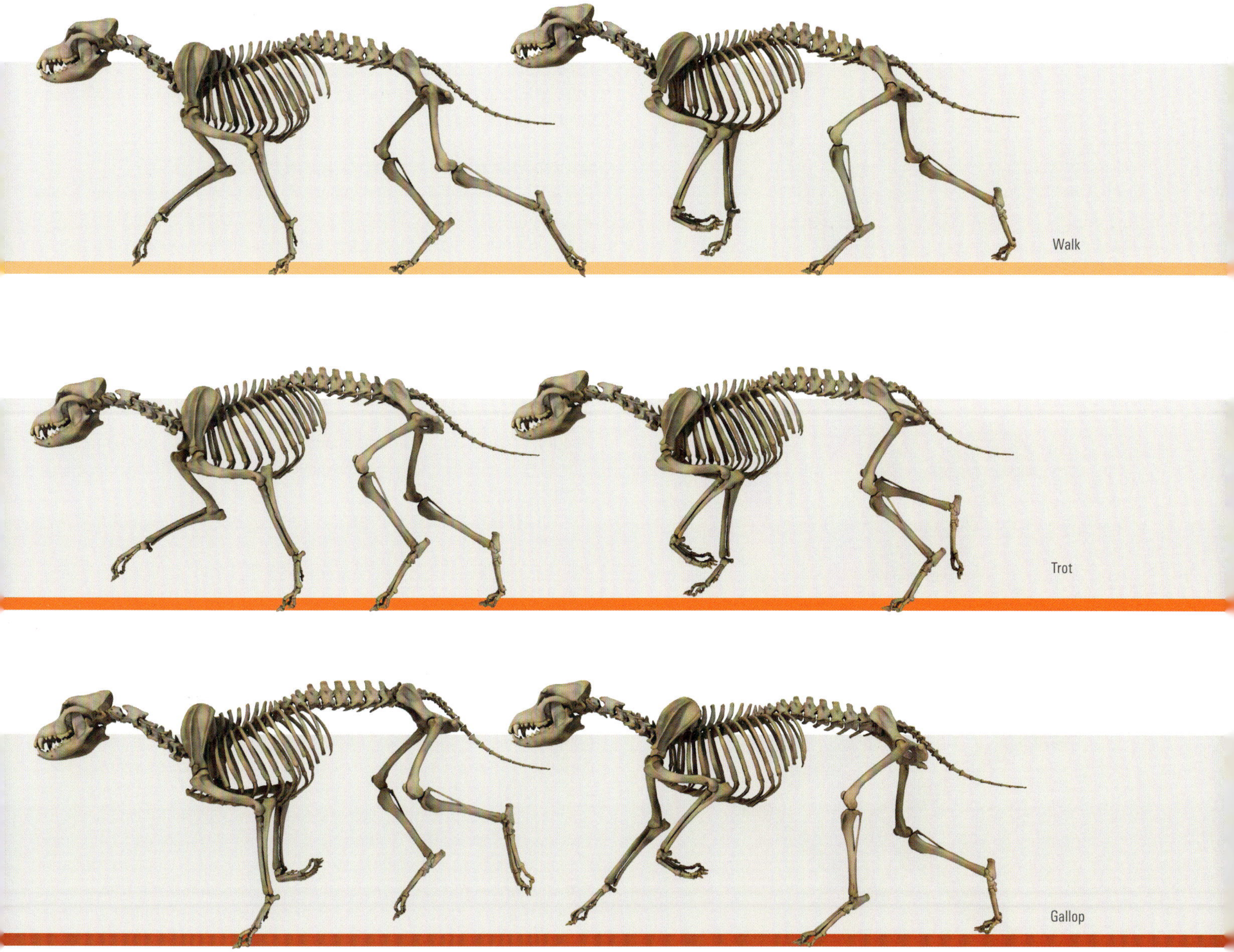

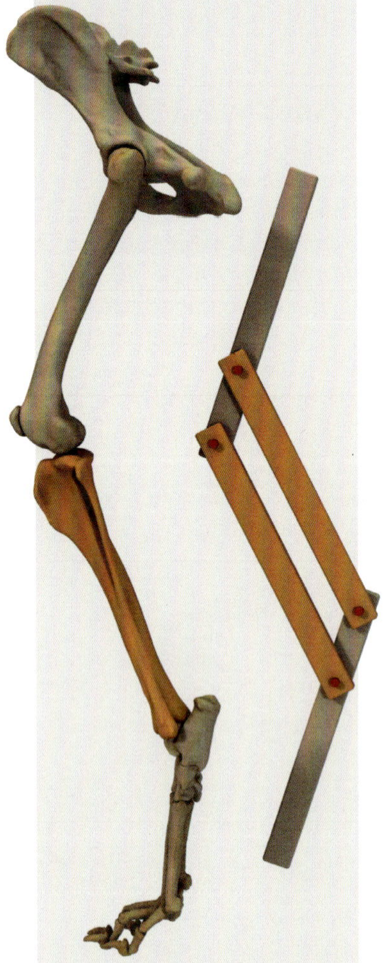

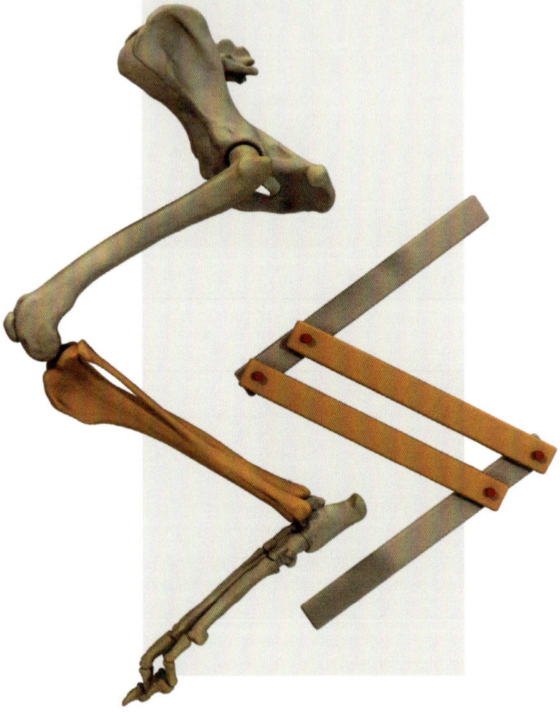

used to create enlargements of existing figures, the excursion of the first element of the limb (shoulder blade or femur) is simply copied, on a larger scale, in the excursion of the third element (forearm, hind pastern). The length and angulation of the middle element determine the degree of enlargement that will take place. As the middle element (upper arm, lower leg) is approximately the same proportional length in all dogs, the excursion of the third element is also the same. The length ratio between the first and the third element influences the force needed to move the third element. The same mechanical principle thus applies to both the fore- and the hindlimb.

Our knowledge of the fact that the first and third element are parallel is helpful in identifying the "hidden" elements shoulder blade and femur. The position of the forearm or hind pastern at touchdown and lift off can provide fairly precise information about the position of the shoulder blade or femur.

The basic motion sequence on which a step cycle is based can be described as the rotation of the fore- and hindlimbs, with the first and third elements moved in parallel, from their touchdown position of between 55° and 65° to a lift off position at or beyond the vertical. The middle segments (upper arm, lower leg/tibia) begin in a relatively steeply inclined position and have moved by lift off to a more (upper arm) or less (lower leg/tibia) horizontal position.

MAXIMUM AND EFFECTIVE AMPLITUDE

The angular displacement of analog elements, e.g. the shoulder blade and femur, are not identical. The amplitude of the movement of the femur is greater than that of the shoulder blade, which is limited by its solely muscular attachment. The upper arm, however, has a greater amplitude than the lower leg, which means that the total of both ranges of motion adds up to the same.

The range of motion can be described in its maximum amplitude or its effective amplitude during the stance and forward swing phase. Effective amplitude is the difference between the angle of touchdown and the angle of lift off. If a joint is highly flexed during the stance phase, for instance, but extended again for lift off, its maximum amplitude is large but its effective amplitude small. Only the effective amplitude contributes to the propulsion of the body during locomotion.

The difference between maximum and effective amplitude shows whether the movement of a joint contributes more to propulsion or to height leveling – the smaller the difference the more effective a joint is in terms of propulsion. The hip joint, like the rotation at the upper edge of the shoulder blade, boasts an effective amplitude of over 90 percent. In dogs, the movement of the hindlimbs comes more or less straight from the hips, with the hip joint able to rotate by 40 - 50°. The effective amplitudes of the knee and ankle joint, on the other hand are, at 10° and 20°, astonishingly low. The main function of the latter two joints is not the contribution to the ground distance traveled by the limb but flexion and extension during the forward swing phase, and height leveling during locomotion over uneven substrates. The topline of a dog moving on uneven terrain remains astoundingly level. The joints responsible for the compensating differences in elevation (in the forelimb the shoulder and, in particular, elbow and in the hindlimb the knee and ankle) act like shock absorbers and compensate for the bumps, allowing the pivot points to remain steady and level with one another.

Contribution of different limb elements to step length in percentage

Walk / Trot	Shoulder blade		Upper arm		Forearm		Forefoot		Pelvis		Femur		Lower leg		Hind foot	
	Walk	Trot	Walk	Trot	Walk	Trot	Walk	Trot	Walk	Trot	Walk	Trot	Walk	Trot	Walk	Trot
Biewer Terrier	69	65	13	15	11	13	8	8	5	6	73	84	12	-6	10	15
Chihuahua	66	63	36	41	-10	-3	7	-1	13	3	77	87	-2	-5	12	16
Dachshund	65	70	13	8	19	15	4	7	12	12	73	66	-2	11	16	11
Pembroke Welsh Corgi	69	-	24	-	1	-	6	-	12	13	73	73	10	10	4	4

SWING PHASE

The speed with which the limbs move is low at the beginning and end of the forward swing phase and high in the middle. The angular velocity of the limb elements does not, then, unlike in the stance phase, remain constant. While touchdown is an instantaneous event, lift off takes place over a "longer" period of several milliseconds. The protraction begins with a forwards rotation of the shoulder blade and femur, which are already engaged in forwards movement while the upper arm and lower leg are still in the process of retraction. The limbs move the quickest during the middle third of the swing phase when, significantly bent, they swing forwards like a short pendulum. In the last quarter of the swing phase the limbs reach their maximum extension. The retraction of the forelimb which, during the swing phase, starts well before touchdown begins in the distal elements. Hindlimb retraction starts only immediately before touch down. Angulation at touchdown is pretty much fixed, varying only minimally in the forelimbs across the different gaits. The forward swing period lasts for an average of 0.3 seconds regardless of gait, but as the distance the limb needs to travel during this phase is longer in the trot, it needs to be moved faster when the dog is trotting. This is achieved via a shortening of the pendulum, which is the distance between the upper pivot point and the toes. The shortening is effected either by moving the swinging limb higher in the air, by lowering the upper pivot point, or by a combination of the two. During the walk, the swinging limb is usually raised, but during the trot it is raised even higher and the pivot point is simultaneously lowered. This shortens the length of the pendulum by 10 – 20 percent.

THE CONTRIBUTION OF A LIMB ELEMENT TO STEP LENGTH

The contribution of a limb element to step length, otherwise known as yield, depends on the height of its pivot point and the angle it describes. A movement of 20° in the hip joint has a different effect on step length than movement through the same angle in the ankle. While the height of the upper pivot points, which are found on the top edge of the scapula and in the hip joint, is basically fixed, the pivot points of the other elements change height during the stance phase. We developed an algorithm to calculate contribution to step length on the basis of the factors mentioned (Fischer & Lehmann 1998), according to which in dogs the movement of the shoulder blade makes up at least 65 percent. In some steps the scapula can contribute as much as 80 percent, independently of gait. In the hindlimbs it is the rotation of the femur in the hip joint that makes the greatest contribution to step length during both the walk and the trot. On average, over 70 percent of the ground distance traveled by the limb comes from the hip joint, more than 5 percent from the pelvis, with the other two elements (lower leg and foot) contributing a mere 25 percent. The yield of the various elements of the forelimb does not change when it comes to the gallop, but the movement of the lower back constitutes what amounts to an additional fourth element in the hindlimb.

The movement of the "pelvis" is an important factor even in the symmetrical gaits (walk and trot). We use the word pelvis in inverted commas because it is not actually the pelvis but the lumbar spine which moves. During the walk and the trot the pelvis describes a movement which combines rotation and tilting. It tilts forwards towards the touchdown limb and then moves upwards and backwards with the leg that is moved backwards. The contribution of the movement of the pelvis to step length is around 15 percent in the walk and the trot. As we did not take X-ray films of the dogs performing the gallop, we can only estimate the contribution of the movement of the back to step length during the gallop. It certainly amounts to at least 50 percent and it can be assumed that this increases significantly when the animal is performing a fast gallop.

Calculating the yield demonstrates the immense importance of the shoulder blade and femur in making up step length during the walk and trot, and that of the back and femur during the gallop. Although the upper limb elements dominate, there are a number of other simple geometric factors which also contribute to step length. The rotation of the scapula, for instance, causes the upper arm to move out of its steep touchdown position into an almost horizontal position. This has the same effect as a connecting rod, causing either the paw to move backwards by around 2/3 of the length of the upper arm or, when the paw is set on the ground, the body to move forwards. Another "trick" used to extend step length without significant joint movement is to flex the limb when the pivot point is low and to extend it when it is higher. As in a pendulum, the length of the distance covered depends on the height of the pivot point, which means that especially in the elbow and ankle, more yield can be attained though the angular displacement remains the same. The exploitation of these mechanical and geometric principles makes it possible to keep the level of movement required to propel the body forwards as low as possible in all the limb joints except the hip.

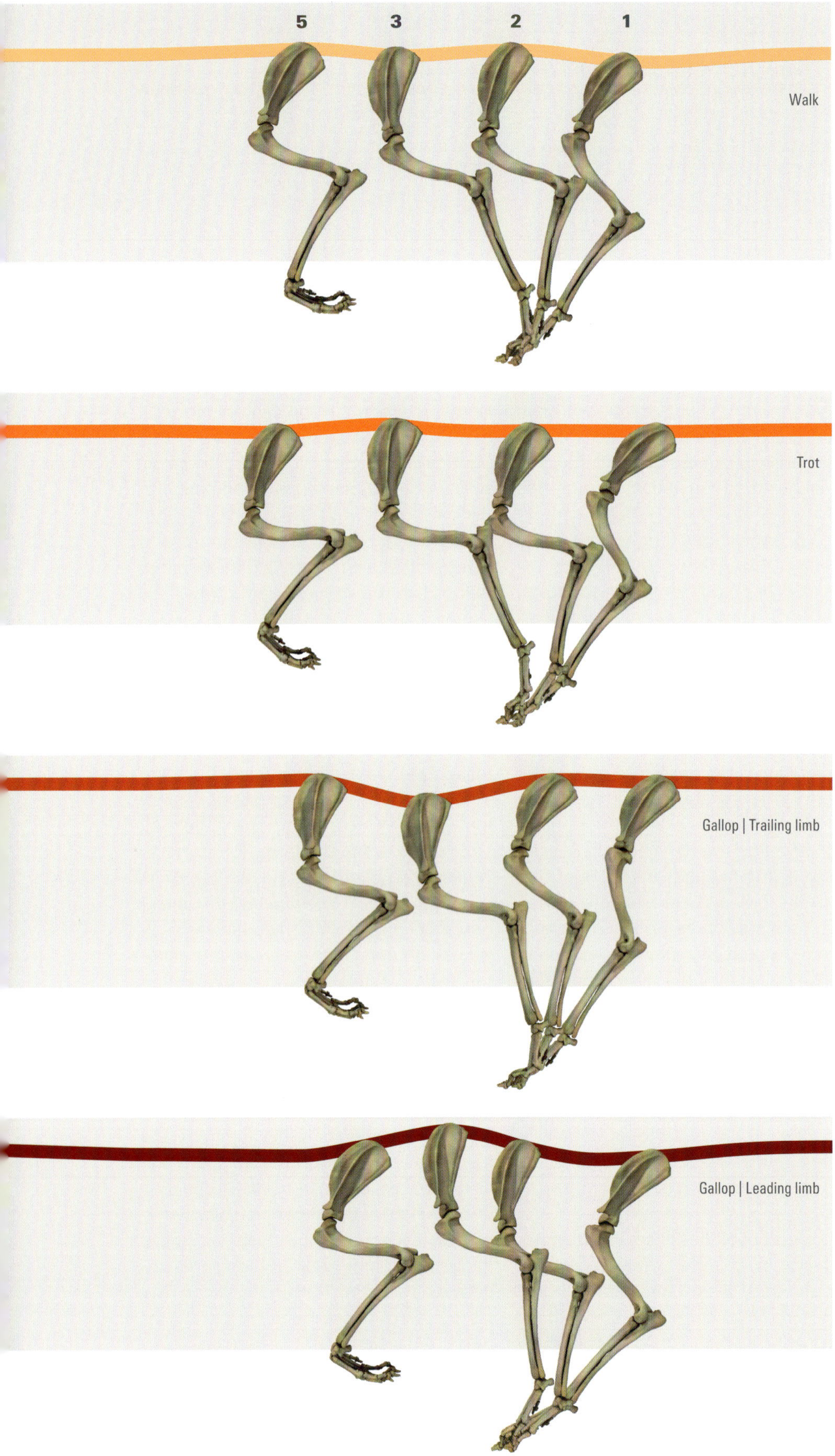

MOTION SEQUENCE OF THE LIMBS

The following description of the sequence of motion (kinematics) performed by the limbs is based on an analysis of X-ray films. Most of the values relate to small breeds, but a comparison with the marker-based data shows that all step cycles essentially follow a very similar pattern. The angles of the limb elements and joints do differ slightly from breed to breed, but the angles of the upper limb elements in particular are largely congruent, even between Dachshunds and Akita (see table).

Stance phase

(I) The front limb touches down perpendicular to the ear. The shoulder blade and forearm are held largely parallel to each other and an almost straight line can be drawn from the forearm to the forefoot. The upper arm is never completely vertical and never extends beyond perpendicular to the shoulder joint, pointing backwards at all times. Unlike in the forelimb, in which the retraction period before touchdown is longer, the retraction of the hindlimb before touchdown only takes place over a small distance, with the angles of maximum extension largely identical to the touchdown angles. The touchdown position is very similar in all gaits except the gallop, in which the position of the trailing limb is steeper.

(II) During the walk, retraction involves less flexion than during the trot. The position of the limbs and their angulation in the middle of the stance phase differ both from breed to breed and between the fore- and hindlimbs, with impaction more clearly observable in the hindlimbs. In the final third of the stance phase the greater flexion which takes place in the trot and gallop is compensated for to such an extent by extension that by the time it comes to lift off, there is no detectable difference between the gaits.

(III) There is almost no difference in lift off angles between the gaits. The scapular, forearm and femur are vertical during lift off or extended by over 100°.

Swing phase

(IV) The swing of the forelimbs is initiated by the raising of the upper arm and is followed by the simultaneous flexion of the wrist and elbow joints and the backwards rotation of the scapular. Lift off in the hindlimbs begins synchronously with the extension (!) of the ankle joint, the forwards movement of the femur, which is caused by flexion in the hip joint, and the tilting forwards of the pelvis.

(V) During the walk, the swinging wrist joint is moved on a level with the stationary wrist joint. During the trot it is moved on a level with the distal forearm. The swinging trailing limb is held higher than the leading limb in the gallop. The swinging leading limb resembles the swinging trotting limb. The elbow is continuously flexed during the forward swing phase until it reaches a position directly underneath the pivot point of the shoulder blade. From the middle of the forward swing phase onwards, all the limb joints begin to extend simultaneously. As in the case of the forelimb, the height of the forward-swinging hindfoot depends on the gait. The foot is only flexed weakly from the ankle and mainly raised by the flexion of the knee and hip joint.

(VI) The forelimb is most extended shortly before touchdown – during touchdown itself it is already engaged in backward movement. The scapula performs a maximum forwards rotation to an angle of up to 50 – 55°; the upper arm is close to vertical, the shoulder is opened to a maximum, the forearm is at its most inclined and only the elbow joint will continue to be extended until touchdown. In the hindlimb, the hip is maximally flexed, putting the femur into its least inclined position of 50 – 60°, while the lower leg is held almost vertically.

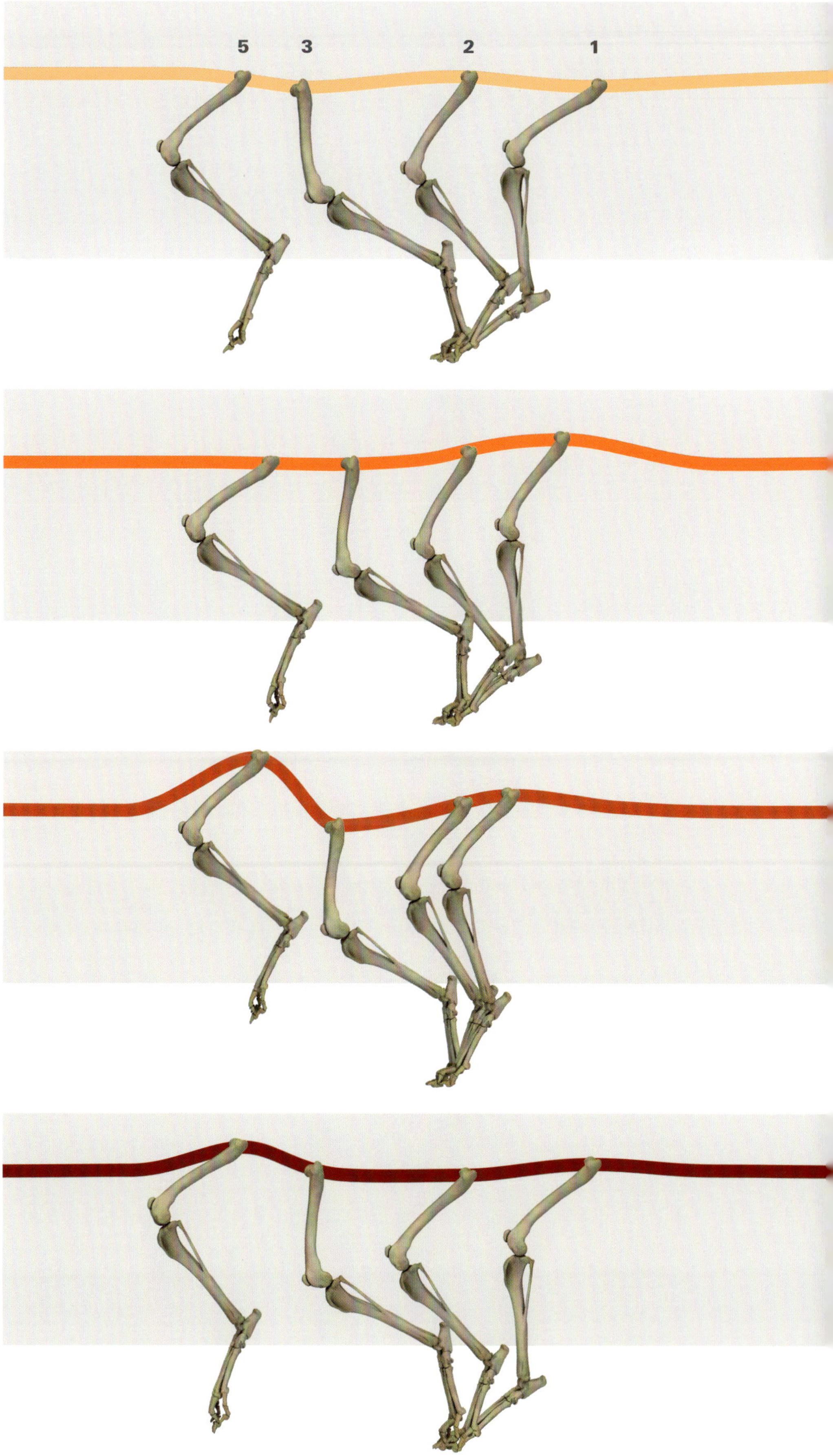

Motion pattern of the forelimb

MV = Median value SD = Standard deviation N = Number of observations	Touch down						Midstance						Lift off						Effective angular displacement						Maximal amplitude					
	MV	SD	N	MV	SD	N	MV	SD	N	MV	SD	N	MV	SD	N	MV	SD	N	MV	SD	N	MV	SD	N	MV	SD	N	MV	SD	N
Shoulder blade																														
Akita	57	2.7	25	54	2.9	25	-	-	-	-	-	-	97	2.4	22	94	4.5	14	40	3.0	19	41	3.3	12	41	2.6	17	45	1.6	12
Biewer Yorkshire Terrier	64	2.8	17	67	2.7	22	69	3.6	12	72	1.7	19	98	4.3	10	100	2.4	28	35	2.6	10	33	2.3	21	39	2.2	8	41	2.7	23
Chihuahua	58	2.5	14	-	-	-	79	2.8	15	-	-	-	94	4.9	10	-	-	-	35	-	10	-	-	-	41	3.9	9	-	-	-
Dachshund	60	1.8	22	63	2.0	22	75	1.6	21	68	2.3	26	97	1.5	4	100	2.8	26	36	3.1	3	37	2.7	20	40	2.9	4	47	3.1	18
Standard Schnauzer	64	2.4	39	60	1.0	16	-	-	-	-	-	-	93	2.7	34	90	1.9	15	30	2.3	33	29	2.2	14	-	-	-	-	-	-
Tibetan Terrier	59	0.4	3	58	1.7	5	-	-	-	-	-	-	88	3.1	4	89	1.9	5	29	5.7	2	31	0.7	4	32	3.3	3	40	2.0	4
Welsh Corgie Pembroke	62	2.1	29	-	-	-	68	2.6	11	-	-	-	96	3.1	22	-	-	-	34	4.0	19	-	-	-	45	2.7	21	-	-	-
Upper arm																														
Akita	98	3.0	15	99	7.3	12	-	-	-	-	-	-	32	3.8	19	22	3.3	8	65	5.1	12	76	7.0	6	69	3.9	14	83	6.8	6
Biewer Yorkshire Terrier	83	3.4	17	79	3.3	26	69	1.8	12	60	3.0	19	36	1.1	8	34	1.5	28	48	3.3	8	45	4.1	26	59	3.8	6	60	3.9	26
Chihuahua	84	2.9	14	-	-	-	37	3.0	15	-	-	-	11	3.1	10	-	-	-	74	-	10	-	-	-	83	5.2	9	-	-	-
Dachshund	78	2.6	22	76	3.2	22	59	1.4	21	63	2.4	26	34	1.0	4	30	2.9	26	44	1.7	3	46	2.6	20	48	5.4	4	56	3.1	18
Standard Schnauzer	-	-	-	-	-	-	-	-	-	-	-	-	-	-	-	-	-	-	-	-	-	-	-	-	-	-	-	-	-	-
Tibetan Terrier	91	3.2	6	90	2.3	9	-	-	-	-	-	-	48	1.8	6	43	1.8	8	44	0.8	2	50	2.1	4	50	4.3	3	66	1.1	4
Welsh Corgie Pembroke	84	4.8	29	86	5.0	6	68	2.5	11	67	4.2	6	34	3.6	23	38	5.0	3	51	3.1	17	42	-	1	61	6.1	17	58	0.7	2
Forearm																														
Akita	-	-	-	-	-	-	-	-	-	-	-	-	-	-	-	-	-	-	-	-	-	-	-	-	-	-	-	-	-	-
Biewer Yorkshire Terrier	59	3.2	12	58	0.8	2	68	2.3	10	64	1.9	16	111	1.7	3	109	2.9	24	52	3.9	4	-	-	-	77	-	1	77	5.3	16
Chihuahua	44	3.6	14	-	-	-	67	2.5	15	-	-	-	93	6.5	7	-	-	-	50	-	7	-	-	-	84	4.9	7	-	-	-
Dachshund	56	2.0	22	51	2.3	22	75	1.3	21	59	1.9	26	115	5.2	4	112	2.1	26	58	4.6	3	61	2.1	20	80	6.4	3	86	2.4	18
Standard Schnauzer	-	-	-	-	-	-	-	-	-	-	-	-	-	-	-	-	-	-	-	-	-	-	-	-	-	-	-	-	-	-
Tibetan Terrier	66	0.5	3	62	0.7	4							106	2.7	3	105	2.8	5	40	-	3	43	-	4	-	-	-	-	-	-
Welsh Corgie Pembroke	58	4.4	8	57	2.7	6	69	3.9	11	68	0.9	6	108	1.5	3	110	1.9	3	49	0.9	2	50	-	1	82	4.1	3	89	2.0	2
Front pastern																														
Akita	-	-	-	-	-	-	-	-	-	-	-	-	-	-	-	-	-	-	-	-	-	-	-	-	-	-	-	-	-	-
Biewer Yorkshire Terrier	-	-	-	-	-	-	-	-	-	-	-	-	-	-	-	-	-	-	-	-	-	-	-	-	-	-	-	-	-	-
Chihuahua	-	-	-	-	-	-	142	3.2	15	-	-	-	87	-	5	-	-	-	-	-	-	-	-	-	110	6.4	9	-	-	-
Dachshund	129	2.2	20	138	1.5	5	115	1.6	21	132	2.4	19	54	5.3	2	57	4.1	14	-	-	-	88	2.0	2	-	-	-	134	2.5	5
Standard Schnauzer	-	-	-	-	-	-	-	-	-	-	-	-	-	-	-	-	-	-	-	-	-	-	-	-	-	-	-	-	-	-
Tibetan Terrier	118	0.3	2	-	-	-	-	-	-	-	-	-	75	0.9	3	75	2.2	4	43	-	2	-	-	-	-	-	-	-	-	-
Welsh Corgie Pembroke	-	-	-	132	3.4	6	-	-	-	130	3.9	6	-	-	-	58	6.4	2	-	-	-	66	-	1	-	-	-	126	3.1	2
Shoulder joint																														
Akita	153	3.2	15	151	7.0	12	-	-	-	-	-	-	128	2.8	19	116	5.5	8	26	3.8	12	32	6.8	7	48	3.5	14	62	7.1	6
Biewer Yorkshire Terrier	147	4.3	17	145	4.5	22	138	3.2	12	132	4.0	19	135	4.9	8	134	2.9	28	12	5.0	8	12	3.1	21	44	4.8	7	48	5.1	20
Chihuahua	143	2.2	14	-	-	-	116	3.7	15	-	-	-	105	7.0	10	-	-	-	38	-	10	-	-	-	58	5.4	9	-	-	-
Dachshund	137	1.5	22	138	2.9	22	134	1.3	21	131	2.3	26	131	0.5	4	129	1.9	25	8	1.9	3	9	2.3	19	34	0.7	4	43	2.9	18
Standard Schnauzer	-	-	-	-	-	-	-	-	-	-	-	-	-	-	-	-	-	-	-	-	-	-	-	-	-	-	-	-	-	-
Tibetan Terrier	152	4.4	3	148	1.7	5	-	-	-	-	-	-	135	2.1	4	130	0.9	5	15	6.5	2	18	2.6	4	40	5.9	3	51	2.4	4
Welsh Corgie Pembroke	145	4.4	29	-	-	-	136	2.1	11	-	-	-	133	7.5	20	-	-	-	18	4.0	15	-	-	-	46	4.2	21	-	-	-
Elbow joint																														
Akita	-	-	-	-	-	-	-	-	-	-	-	-	-	-	-	-	-	-	-	-	-	-	-	-	-	-	-	-	-	-
Biewer Yorkshire Terrier	142	3.9	13	136	2.6	2	137	3.5	10	123	3.3	16	147	1.1	3	142	2.9	24	5	1.8	3	-	-	-	68	-	1	67	4.6	23
Chihuahua	128	4.0	14	-	-	-	104	4.7	15	-	-	-	104	7.6	7	-	-	-	25	-	7	-	-	-	71	6.4	9	-	-	-
Dachshund	134	2.4	22	127	3.5	22	134	2.5	21	122	3.9	26	149	1.9	4	141	4.0	25	15	5.5	3	15	2.5	19	64	8.7	3	69	2.8	18
Standard Schnauzer	-	-	-	-	-	-	-	-	-	-	-	-	-	-	-	-	-	-	-	-	-	-	-	-	-	-	-	-	-	-
Tibetan Terrier	155	2.0	3	152	1.6	4							156	0.6	2	153	1.3	3	2	-	2	1	-	3	-	-	-	-	-	-
Welsh Corgie Pembroke	140	7.6	8	143	3.5	6	138	5.9	11	136	4.8	6	146	2.4	4	148	3.8	3	3	0.3	2	8	-	1	65	0.8	2	65	5.9	2
Carpal joint																														
Akita	-	-	-	-	-	-	-	-	-	-	-	-	-	-	-	-	-	-	-	-	-	-	-	-	-	-	-	-	-	-
Biewer Yorkshire Terrier	-	-	-	-	-	-	-	-	-	-	-	-	-	-	-	-	-	-	-	-	-	-	-	-	-	-	-	-	-	-
Chihuahua	-	-	-	-	-	-	208	4.7	15	-	-	-	179	-	5	-	-	-	-	-	-	-	-	-	104	5.8	7	-	-	-
Dachshund	185	2.1	20	190	3.1	4	190	1.6	21	192	2.9	13	177	3.8	2	170	5.1	13	-	-	-	25	2.5	2	-	-	-	117	6.2	6
Standard Schnauzer	-	-	-	-	-	-	-	-	-	-	-	-	-	-	-	-	-	-	-	-	-	-	-	-	-	-	-	-	-	-
Tibetan Terrier	184	0.2	2	-	-	-	-	-	-	-	-	-	182	0.9	3	179	1.1	4	2	-	2	-	-	-	-	-	-	-	-	-
Welsh Corgie Pembroke	-	-	-	189	2.8	6	-	-	-	199	4.2	6	-	-	-	168	3.9	2	-	-	-	17	-	1	-	-	-	114	0.4	2

Motion pattern of the hindlimb

MV = Median value
SD = Standard deviation
N = Number of observations

	Touch down						Midstance						Lift off						Effective angular displacement						Maximal amplitude					
	MV	SD	N	MV	SD	N	MV	SD	N	MV	SD	N	MV	SD	N	MV	SD	N	MV	SD	N	MV	SD	N	MV	SD	N	MV	SD	N
Pelvis																														
Akita	56	1.8	8	54	0.7	6	-	-	-	-	-	-	-	-	-	52	1.5	8	-	-	-	2	1.2	6	-	-	-	8	1.3	6
Biewer Yorkshire Terrier	68	3.0	10	65	1.6	3	-	-	-	-	-	-	66	2.2	9	64	0.0	2	4	1.4	9	2	0.8	2	8	2.0	8	3	2.3	2
Chihuahua	65	1.6	23	-	-	-	58	1.8	22	-	-	-	61	2.4	15	-	-	-	5	-	15	-	-	-	9	1.7	15	-	-	-
Dachshund	-	-	-	-	-	-	-	-	-	-	-	-	-	-	-	-	-	-	-	-	-	-	-	-	-	-	-	-	-	-
Standard Schnauzer	-	-	-	-	-	-	-	-	-	-	-	-	-	-	-	-	-	-	-	-	-	-	-	-	-	-	-	-	-	-
Tibetan Terrier	59	1.0	3	56	2.4	5	-	-	-	-	-	-	56	1.6	2	54	1.2	5	6	-	1	3	2.9	4	6	0.3	2	6	1.5	4
Welsh Corgie Pembroke	56	4.2	17	51	1.9	15	47	1.0	2	-	-	-	47	2.4	20	44	1.5	14	7	2.1	13	7	2.5	14	9	1.9	13	9	1.6	14
Femur																														
Akita	50	1.6	6	57	2.1	12	-	-	-	-	-	-	-	-	-	110	1.9	13	-	-	-	54	1.7	12	-	-	-	73	1.9	11
Biewer Yorkshire Terrier	61	2.8	25	63	2.3	17	74	1.4	15	69	1.5	13	103	5.6	25	105	1.7	15	41	7.2	22	42	3.3	14	46	3.4	20	52	2.8	13
Chihuahua	55	3.5	23	-	-	-	75	3.0	22	-	-	-	107	1.9	15	-	-	-	52	-	15	-	-	-	72	2.7	15	-	-	-
Dachshund	66	2.0	21	69	1.7	35	82	2.1	21	81	1.5	39	117	2.3	18	113	1.4	19	-	-	-	-	-	-	63	3.1	18	60	2.6	15
Standard Schnauzer	63	2.0	13	-	-	-	-	-	-	-	-	-	107	3.1	11	-	-	-	45	1.8	10	-	-	-	49	3.3	10	-	-	-
Tibetan Terrier	60	1.3	9	63	1.3	9	-	-	-	-	-	-	102	0.4	2	103	3.8	5	41	-	1	39	4.8	4	45	3.8	2	54	0.7	4
Welsh Corgie Pembroke	59	4.4	20	64	2.1	15	74	2.5	3	-	-	-	107	8.5	19	113	4.1	14	49	5.0	12	52	4.0	25	55	5.8	15	59	4.3	14
Lower leg																														
Akita	-	-	-	-	-	-	-	-	-	-	-	-	-	-	-	-	-	-	-	-	-	-	-	-	-	-	-	-	-	-
Biewer Yorkshire Terrier	91	1.7	11	80	1.7	13	72	1.4	15	66	2.5	13	46	2.1	16	44	2.9	11	45	2.0	11	37	2.6	10	56	2.7	11	59	2.7	11
Chihuahua	78	3.8	23	-	-	-	49	2.9	22	-	-	-	36	2.5	15	-	-	-	42	-	15	-	-	-	77	2.0	15	-	-	-
Dachshund	85	2.2	20	87	2.8	35	59	1.9	21	56	1.7	39	37	1.7	18	38	1.3	19	-	-	-	-	-	-	68	2.5	18	75	1.8	15
Standard Schnauzer	-	-	-	-	-	-	-	-	-	-	-	-	-	-	-	-	-	-	-	-	-	-	-	-	-	-	-	-	-	-
Tibetan Terrier	92	0.8	6	82	2.2	4	-	-	-	-	-	-	44	1.7	3	-	-	-	48	-	3	-	-	-	-	-	-	-	-	-
Welsh Corgie Pembroke	94	2.7	23	91	4.0	38	71	1.1	3	54	3.4	10	41	1.7	17	39	2.2	23	53	3.1	13	55	3.4	16	73	3.3	13	76	3.1	24
Hind pastern																														
Akita	-	-	-	-	-	-	-	-	-	-	-	-	-	-	-	-	-	-	-	-	-	-	-	-	-	-	-	-	-	-
Biewer Yorkshire Terrier	-	-	-	-	-	-	-	-	-	-	-	-	-	-	-	-	-	-	-	-	-	-	-	-	-	-	-	-	-	-
Chihuahua	133	3.6	23	-	-	-	122	3.1	22	-	-	-	67	3.3	14	-	-	-	66	-	14	-	-	-	91	2.0	15	-	-	-
Dachshund	137	3.9	4	131	2.8	28	115	1.7	20	116	2.1	38	62	3.7	15	63	2.4	13	-	-	-	-	-		89	-	1	82	3.5	6
Standard Schnauzer	-	-	-	-	-	-	-	-	-	-	-	-	-	-	-	-	-	-	-	-	-	-	-	-	-	-	-	-	-	-
Tibetan Terrier	129	2.7	5	-	-	-	-	-	-	-	-	-	80	3.1	3	-	-	-	50	-	3	-	-	-	-	-	-	-	-	-
Welsh Corgie Pembroke	-	-	-	134	5.8	6	-	-	-	113	3.7	10	76	2.8	9	68	1.6	9	-	-	-	-	-	-	-	-	-	76	-	1
Hip joint																														
Akita	106	3.3	6	113	2.0	6	-	-	-	-	-	-	-	-	-	164	2.2	8	-	-	-	-	-	-	-	-	-	-	-	-
Biewer Yorkshire Terrier	125	4.6	10	126	3.7	3	-	-	-	-	-	-	174	5.8	8	169	4.0	2	45	3.6	6	42	8.0	2	54	5.3	6	52	4.2	2
Chihuahua	120	3.8	23	-	-	-	133	3.2	22	-	-	-	168	3.2	15	-	-	-	48	-	15	-	-	-	70	2.3	14	-	-	-
Dachshund	-	-	-	-	-	-	-	-	-	-	-	-	-	-	-	-	-	-	-	-	-	-	-	-	-	-	-	-	-	-
Standard Schnauzer	-	-	-	-	-	-	-	-	-	-	-	-	-	-	-	-	-	-	-	-	-	-	-	-	-	-	-	-	-	-
Tibetan Terrier	120	1.2	3	120	2.6	5	-	-	-	-	-	-	158	2.0	2	158	2.2	4	36	-	1	41	4.2	3	44	5.8	2	53	2.0	4
Welsh Corgie Pembroke	115	3.9	17	114	2.7	15	120	0.5	3	-	-	-	153	7.7	19	157	4.0	14	41	4.7	11	43	3.4	14	49	5.4	13	54	5.0	14
Knee joint																														
Akita	-	-	-	-	-	-	-	-	-	-	-	-	-	-	-	-	-	-	-	-	-	-	-	-	-	-	-	-	-	-
Biewer Yorkshire Terrier	155	2.2	11	143	2.6	13	147	2.7	15	135	3.9	13	146	1.9	16	149	2.4	11	9	1.9	11	5	2.3	10	39	2.8	11	52	2.4	11
Chihuahua	133	5.2	23	-	-	-	124	5.7	22	-	-	-	143	3.2	15	-	-	-	10	-	15	-	-	-	66	3.5	15	-	-	-
Dachshund	151	2.8	20	156	2.7	35	142	3.0	21	137	2.5	39	154	3.1	18	151	2.2	19	-	-	-	-	-	-	56	3.8	18	67	3.1	15
Standard Schnauzer	-	-	-	-	-	-	-	-	-	-	-	-	-	-	-	-	-	-	-	-	-	-	-	-	-	-	-	-	-	-
Tibetan Terrier	152	1.6	6	145	2.4	4	-	-	-	-	-	-	-	-	-	-	-	-	-	-	-	-	-	-	-	-	-	-	-	-
Welsh Corgie Pembroke	153	3.3	17	155	3.4	15	145	2.5	3	-	-	-	145	7.8	15	148	0.6	3	9	6.2	7	8	5.6	3	54	4.2	10	58	4.7	14
Ankle joint																														
Akita	-	-	-	-	-	-	-	-	-	-	-	-	-	-	-	-	-	-	-	-	-	-	-	-	-	-	-	-	-	-
Biewer Yorkshire Terrier	-	-	-	-	-	-	-	-	-	-	-	-	-	-	-	-	-	-	-	-	-	-	-	-	-	-	-	-	-	-
Chihuahua	126	5.3	22	-	-	-	108	5.2	22	-	-	-	149	5.1	14	-	-	-	23	-	14	-	-	-	69	2.6	13	-	-	-
Dachshund	127	5.4	4	136	3.6	28	124	3.0	20	119	3.0	38	155	3.5	15	155	1.7	13	-	-	-	-	-	-	55	3.5	16	63	3.5	10
Standard Schnauzer	-	-	-	-	-	-	-	-	-	-	-	-	-	-	-	-	-	-	-	-	-	-	-	-	-	-	-	-	-	-
Tibetan Terrier	143	2.8	5	-	-	-	-	-	-	-	-	-	145	1.4	3	-	-	-	2	-	3	-	-	-	-	-	-	-	-	-
Welsh Corgie Pembroke	-	-	-	134	3.1	6	-	-	-	121	6.2	10	146	1.7	9	153	1.6	9	-	-	-	-	-	-	32	4.4	8	47	5.1	4

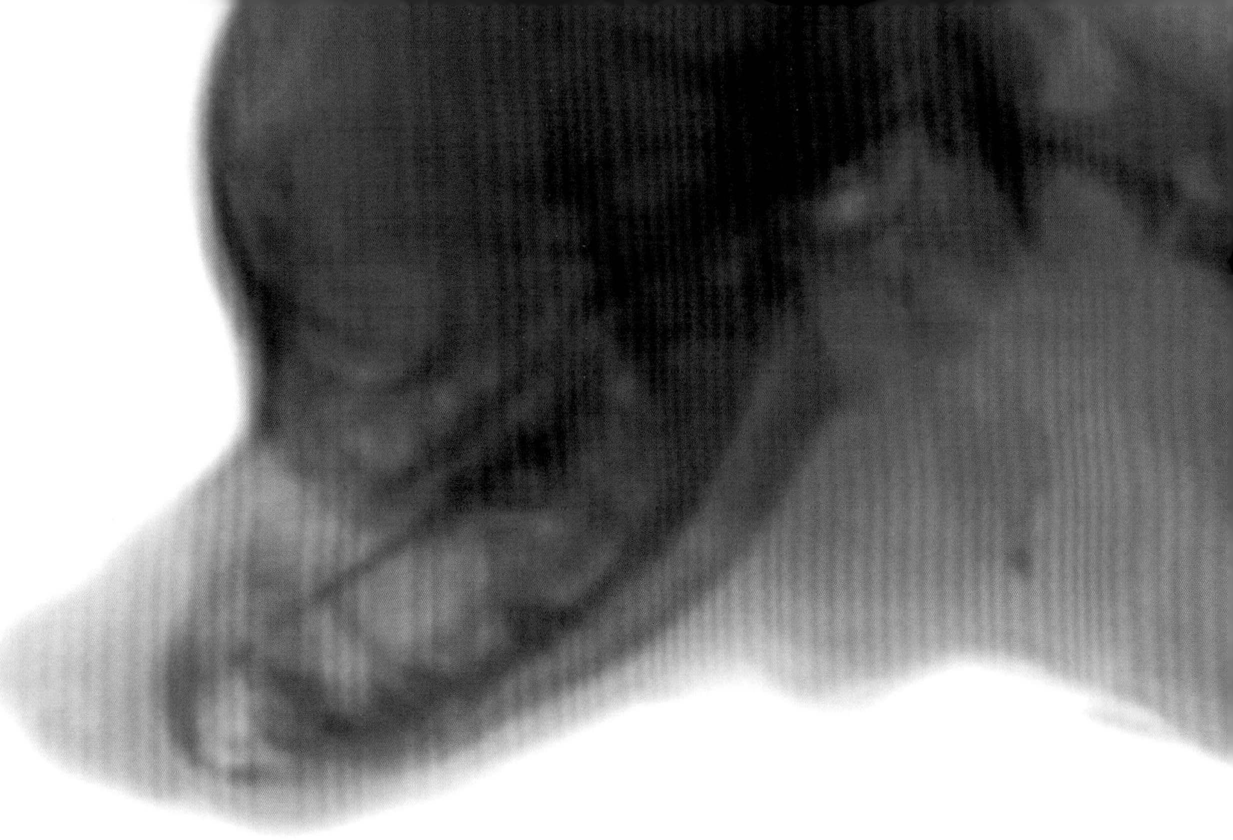

PREVIOUS STUDIES

As the data from this study are the first to have been obtained for all joints quantitatively on the basis of high-frequency X-ray films (though Tashman & Anderst (2003) did study the movements of the knee using biplanar X-ray technology), it is difficult to compare them with other data obtained under different conditions and usually only at 60 frames per second. Furthermore, many authors only mention the maximum amplitude of an entire step cycle (e.g. Gillette & Zebas 1999) and do not differentiate between the stance phase and the swing phase, meaning that we were only able to approximate values for effective joint movement from the angular curves depicted. Previous studies are therefore only presented here as a means of completing the picture.

The [range of] movement of the shoulder blade is only mentioned by Roy (1971), who estimates it to be a mere 15°. Inspired by one of the pioneers of motion research, Ted Goslow (Goslow et al. 1981), Brown (1986) attempted, using a plastic model and video films, to determine the [range of] movement of the shoulder blade. Although he/she came close, he/she still underestimated it.

For reasons connected with the measuring techniques used (video films, marker-based motion analysis), the movements of the hip joint are often dramatically underestimated in all the studies that have been carried out to date, and are given at between just 20° and 30° (e.g. Allen et al. 1994, Bockstahler 2007, Colborne 2008, DeCamp et al. 1993, Hottinger et al. 1996, Schaeffer et al. 1998, Unkel-Mohrmann 1999). Only Goslow et al. (1981) and Carrier et al. (1998) come up with values of 45° in the trot and over 60° in the gallop.

The values provided (or implied by illustrations) for the other limb joints lie in the region of the limb excursion measured in the present study.

One thing that earlier investigations do confirm is the basic observation that the angles of the elbow, hip and ankle joint remain more or less the same even when speed increases and the gait changes (Arshavskii et al. 1965). The angle of the wrist increases by a factor of 1.2, that of the shoulder by a factor of 1.3 and that of the knee by a factor of 1.7. Hottinger et al. (1996) also observed that the movements of the joints of the forelimb are only minimal during the stance phase, while those of the hindlimb are slightly larger.

Williams et al. (2009) describe how the initial differences in the motion sequences of trailing and leading hindlimbs become less and less pronounced as acceleration increases and the limbs move through larger and larger angles.

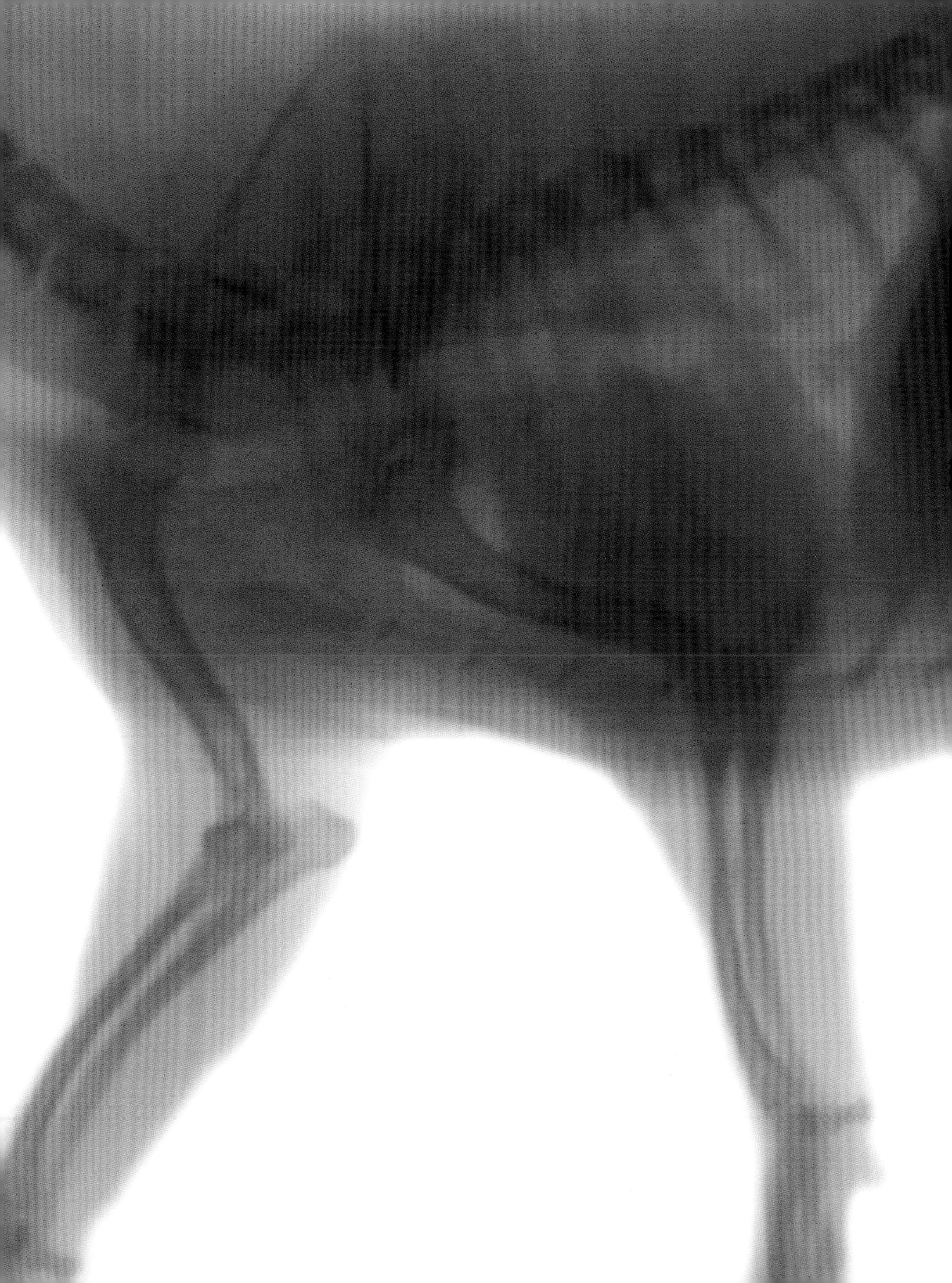

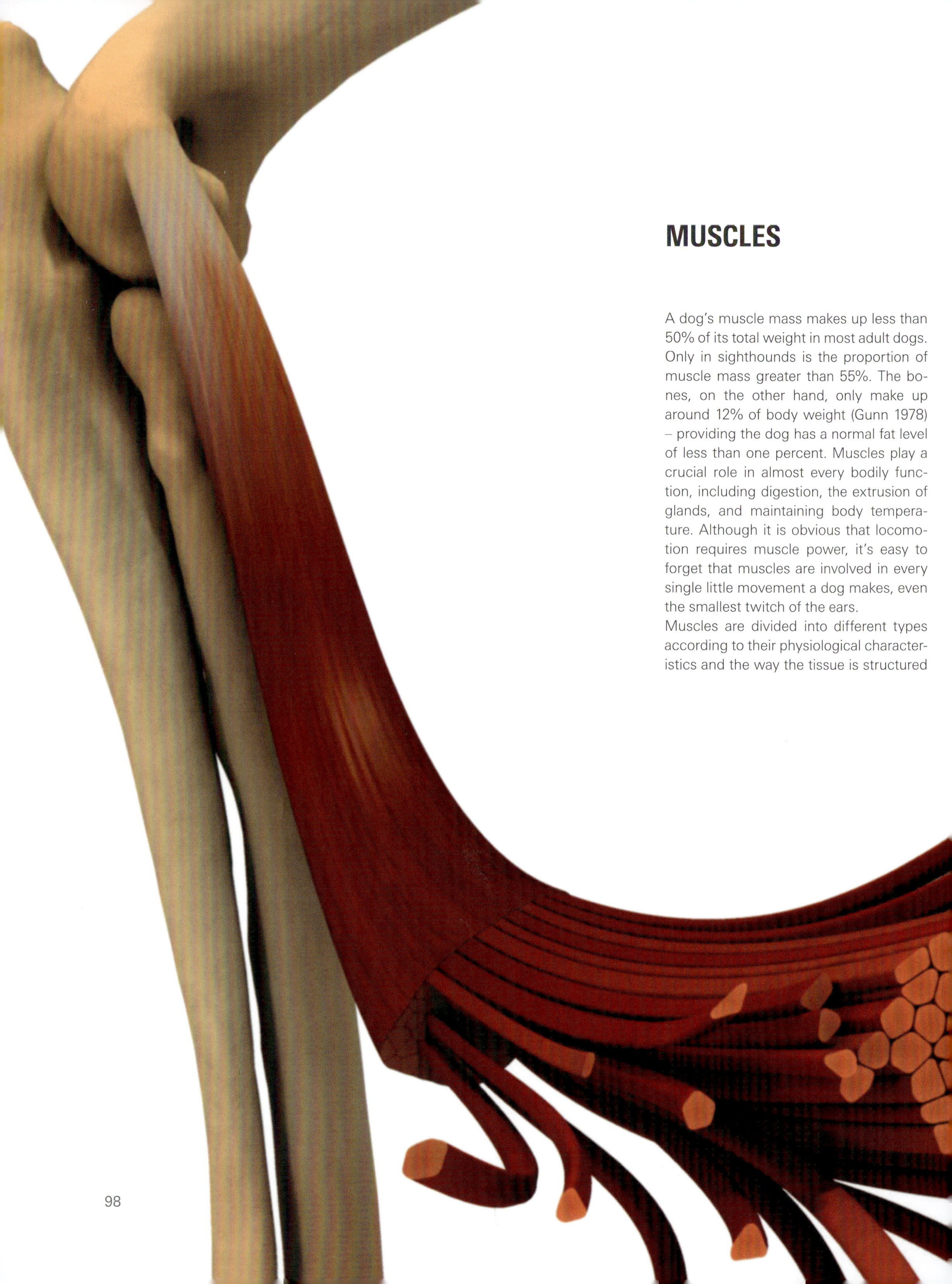

MUSCLES

A dog's muscle mass makes up less than 50% of its total weight in most adult dogs. Only in sighthounds is the proportion of muscle mass greater than 55%. The bones, on the other hand, only make up around 12% of body weight (Gunn 1978) – providing the dog has a normal fat level of less than one percent. Muscles play a crucial role in almost every bodily function, including digestion, the extrusion of glands, and maintaining body temperature. Although it is obvious that locomotion requires muscle power, it's easy to forget that muscles are involved in every single little movement a dog makes, even the smallest twitch of the ears.

Muscles are divided into different types according to their physiological characteristics and the way the tissue is structured

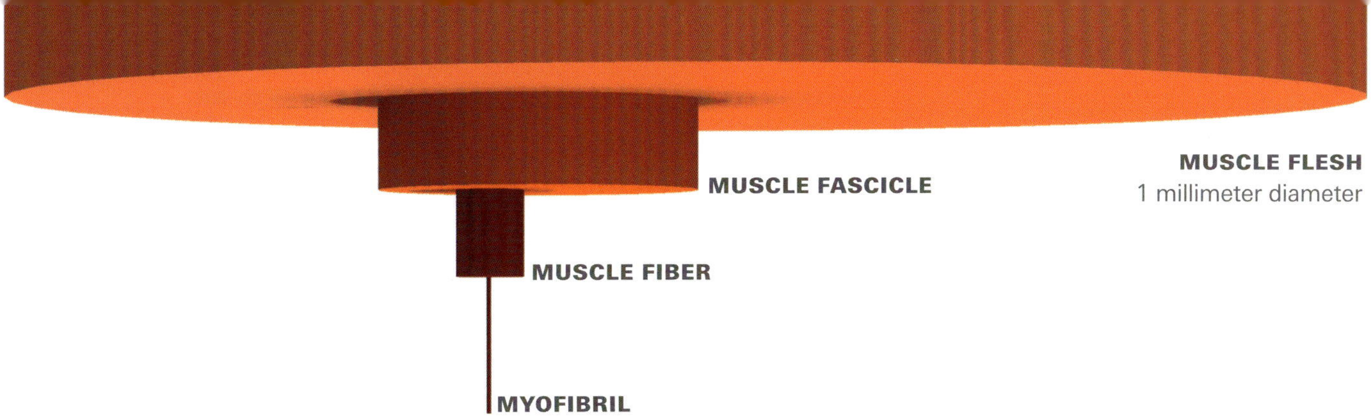

(histology). The skin and the digestive tract are made up of "smooth" muscle, the heart is made of a specialized type of striated muscle called cardiac muscle, and then there is skeletal muscle. This book only deals with the musculature involved in locomotion, which is the skeletal musculature, so when we use the term "muscle", this is what we are referring to.

When a muscle contracts in order to effect a movement, chemical energy is converted into mechanical work. A muscle can only actively contract – stretching is a passive process caused by the contraction of other muscles or by gravity. Muscles work on the agonist-antagonist principle.

While the processes which generate force take place on an invisible, molecular level, the bulging of a muscle is easily visible to the naked eye. This bridging of scale is achieved through the serial cascade of vast numbers of identical units. Force is generated by contractile elements, called sarcomeres, that are just 1/100,000 of a millimeter in size. Thousands of these protein complexes together form a myofilament, and a large number of myofilaments bunched together form a myofibril. Hundreds of myofibrils together form a muscle fiber. Muscle fibers and muscle cells are the same thing. Groups of muscle fibers form muscle fascicles. Varying numbers of muscle fascicles form the muscle proper (the "meat"), the tissue that we can see with the naked eye. Leg muscles contain around 300 muscle fibers per square millimeter, which extrapolated to the muscle as a whole makes for around 50,000 – 90,000 muscle fibers.

The myofilaments in the myofibrils are arranged in a specific pattern which is only visible under a microscope and has a special functional significance. A myofibril consists of repeating strings of sarcomeres, a Greek word meaning "flesh parts". A sarcomere is the smallest functional unit of a muscle. The contraction and visible bulging of the muscle associated with it is the result of the serial action of many sarcomeres: in one muscle fiber 10,000 serially connected sarcomeres effect a muscle contraction of just 4 millimeters.

Sarcomere – myofibril – muscle fiber – muscle fascicle. These are the four functional elements of which muscle tissue consists.

MUSCLE FLESH – MUSCLE FASCICLE – MUSCLE FIBER

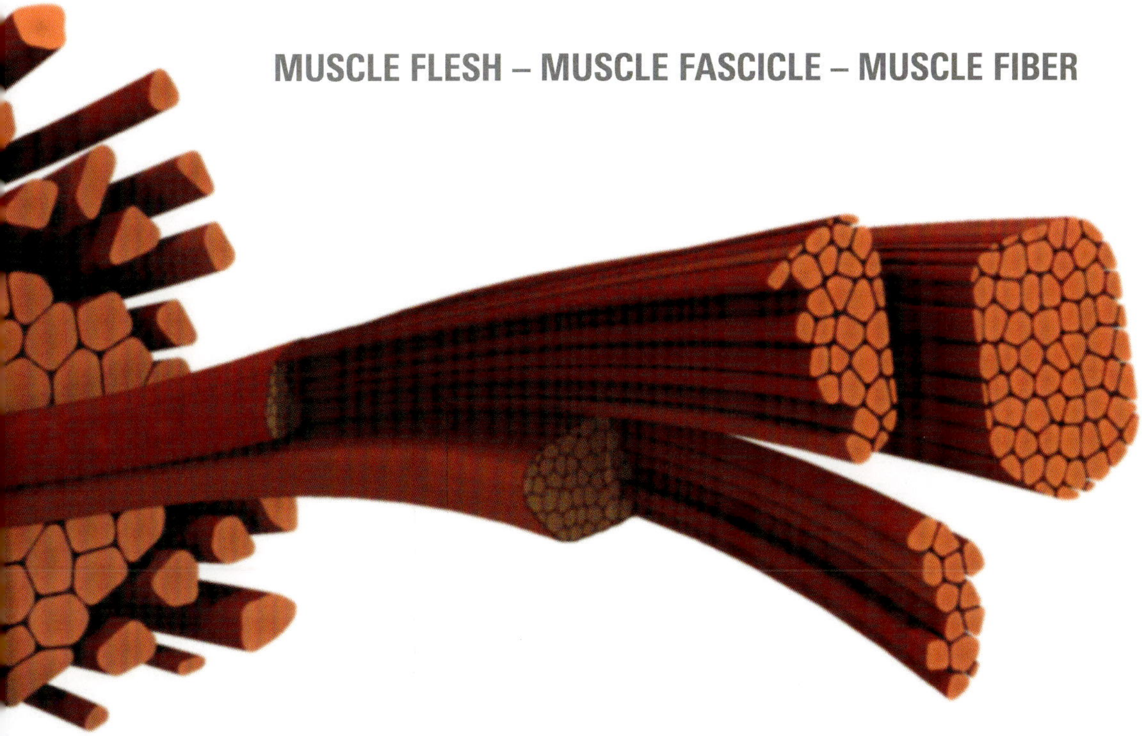

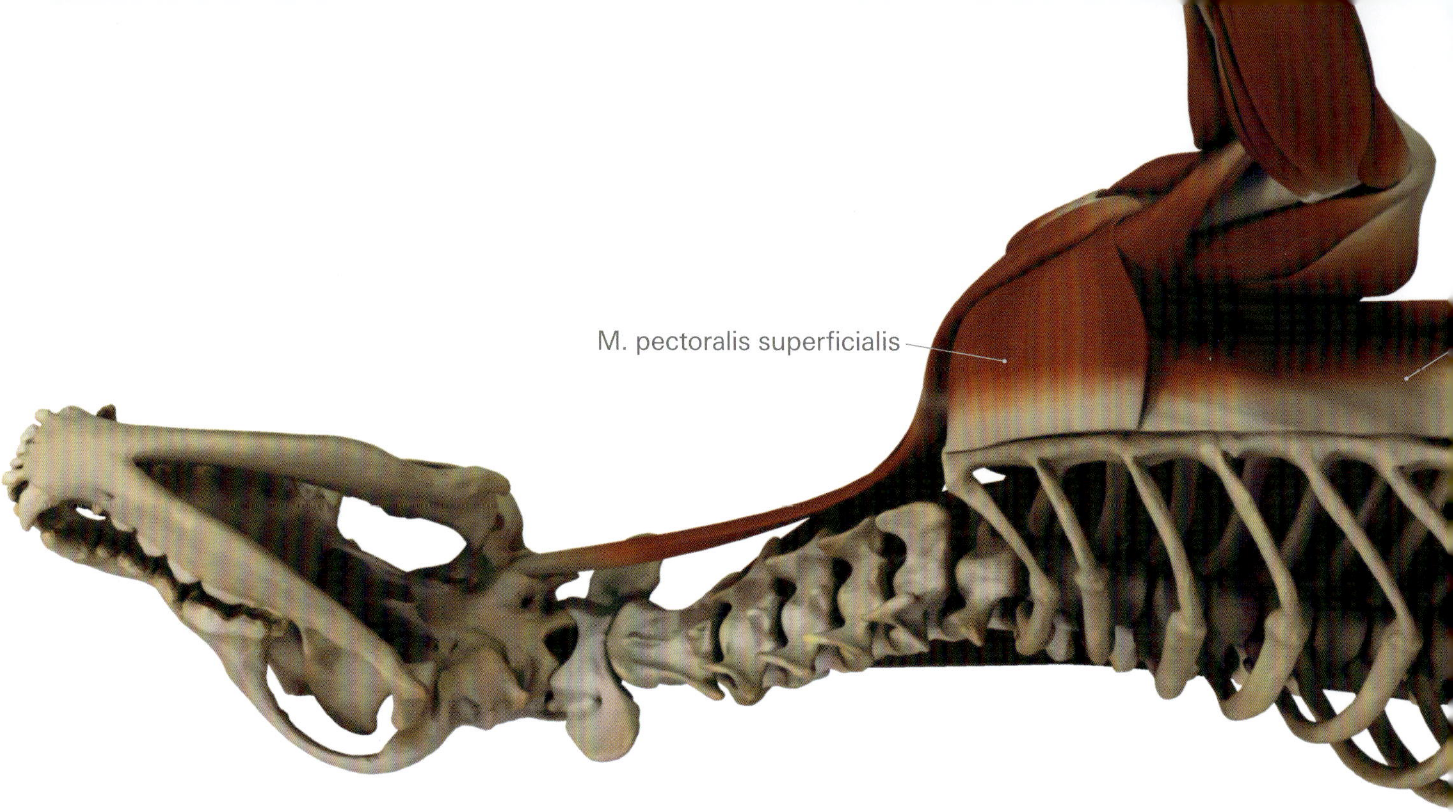

THE MUSCLES OF THE FORE - AND HINDLIMBS

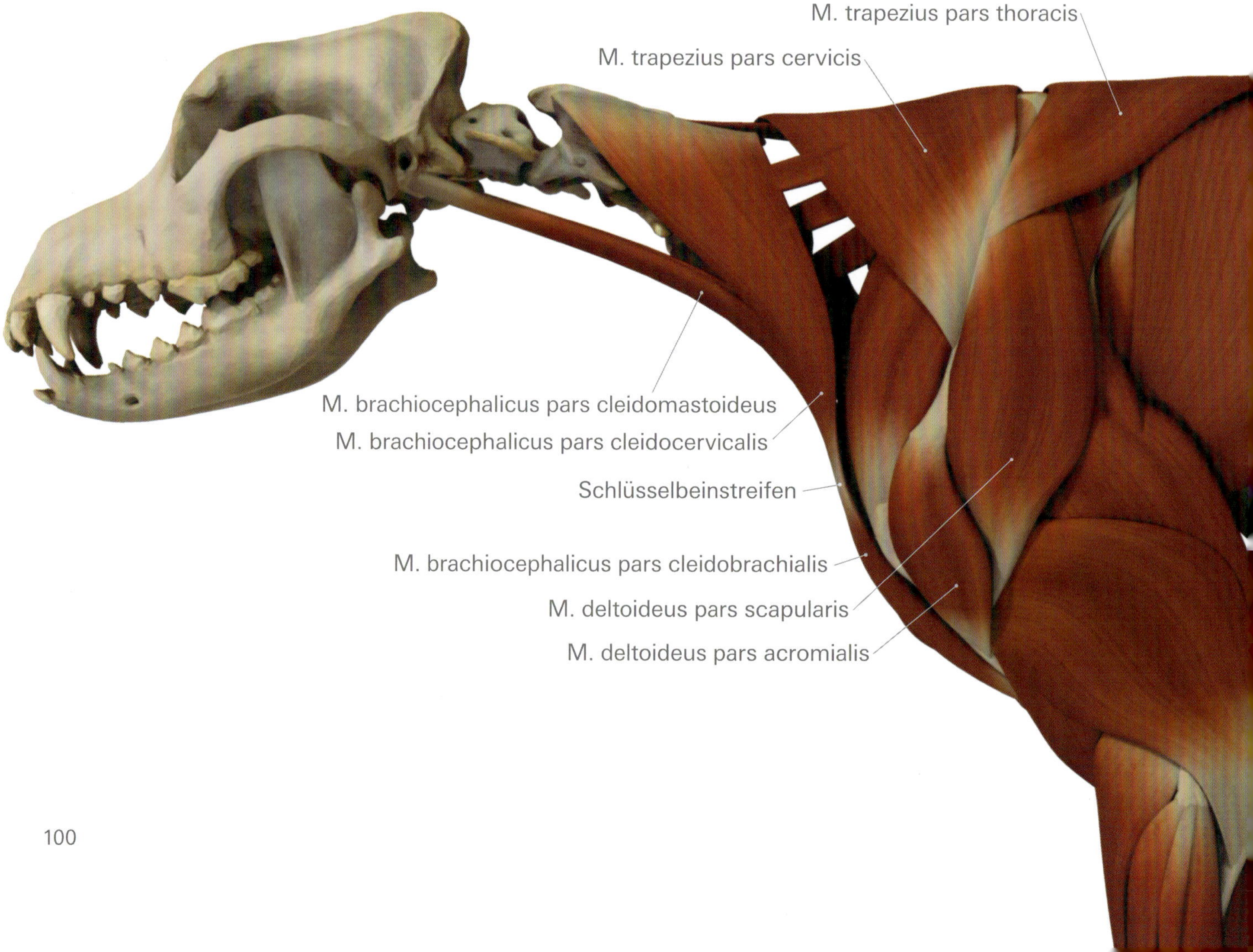

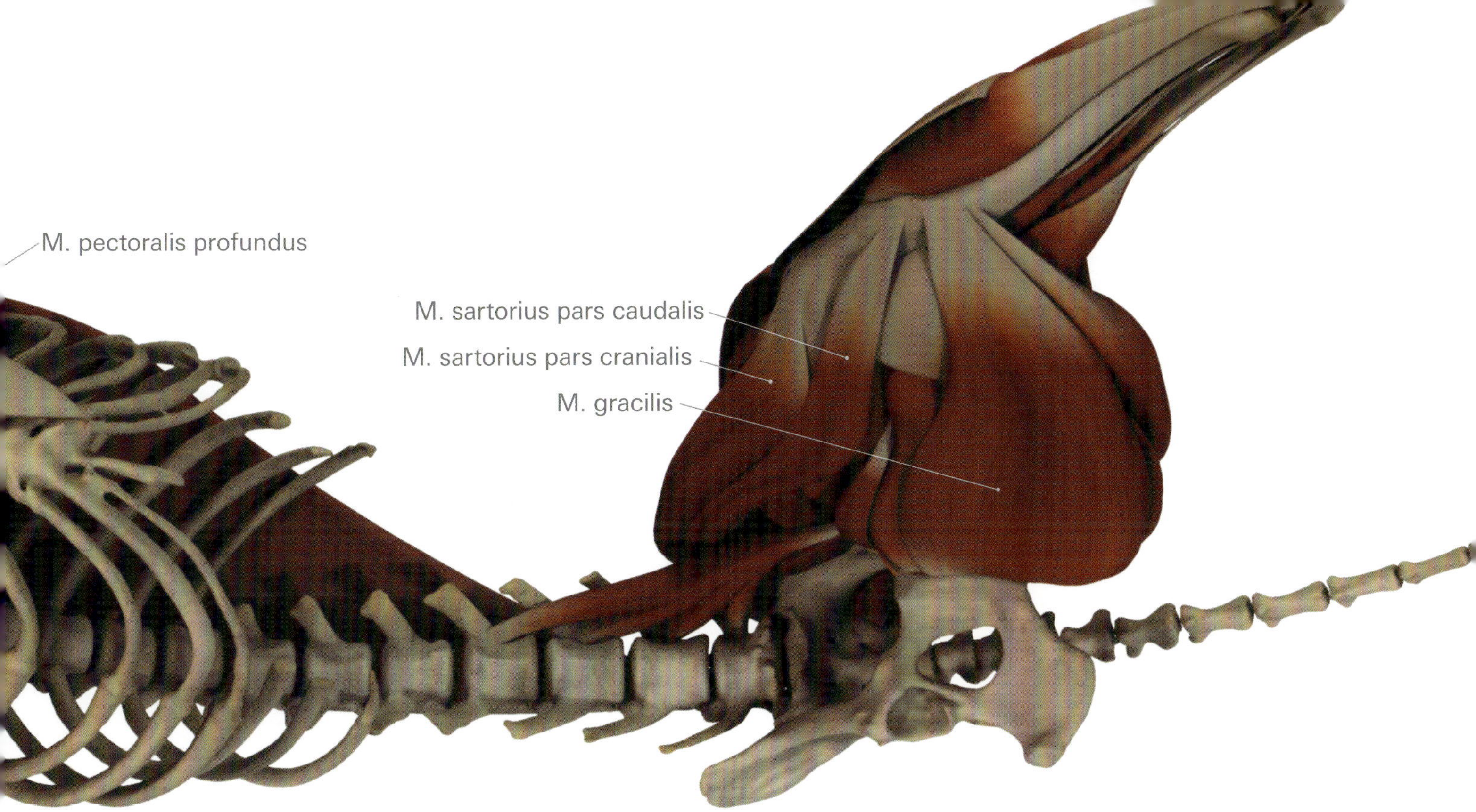
M. pectoralis profundus
M. sartorius pars caudalis
M. sartorius pars cranialis
M. gracilis

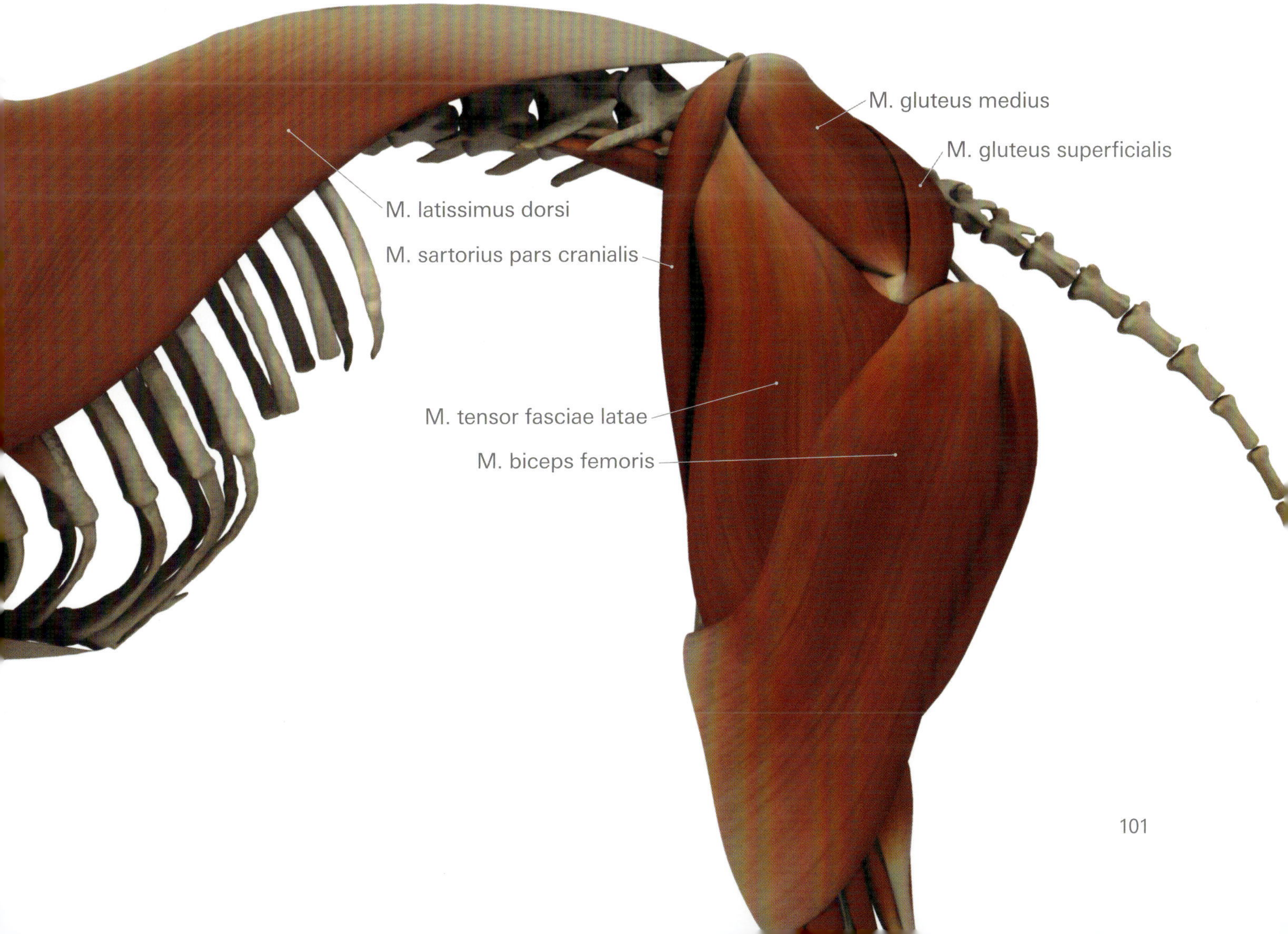
M. gluteus medius
M. gluteus superficialis
M. latissimus dorsi
M. sartorius pars cranialis
M. tensor fasciae latae
M. biceps femoris

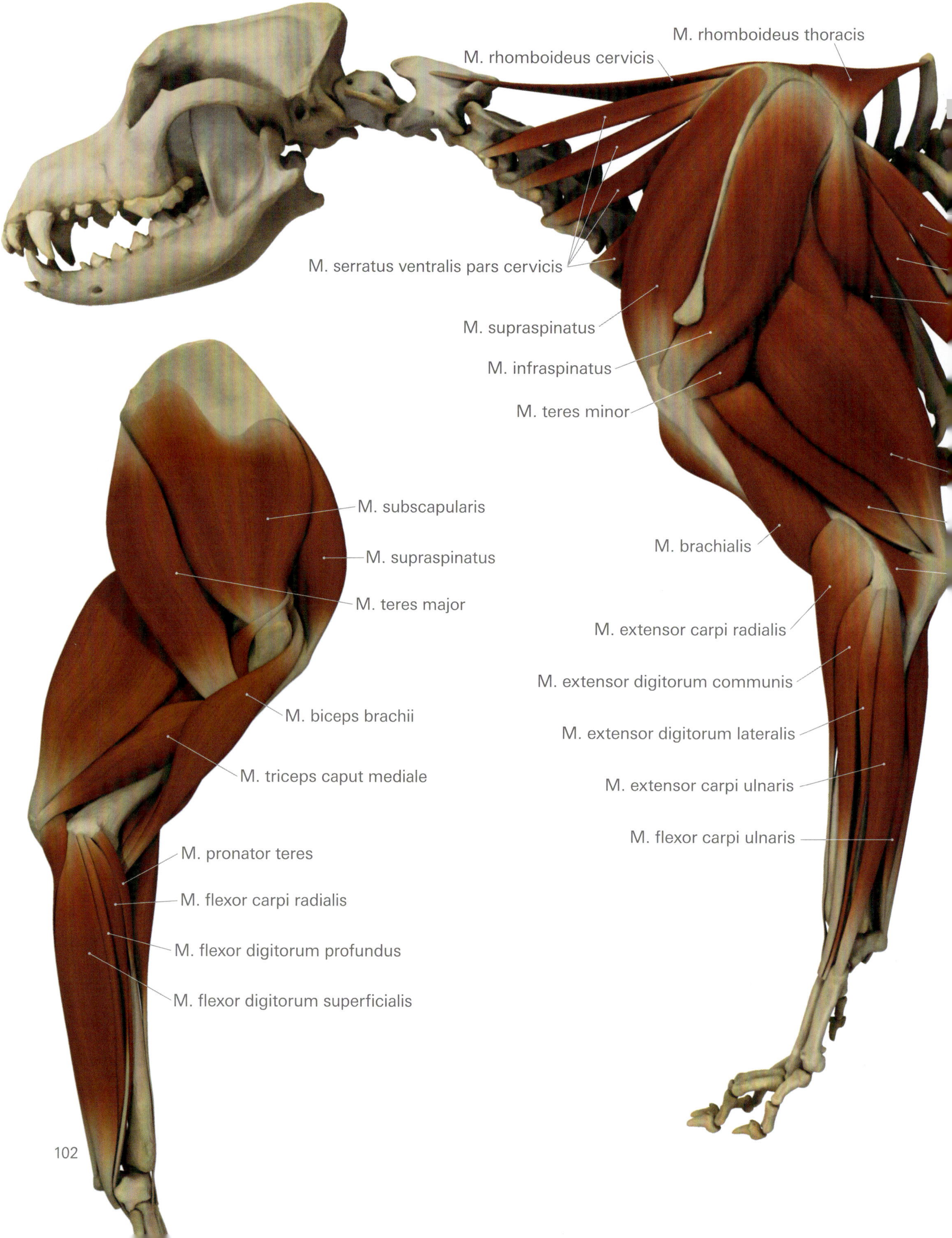
M. rhomboideus thoracis
M. rhomboideus cervicis
M. serratus ventralis pars cervicis
M. supraspinatus
M. infraspinatus
M. teres minor
M. subscapularis
M. supraspinatus
M. teres major
M. brachialis
M. extensor carpi radialis
M. extensor digitorum communis
M. biceps brachii
M. extensor digitorum lateralis
M. triceps caput mediale
M. extensor carpi ulnaris
M. flexor carpi ulnaris
M. pronator teres
M. flexor carpi radialis
M. flexor digitorum profundus
M. flexor digitorum superficialis

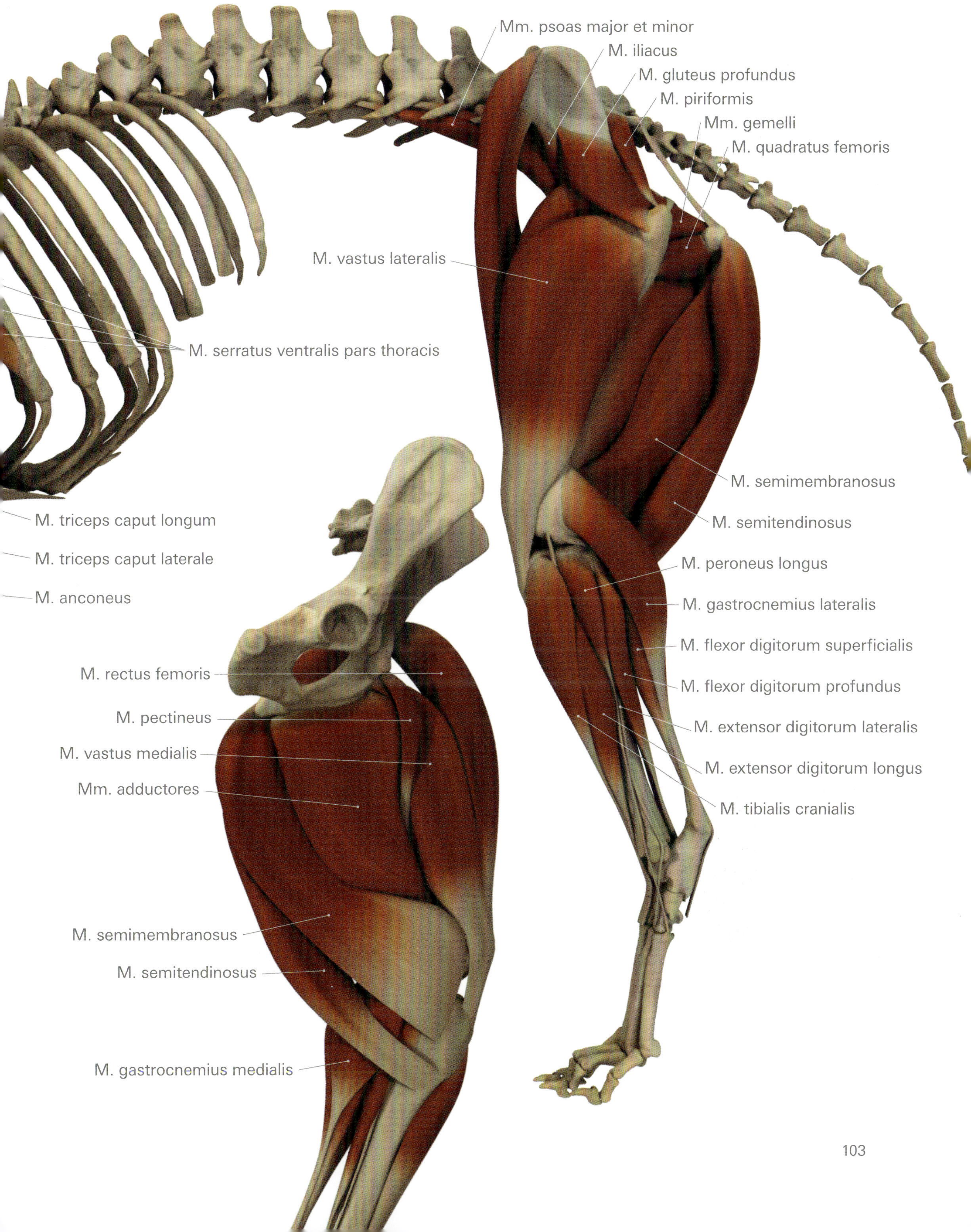
Mm. psoas major et minor
M. iliacus
M. gluteus profundus
M. piriformis
Mm. gemelli
M. quadratus femoris
M. vastus lateralis
M. serratus ventralis pars thoracis
M. semimembranosus
M. semitendinosus
M. triceps caput longum
M. triceps caput laterale
M. anconeus
M. peroneus longus
M. gastrocnemius lateralis
M. flexor digitorum superficialis
M. rectus femoris
M. flexor digitorum profundus
M. pectineus
M. extensor digitorum lateralis
M. vastus medialis
M. extensor digitorum longus
Mm. adductores
M. tibialis cranialis
M. semimembranosus
M. semitendinosus
M. gastrocnemius medialis

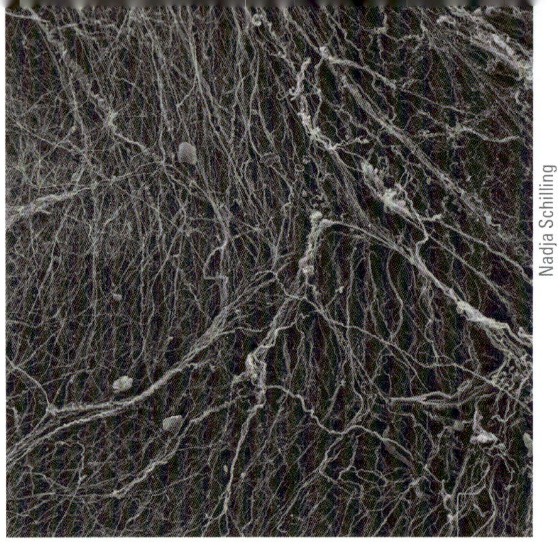

Epimysium 200 micrometers

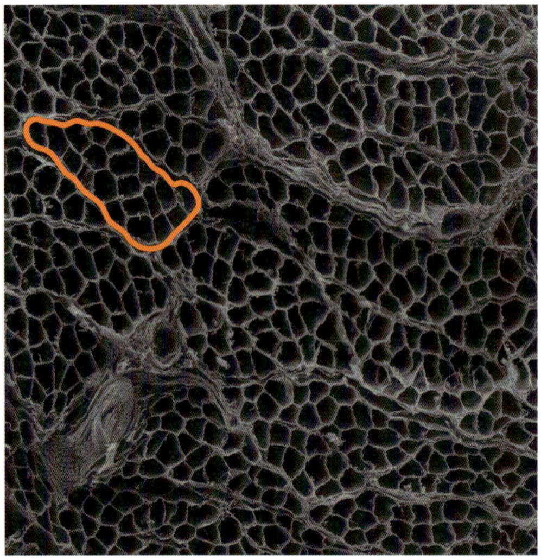

Perimysium 200 micrometers

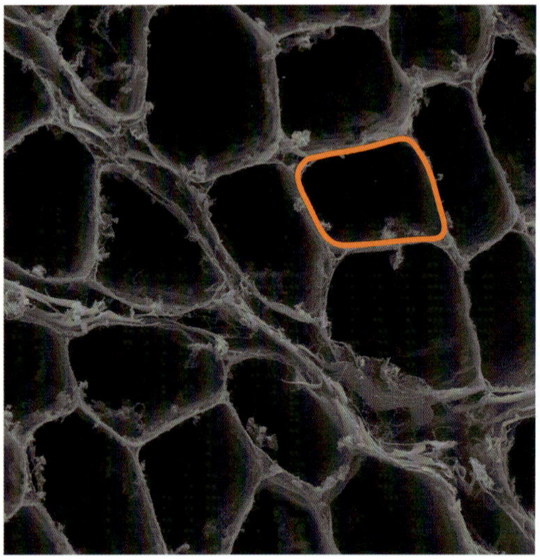

Endomysium 20 micrometers

Connective tissue structure: the epimysium surrounds the muscle, the perimysium the muscle fascicle and the endomysium the muscle fiber.

STRUCTURE

Connective tissue

All muscles consist of two types of tissue: connective tissue and muscle tissue. Muscle tissue generates forces, and connective tissue, which is highly tensile, transmits them. Connective tissue encloses the various functional elements which make up a muscle and fixes the muscle to the bone.

The connective tissue which covers the muscle as a whole is called the epimysium. Muscle fascicles are covered by another layer of connective tissue, the perimysium, which divides the muscle into distinct individual units. This is what the "stringy" in "stringy meat" describes. A third layer of what is usually loose connective tissue is found between the individual muscle fibers, and this is known as the endomysium. Finally, every muscle fiber is enclosed by a tubular sheath, the sarcolemma, or cell membrane of the muscle cell. Because the serial structure of the connective tissue reflects the distinct structural levels of the muscle, its configuration also reveals the direction in which the muscle fibers run.

Muscles are attached to the bone either by flat, taut, plexiform aponeuroses or by tendons – tough bands of connective tissue made up of parallel fibers. Tendons are given their characteristic structure by the mechanical strain they are subjected to in a preferred direction. When the tendon is stretched and taut, the connective tissue fibers are closely packed together. Additional, elastic fibers then return them to their original position.

Muscle topography

The study of the position and arrangement of muscles in the body is known as topographical myology. When Renaissance scientists began dissecting the human body, they described the musculature as a landscape and gave all the muscles they could distinguish different names. The nomenclature they used would have been comprehensible to the small educated class of the time ("Musculus cuccularis", a muscle in the back, is so called because it looks, in humans, like a monk's hood, for instance), but nowadays it can be extremely puzzling. The fact that the muscles of a dog are described using the names coined for the human anatomy makes it even more confusing. Eventually, the musculature in both humans and animals was described in full. Astonishingly, almost five hundred years later, gaps are still being discovered in our knowledge of topographical myology, though what we do know is becoming more and more detailed.

Basically, muscles are described on the basis of their origin and insertion and the nerves that supply them. The proximal end is classed as the origin. Every muscle is given the Latin name it bears in humans (if it is present in humans) and which it was given on the basis of the human anatomy. Muscles are introduced by "Musculus" and abbreviated to M., the plural is "Musculi" which is abbreviated to Mm. Even where a muscle is different in appearance in dogs, has a different origin or insertion or is modified in some other way, it is "homologized" with the human muscle. The decisive criterion is its position and nerve supply. In humans, for example, the M. biceps brachii has, as the Latin name indicates, two heads. In dogs, on the other hand, it only has one. M. triceps brachii has four heads in dogs, not three, which illustrates how misleading the anatomical

names can be. The international convention "Nomina anatomica veterinaria" does not permit the renaming of muscles, so the Latin names are used and understood throughout the world.
The function of a muscle can often be deduced just from knowing its origin and insertion. By linking the two points and imagining what would happen if the distance between them was shortened, it is possible to work out the potential effect of the contraction of the muscle in question. A muscle's function may be to bend (in the case of a flexor), stretch (in the case of an extensor), pull up (in the case of an adductor), move to the side (in the case of an abductor), lower (in the case of a depressor) or raise (in the case of a levator) a body part. When a muscle contracts, however, the insertion is pulled towards the origin just as much as the origin is pulled towards the insertion. For the forces to be directed effectively, then, other muscles or designated mechanisms at the joint need to fix one of the two points so that one (the movable point) is moved to the other (the fixed point), rather than both points moving towards each other. Sometimes the fact that one point (usually the origin, which is situated closer to the body) has a much greater mass than the other is enough to prevent it from moving significantly. The role of a muscle can often be just to prevent the movement of a joint, e.g. to stabilize a joint against the effect of gravity or to provide controlled resistance to gravity. The function of a leg muscle during the swing phase, when the leg does not have contact to the ground, can be different from that during the stance phase, when it does.
A muscle, then, is a topographical structure which, as an anatomically distinct unit, has its own name, and which performs work in a preferred direction. Muscles of the same name are supplied by the same nerve or nerves in all mammals (and often in all vertebrates). The way in which a muscle works cannot be determined through knowledge of its position alone and often only becomes clear from dissecting the tissue into its functional components.

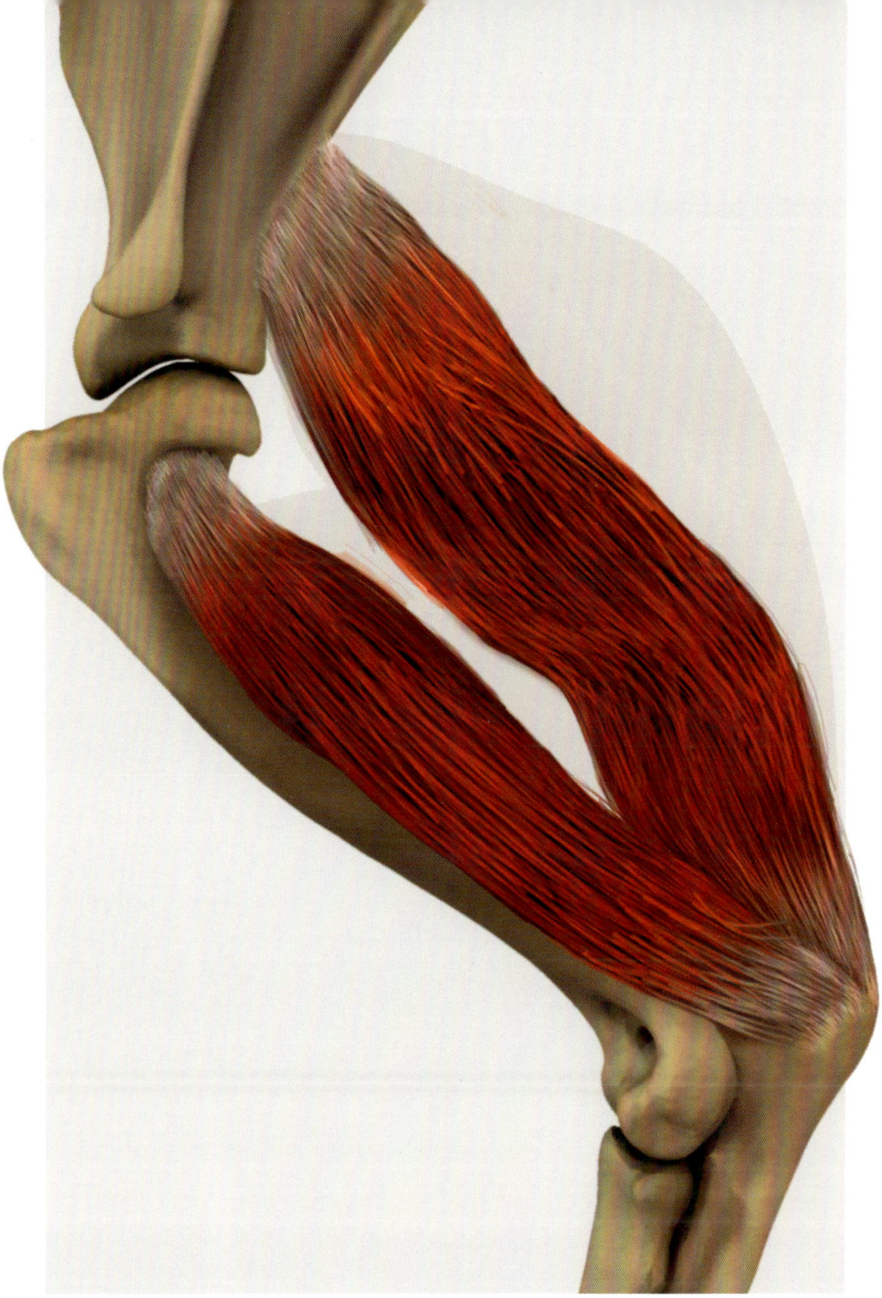

Arrangement of muscle fascicles in the M. triceps brachii.

Muscle fascicles

The muscle fascicle determines the direction, or more precisely the vector, of force transmission. One muscle fascicle, or muscle fiber bundle, contains between 20 and 50 muscle fibers, which all run parallel to each other and thus define the direction of force development. The individual fascicles within a muscle, on the other hand, might well be oriented in different directions. Only in exceptional cases are they all arranged in the direction of pull of the muscle. This is because only very few muscles originate from or insert onto a single point. Most of them need to transmit forces in various directions to tendons or bones. The fascicles in many muscles run diagonally to longitudinal, like the fibers in the feathers of a bird. This is referred to simplistically as pennation. In actual fact the arrangement of muscle fascicles is extremely complex. Working out the main direction of force transmission in any given muscle is made even more difficult by the fact that the fascicles do not just contract when the muscle contracts, but change their angle of pennation (Stark 2007).

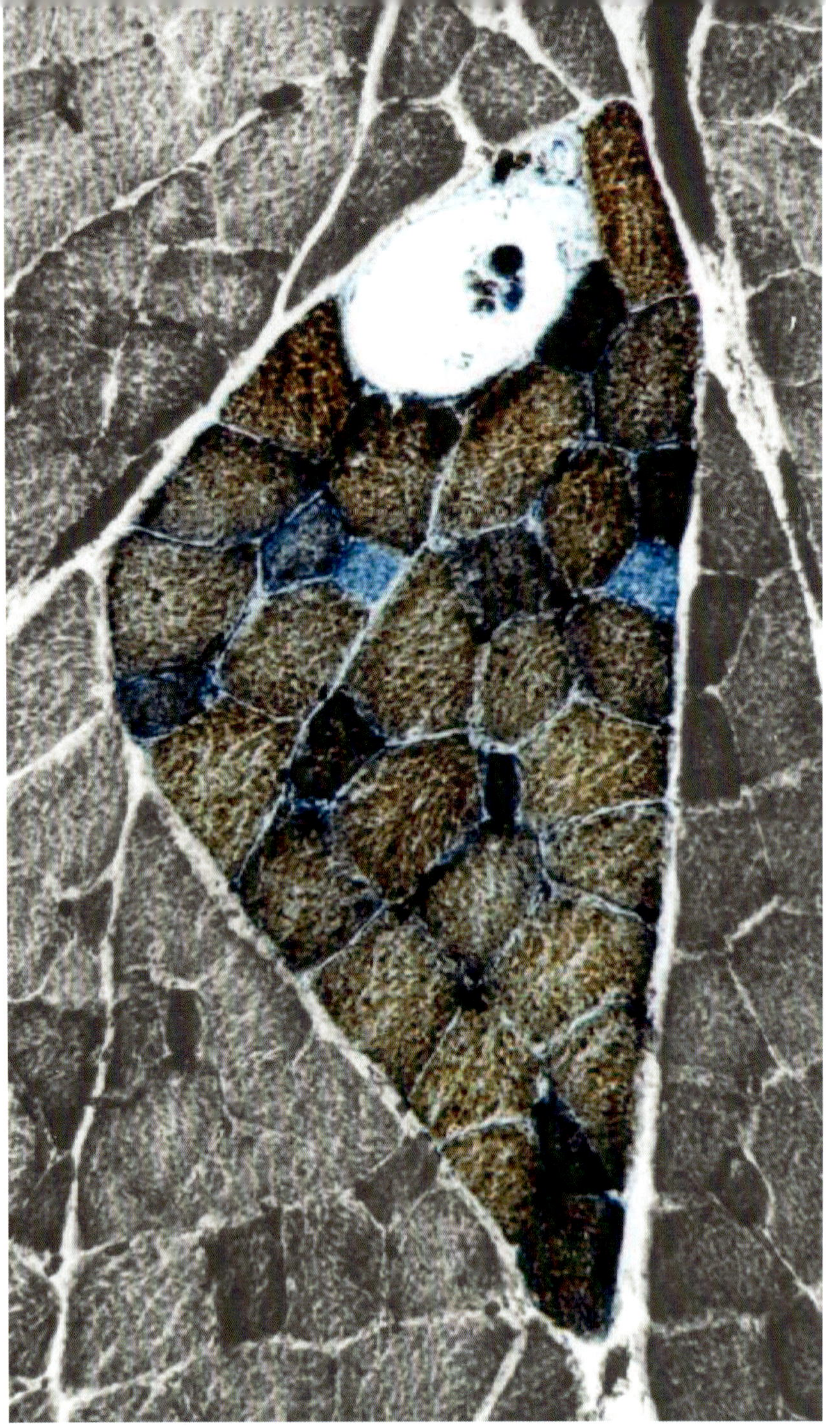

Using an enzyme reaction to visualize muscle fibers. The colored region corresponds to one muscle fascicle.

Muscle fiber types in dogs		
Red muscle fibers	White muscle fibers	
Type I fibers	Type II fibers	
SO fibers	FOG fibers	FX fibers
Oxidative	Oxidative / glycolytic	Oxidative / glycolytic
Fatigue-resistant	Fatigue fast	
Contract slowly	Contract rapidly	

Muscle fibers

Muscle fibers reflect the cellular level of the musculature. They work and are structured in a variety of different ways, with the result that they are divided into various different types. Within vertebrates, a distinction is made between tonic and phasic muscle fibers. As is the case in all mammals, the skeletal musculature is made up exclusively of phasic fibers, also known as twitch fibers. Accordingly, the following text only deals with this type.

Phasic muscle fibers possess thousands of cell nuclei. They can reach lengths of up to 10 centimeters and are usually around 1/20th of a millimeter (45 – 60 micrometers) in diameter. Interestingly, the diameter of muscle fibers does not depend on body size but on various factors such as the distance over which oxygen and nutrients need to be transported, the ratio of fiber volume to fiber surface area and other things (Goldspink 1977, Kimura 1993, Marx et al. 2006). The theory put forward by Kimura (1993), that dogs have strikingly thin muscle fibers in comparison to other mammals, has never been confirmed. The cross-sectional area of a muscle fiber depends on the type of fiber, its position in the muscle and the muscle itself. It is generally between 0.0015 mm^2 (1.5 μm^2) and 0.0019 mm^2 (1.9 μm^2) (Marx et al. 2006). The animal's state of fitness also plays a role. The cross-sectional area of a muscle fiber can be increased through training, but the number of fibers cannot. Training and nutrition have the greatest influence on the size of muscle fibers (Hoppeler Flück 2002). In a study of Arctic sledge dogs, Gerth et al (2009) showed that in summer, when the dogs are much less active, fed smaller amounts and thus lose up to 30 per cent of their body weight, muscle atrophy sets in. The diameter of their muscle fibers reduces and the myofibrils begin to die off.

Phasic muscle fibers are, in turn, divided into two subtypes which differ in appearance, metabolism, structure, contractile behavior and nerve supply. Since Ranvier (1873), a distinction has been made between red and white muscle fibers which, when large numbers occur together in one muscle, can be distinguished with the naked eye. Chicken and trout con-

In type I muscle fibers the motoaxon innervates on one side and branches off parallel to the muscle.

sists mainly of white muscle fibers, partridge and tuna of red muscle fibers. The coloring is caused by a particular protein which is present in different amounts. Red muscle fibers contain a large amount of myoglobin, a protein which (like the hemoglobin in red blood cells) can bind and store oxygen. Red muscle fibers also have more mitochondria, which use oxygen to generate the energy necessary for mechanical work. This form of metabolism is known as oxidative, or aerobic metabolism. White muscle fibers, which have significantly less oxygen at their disposal due to their lower myoglobin content, generate energy by breaking down carbohydrates - generally a particular type of sugar (glucose). Metabolism in the absence of oxygen is described as glycolytic, or anaerobic.

Red and white fibers differ not just in their methods of metabolism but also in the way in which they contract. This is determined by the type of contractile proteins they contain. Red muscle fibers are made up of contractile proteins which work slowly and enduringly. The proteins responsible for contraction in white muscle fibers, on the other hand, work fast and fatigue easily.

Muscle fibers are conventionally divided into type I and type II fibers according to their metabolic strategy, the presence or absence of various myoglobin proteins and their contractile behavior. Type I fibers are oxidative, slow and for sustained work; type II fibers are glycolytic, fatigue more quickly and contract rapidly.

Force development depends on the diameter of muscle fibers. Type II fibers are often, but not always (Armstrong et al. 1982, Rodriguez-Barbudo et al. 1984, Aguera et al. 1990), larger in diameter than type I fibers (1.9-3.1 μm^2 as opposed to 1.1-2.4 μm^2, Newsholme et al. 1988), so that in absolute terms, type II fibers often generate greater forces than type I fibers. It is, however, incorrect to say that type II fibers always generate greater forces than type I fibers (Marx et al. 2006), because when they are equal in diameter, both types are capable of generating the same amount of force.

The two muscle fiber types differ in their nerve supply, too. Type I fibers are supplied by a nerve whose branches only enter the muscle from one side, run parallel to the direction of the fibers and, at regular intervals, project offshoots vertically into the muscle fibers. In the case of type II fibers, the nerve branches out like a tree vertically to the direction of the muscle fibers on both sides.

The white, rapidly contracting and quickly tiring glycolytic type II fibers can be divided into two subtypes, type IIA and type IIB. Type IIA muscle fibers have a metabolism which is a cross between oxiadative and glycolytic and which is known as FOG (fast oxidative glycolytic). Type IIB muscle fibers are purely glycolytic (FG: fast glycolytic).

In type II muscle fibers the motoaxon innervates centrally and branches off in all directions.

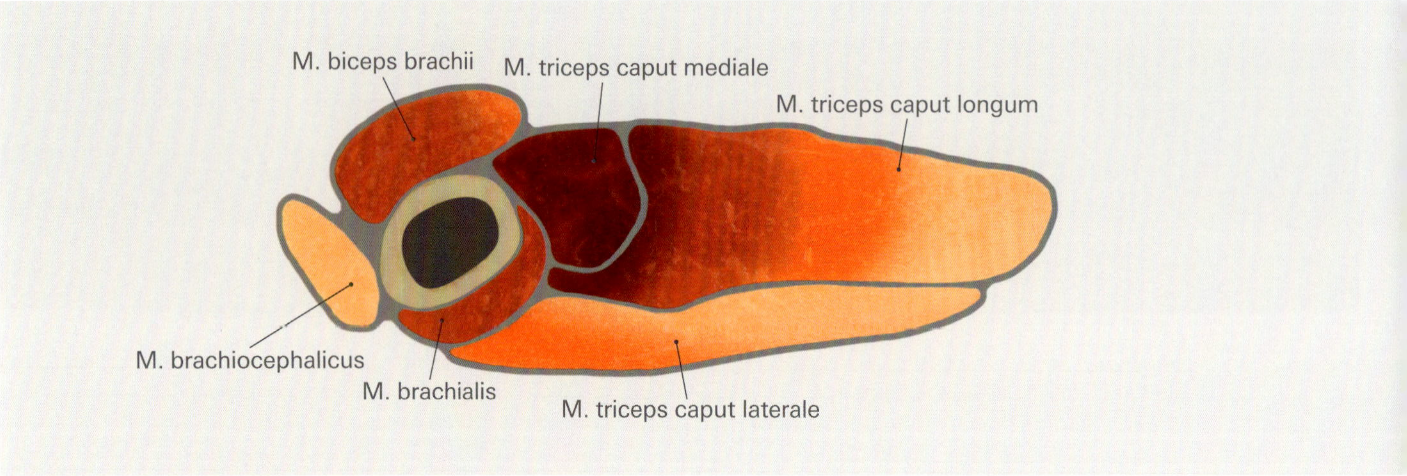

Proportion of oxidative type I muscle fibers in a cross-section of the upper arm. The density of oxidative fibers increases towards the bone (based on Armstrong et al. 1990 b).

One peculiarity of dogs is their lack of purely glycolytic type IIB fibers. This fact is sometimes disputed in the literature but has been amply proven (Maxwell et al. 1977, Braund et al. 1978, Gunn 1978, Snow et al. 1982, Latorre et al. 1993, Strbenc et al. 2004, Acevedo Rivera 2006). Instead of type IIB fibers, dogs possess another mixed fiber type known as type IIA/X, or IIX for short. This type has been found in the leg muscles of various dogs, but was not found to be present in the back musculature of a beagle (Steuernagel 2009). Type IIX fibers have also been found in isolated muscles in other large mammals (Smerdu et al. 2009).

The occurrence of type IIX fibers rather than purely glycolytic type IIB fibers is a key factor in the extraordinary levels of stamina often displayed by dogs. Unlike almost all other mammals, dogs can keep all their muscle fibers working over long periods of time. Furthermore, the presence of numerous capillaries ensures that the fibers are well supplied with blood.

The dog's legendary stamina is not just a myth propagated by dog owners who are less than fit themselves. Dill et al. (1932) tested canine stamina when outside temperatures were low, oxygen supply good and the animals well fed and watered and were forced to conclude that under these conditions, it is almost impossible to tire a dog out. One dog ran for 17 consecutive hours with just five minutes' break every half hour, covering a distance of 132 km. The experiment is likely to have been much more tiring for the scientists than for the dogs involved. The Iditarod race is an even better example of stamina. The race is 1850 km and the race record is 219 hours. The lab example is an average speed of 7.76 km/hr for 17 hours. The race record average speed is 8.45 km/hr for 219 hours. 200 km a day for 9 days! There was a team of dogs that did that distance and they pulled a sled and human the whole way. For this reason, Snow et al. (1982) respectfully refer to dogs in their publications as "perfect running machines".

Distribution of muscle fiber types

The distribution of the different fiber types varies enormously from muscle to muscle. Generally speaking, the muscles of the forelimbs contain a higher proportion of type I fibers than those of the hindlimbs. However, no two muscles in the body boast exactly the same distribution of fiber types. Armstrong et al. (1982) determined the proportion of each fiber type in all the fore and hindlimb muscles of three mongrels. The proportion of type I fibers ranged between 14 and 100 %.

The way in which the fiber types are arranged within the muscles is far also from uniform. In almost all the muscles investigated so far, the proportion of red fibers increases from the outer edges to the interior of a muscle. Significantly more type I fibers are found close to the bone than in areas of the muscle further away from the bone. The fibers further from the bone have a more favorable angle of attack with regard to the direction of pull than those near the bone, with the result that the more rapidly contracting type II fibers tend to be located in the outer regions of a muscle.

There is not much variation in fiber type distribution between individuals of the same breed, and no significant differences have yet been discovered between breeds, either (Newsholme et al. 1988,

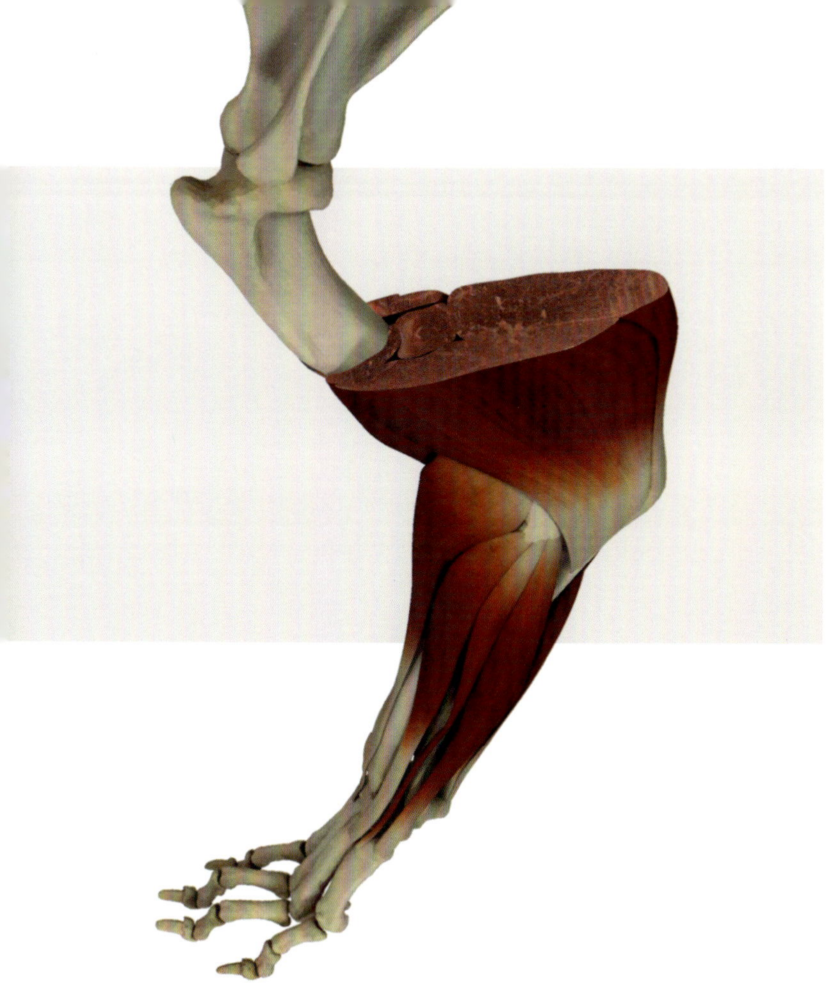

MUSCLE TRAINING

Muscle performance can be increased through training, and too little movement reduces the ability of a muscle to generate force. Training generates additional myofibrils inside the muscle fibers, thus increasing their overall diameter. Generally, no new fibers are created unless special satellite cells are activated for the purposes of regeneration after an injury. Training also improves the ability of the muscle cells to absorb oxygen, increases cardiac output and blood supply to the muscles, and sometimes promotes the formation of new capillaries in the muscles.
Interval training, which is a type of training practiced by humans to increase the capacity of type IIB fibers and the body's tolerance of high levels of lactic acid in the blood, is of little benefit to dogs, as they do not possess this kind of muscle fiber.

Kuzon et al. 1989, Aguera et al. 1990, Höche 2008). A notable exception is formed by the greyhounds, which boast a higher proportion of fast contracting type II fibers than other breeds (Gunn et al. 1978, Guy Snow 1991, Rodriguez-Barbudo et al. 1984).
90-95 percent of the muscle fibers of newborn pups are still undifferentiated. The few differentiated fibers to be found are very large red fibers which disappear again after 4-5 weeks. A gradual differentiation into red and white fibers takes place gradually in the first four weeks of life. By week 12, the distribution pattern found in the adult dog has formed. The muscle fibers in the different muscles all develop at approximately the same rate. It is not true that puppies exhibit lots of white fibers and adult dogs lots of red fibers (as claimed in Brown 1986, p. 70, for example). Type I fibers cannot change into type II fibers or vice versa.

Energy for muscle contraction
Energy in the form of ATP, the body's "energy currency", can be obtained in two different ways, one involving oxygen, the other not. In their resting state, muscles consume less than 20 percent of the body's total energy; when working hard or at maximum capacity this can increase to 90 percent (Lindstedt et al. 1985). If a very large amount of energy is needed, the body can no longer provide the muscles with enough oxygen to obtain it, mainly because only small amounts of oxygen can be stored in the cells. Sugar is burned by both anaerobic and aerobic pathways. It can be deposited in the cells for future use in its storage form glycogen. These stores are used for energy retrieval during times of extreme exertion.
While the process by which energy is obtained with the help of oxygen (cell respiration or aerobic energy metabolism) is slow but very efficient, glycogenic metabolism (glycolysis or anaerobic energy metabolism) is quick but costly. During cell respiration, 37 ATP molecules are obtained from one glucose 6-phosphate molecule (the normal form of sugar in a cell), while during glycolysis only two ATP molecules are obtained.
Explained simply, during cell respiration the nutrients transported into the cell are broken down in the mitochondria into carbon dioxide ($CO2$) and oxygen ($O2$). Approximately 50 percent of the energy obtained from carbohydrates and fats can be converted into ATP. The rest is not wasted but is converted into heat, thus helping to maintain the operating temperature. During glycolysis on the other hand, the deposits in the cells are made use of and broken down into pyruvic acid. When the oxygen capacity of the mitochondria is exhausted, the pyruvic acid is converted into lactic acid. Lactic acid and other processes cause the muscle to become too acidic. As the acidity in the vicinity of a muscle fiber increases, the ability of the myosin to bind to actin decreases and the muscle tires or, worse, becomes exhausted and hard.
In the case of dogs, whose stamina is legendary, it should thus be clear that the absence of purely glycolytic type IIB fibers significantly reduces the danger of muscle over-acidification. Even when dogs are on the go for a very long time, their muscles always operate using oxygen and do not get tired under normal outdoor conditions, except during extremely fast sprints.

THE GREYHOUND – THE SPRINTER AMONG DOGS

The speed and sprinting prowess of Greyhounds has long fascinated not only scientists, which goes some way to explaining the sheer number of studies dedicated to this breed. A Greyhound is capable of speeds of up to 18.5 meters per second (67 km/h) and can run an 800 meter race in less than 60 seconds. A comparison of the speeds recorded in the three main Greyhound races over the years reveals no increase in the last 40 years (Denny 2008). When running at maximum stride frequency, Greyhounds can take up to four strides per second (Jayes & Alexander 1982). By way of comparison, a cheetah can run 29 meters per second, and the world record sprinter Usain Bolt 10.4 meters per second (37.5 km/h).

The cross-sectional area of the muscle fibers in the leg muscles of Greyhounds is greater than in other dogs, and the proportion of type II fibers is significantly higher (Gunn 1978, Guy & Snow 1981, Rodriguez-Barbudo et al. 1984). Dobson et al. (1988) investigated the metabolic characteristics of three hindlimb muscles (M. biceps femoris, M. vastus lateralis, M. gastrocnemius) which in the Greyhound consist of over 90% rapidly contracting type II fibers. After a sprint, the lactic acid content in the blood plasma was 16 times higher than resting levels and the pH value of the blood had fallen from 7.4 to 6.9. In the muscle fibers themselves the lactic acid level was six to seven times higher and the pH value fell from 6.9 to 6.5, indicating acidity.

Myofibrils and sarcomeres: the transformation of chemical energy into mechanical

Myofibrils and sarcomeres are the smallest components of a muscle. A myofibril consists of three main proteins, actin, myosin and titin, and is serially divided along its length into sarcomeres. Sarcomeres are between one and a half and three micrometers long (Sharir et al. 2006) and form the contractile element of a muscle. Sarcomeres are connected with each other along the length of the muscle by disc-like structures known as z lines. Z lines consist of at least 30 different proteins about whose function not much is known. Each actin filament is fixed to a z line. The actin filaments are interspersed with myosin filaments, with which they are sporadically connected via myosin heads. Each myosin filament is surrounded by 6 actin filaments, but the overall ratio of myosin to actin filaments is 1:2. A sarcomere, then, consists of approximately 2000 thin actin filaments and around 1000 myosin filaments. Myosin filaments are 0.01 micrometers in diameter and 1.6 micrometers long. Actin filaments are half as thick and two thirds as long.

Contraction takes place on the level of the myofibrils. The myosin filaments pull the actin filaments towards the middle of the sarcomere. The filaments are responsible for the direction and accumulation of the movements of the individual molecules. Because the sarcomeres which make up the myofibrils are arranged in series, the contractions of the individual sarcomeres add up. Myofibrils run through the whole length of the muscle and can be up to 20 centimeters long.

The pulling of the filaments past each other is triggered by the release of calcium in the muscle fiber; the energy for the movement is generated by the splitting of the energy molecule ATP to produce ADP. Chemical energy (ATP) is transformed into mechanical work by the pulling of the filaments (= contraction). During contraction, a sarcomere shortens by 1/40,000 millimeter, or 0.4 micrometers. To make it easier to imagine this process, we talk about a "molecular oar stroke". The oars are the tiny mobile heads on the myosin. The oar stroke begins with the binding of the myosin head to a particular point on the actin. This requires calcium. The head then "nods" (a movement of just 5 nanometers) and the actin is pulled past it. The energy for this work comes from the breaking down of ATP into ADP. ATP then binds with myosin again, the head is released from the actin filament, ATP is split and the energy produced leads to another nodding movement of the head. Take up of energy – docking – nodding – release of energy – release of head... This oar stroke cycle can be repeated up to 100 times a second, and billions of these oar strokes are needed for a dog to take a single step – yet only a fraction of the available myosin molecules are activated at any one time.

A sarcomere and its three most important myofilaments. Right: this section is fully encased by actin filaments. Center: some of the actin filaments have been removed to reveal the myosin filaments in the interior. Left: titin.

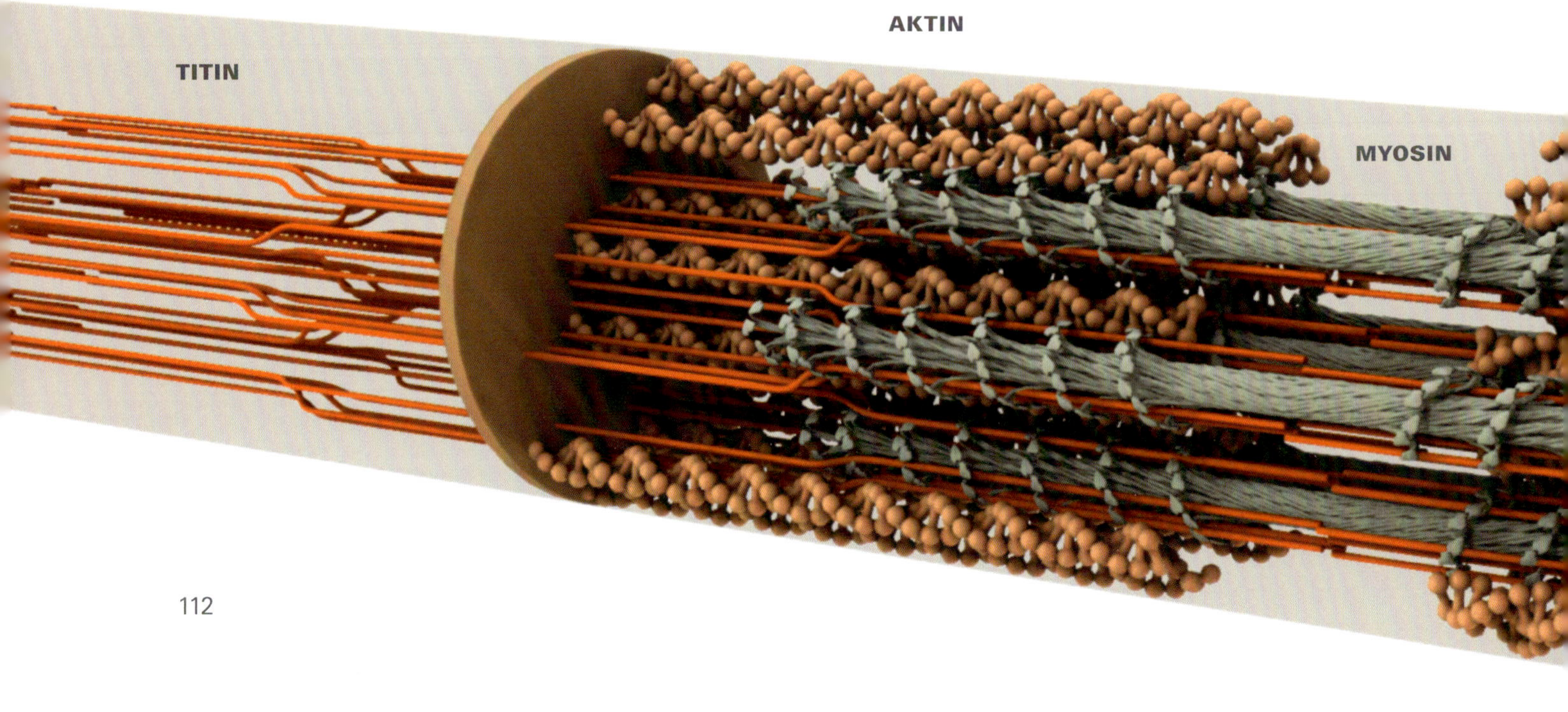

Elasticity

The actin-myosin-sarcomere model explains the molecular mechanism which leads to the contraction and active shortening of muscles. In order for them to return to their starting position, however, a spring mechanism is also necessary. This role is assumed by the protein titin, which forms a type of muscle cell skeleton which centralizes actin and myosin filaments and is responsible for the resting tension of muscles. Titin is the length of half a sarcomere, around 1 micrometer, and is the largest protein in the mammalian body. Its most important characteristic is its elasticity. Titin needs to be elastic so as not to hinder contraction and in order to be able to return the myofilaments to their original position after contraction. The many titin molecules which occur in a sarcomere behave like serially connected springs, and give muscles spring-like properties. The limb stiffness and compliance observed during locomotion is thus not just down to the elastic behavior of tendons but is also caused by the elasticity of the muscles.

Muscles do not just work actively, they also work passively. When a dog makes contact with the ground during locomotion, gravity ensures that its extensor muscles are passively stretched simply by its body weight. This stretching takes place initially on the level of the tendons, with long tendons able to absorb more elastic energy than short ones. The titin in the muscles stretches like a molecular spring. The length of a muscle can vary according to its state of contraction or extension. A contraction is most effective when the length of the muscle at the start of the contraction is 120 percent of its resting length. The passive stretching of the extensor muscles caused by the absorption of body weight brings them to precisely this length.

The elastic energy stored in the tendons and muscles is recovered at the end of the stance phase, which saves both active work and energy (see page 57). The energy costs of locomotion differ between breeds. Small dogs require significantly more energy than large ones to produce the same force. Large dogs benefit from the fact that their greater mass allows them to recover more energy from the spring-like properties of the tendons and muscles, especially during the trot.

A SARCOMERE IN ITS RELAXED STATE AND IN ITS CONTRACTED STATE.

During a contraction, the myosin heads consume energy and "nod" and the titin changes its structure.

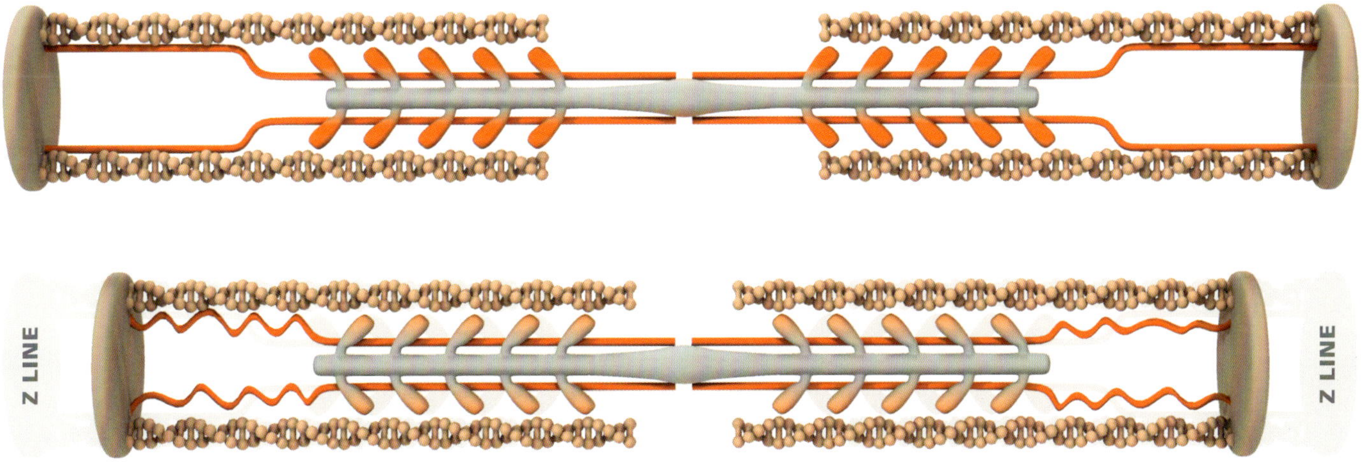

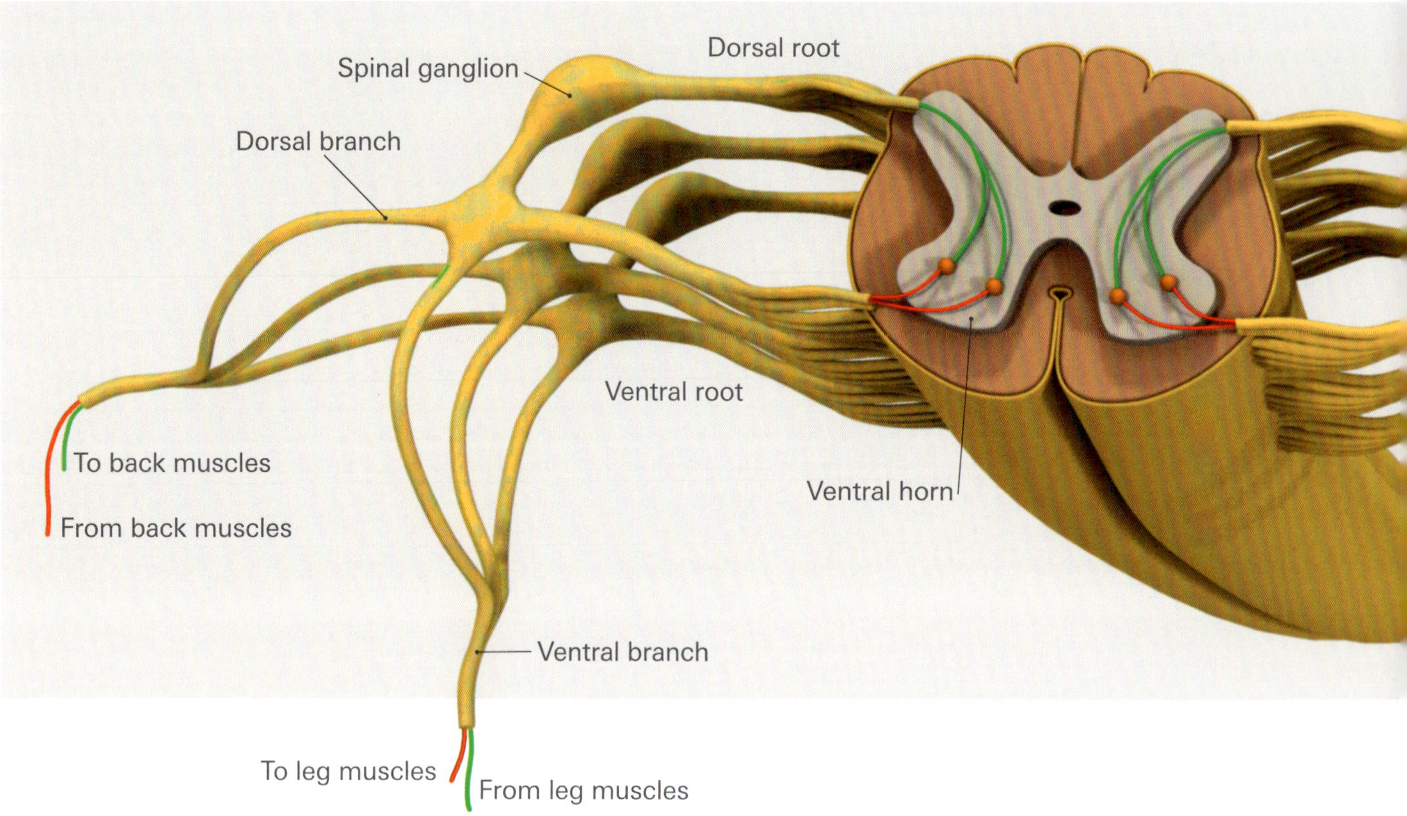

CONTROL

Muscle control

Muscle control is as unconscious a process in dogs as it is in humans. The brain can neither send direct commands to the muscles nor receive direct sensory information from them. Both motor and sensory pathways have one or several switch points in the spinal cord where the information from one nerve cell is conveyed to another, either upwards in the direction of the brain or downwards in the direction of the muscles. In experiments involving cats, the spinal cord was severed at various different levels, thus "turning off" specific parts of the brain, or parts of the brain were actually removed. When the cats were hung by the trunk over a treadmill (this was necessary due to the loss of their sense of balance), they were able to perform the walk, trot and gallop. The critical impulse or initial command from the brain which triggered the first step was provided passively by the turning on of the treadmill and the retraction of the limbs.

Once cyclic locomotion - constant rhythmic locomotion with no external interference - is underway, no signals or commands from the brain are necessary. Cyclic locomotion is probably controlled by certain rhythm generators in the spinal cord (central pattern generators, or CPGs, Grillner 1975), although no anatomical evidence of these has yet been found in mammals. CPGs are thought to be responsible for the simultaneous touchdown of the diagonal limb pairs during the trot. Once they receive information from one limb, they can adjust the rhythm of the other limbs (Buxton Perk 1990).

Every command relating to the movement of a muscle emanates from the ventral horn of the spinal cord. This is where the commands are received from the superordinate centers of the brain that are responsible for consciously triggered movement, or voluntary motor function. The nerve cells (upper motor neurons) are located in the motor cortex of the brain. The fibers which branch off them (axons), which also form part of the cell, form structures called pyramidal tracts. The axons do not run directly into the muscle but into the "lower motor neurons" in the spinal cord. There are two types of lower motor neuron: alpha and gamma. The first innervate the muscles, the second the muscle sensors, or muscle spindles. Muscle spindles register the level of contraction of the muscle. Should the desired posture or movement require a change in the strength of the contraction, the muscle spindles forward this information on. If a muscle needs to be contracted, the alpha and gamma motor neurons are activated at the same time. The activation of the alpha motor neurons enables the movement to be rapidly initiated, while the activation of the gamma motor neurons leads to the re-setting of the reference value of the muscle spindle control loop.

Alpha motor neurons are organized into groups of cells which supply muscle groups that have a similar function. The axons of the alpha motor neurons leave the spinal cord together via the ventral root and split into an upper (dorsal) and

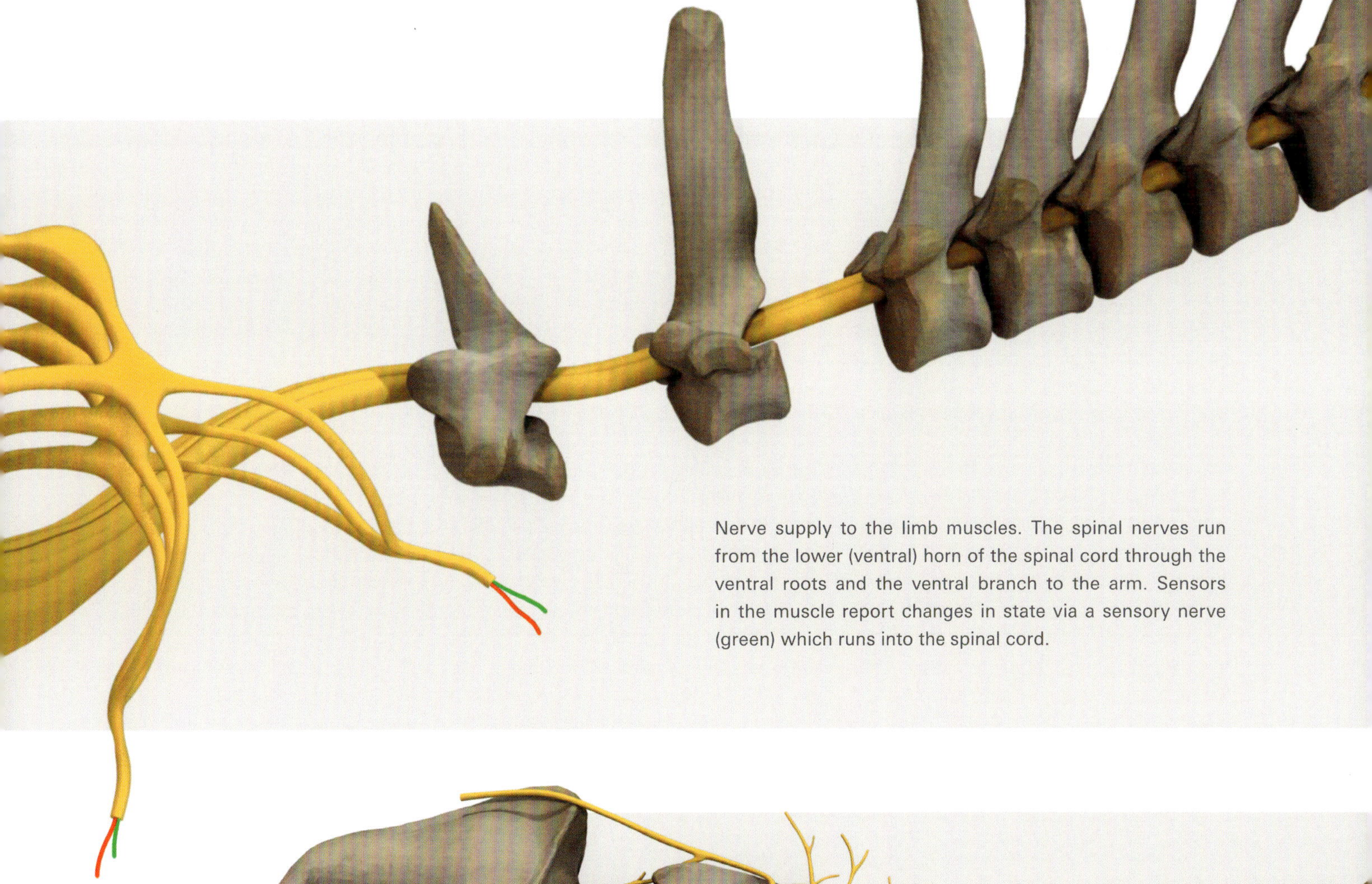

Nerve supply to the limb muscles. The spinal nerves run from the lower (ventral) horn of the spinal cord through the ventral roots and the ventral branch to the arm. Sensors in the muscle report changes in state via a sensory nerve (green) which runs into the spinal cord.

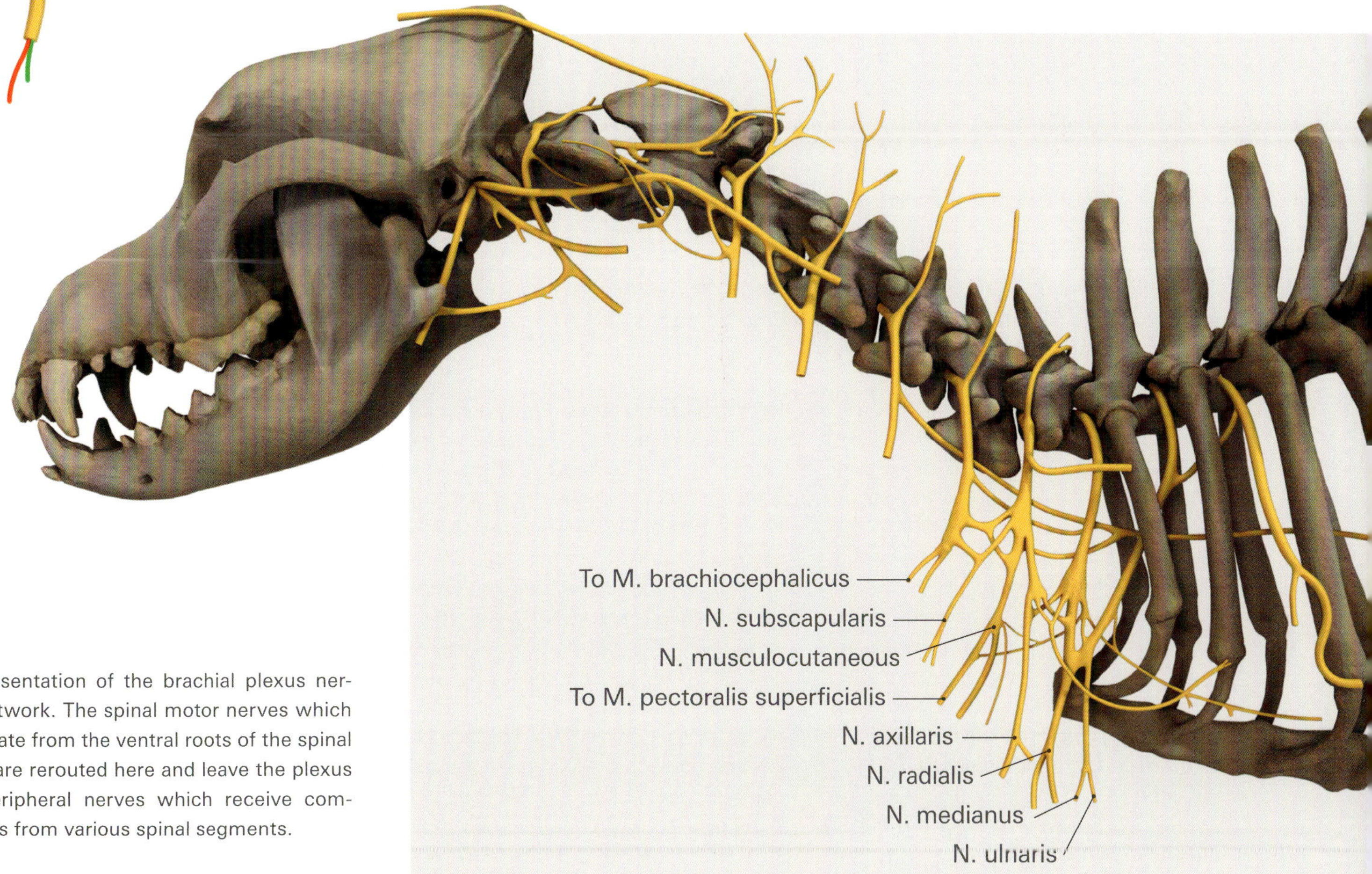

Representation of the brachial plexus nerve network. The spinal motor nerves which emanate from the ventral roots of the spinal cord are rerouted here and leave the plexus as peripheral nerves which receive commands from various spinal segments.

lower (ventral) spinal cord nerve, also known as the spinal nerve. The dorsal spinal nerve supplies the muscles of the back, the ventral nerve all other muscles in the trunk and limbs. Dogs have 36 pairs of spinal nerves, eight in the region of the neck, thirteen in the thoracic region, seven in the lumbar region, three in the sacral region and five in the tail region (Flechter Kitchell 1966). Spinal nerves emerge from the spinal cord at every vertebra in correspondence with the originally segmental nature of the musculature. This is particularly evident in fish. Fish fillets consist of a series of consecutive muscle lamellae, with each lamella associated with a spinal nerve. The trunk region is segmental in all vertebrates. Segments in this context are defined as serially arranged, identical elements which structure the nerves, muscles and certain skeletal elements (e.g. the vertebrae) in the trunk region.

In the course of the formation of the limbs during the evolution of land vertebrates, the strict segmental arrangement of the muscles disappeared, but that of the nerves remained. Almost every limb muscle is supplied by parts of two to three spinal nerves. After they have emerged from the spinal cord and before they run to the muscles, the spinal nerves are rerouted in a way which can be compared to the switching of railway points. The points are located in the nerve networks of the fore- and hindlimbs, the brachial plexus and the lumbosacral plexus. Only after their rerouting within the plexus do the spinal nerves become the peripheral nerves of the arms and legs, each of which is made up of various spinal nerves. The rerouting which takes place in the plexus can vary greatly from individual to individual even within the same species, and this certainly applies to dogs (Fletcher 1970). The muscles of the hindlimbs generally originate from five segments which correspond to the 5th to the 7th lumbar vertebrae (L5-L7) and the first two sacral vertebrae (S1-S2) (Ziegler 1934). The spinal nerve of L6 is involved in the nerve supply of almost all the hindlimb muscles. The muscles of the arms originate in the last two cervical (neck) segments and the first two thoracic (chest) segments (C6-Th2) (Sharp et al. 1991).

The motor unit

Muscle contractions are triggered by nerve impulses. Nerves supply groups of muscle fibers rather than individual fibers, which simplifies muscle control significantly. All the muscle fibers that are supplied by the same nerve form a motor unit. All the muscle fibers in a unit contract at the same time. One nerve fiber can be responsible for the activation of up to a thousand muscle fibers. The peripheral nerves which run from the spinal nerves to the muscles form increasingly fine ramifications until they eventually reach the muscle fibers. The muscle fibers are activated indirectly through the release from one of the bulbous extensions leading off the nerve endings (the motor end-plates) of a neurotransmitter, acetylcholine, which is absorbed by the muscle cell receptors, thus initiating the depolarization of the muscle fibers.

All the fibers of a motor unit are the same fiber type. The motor unit is not to be confused with the muscle fascicle, which can consist of different types of muscle fibers. Muscle fascicles are thus made up of the fibers of various motor units. The size of a motor unit depends on how precise the movements in which the muscles are involved have to be. If the muscles are needed for very precise movements, only a few hundred fibers of one type are innervated. If the muscles are responsible for gross movements, a motor unit can contain up to 2,000 muscle fibers.

The strength with which a muscle contracts depends on the number of activated motor units. This is known as recruitment. Firstly, the slow, red muscle fibers are activated, followed gradually by the fast white fibers. Once the nerve impulse has arrived at the muscle fiber it takes approximately 40 milliseconds before force develops. In other words, there is an electromechanical delay. Activation takes place via a contact point between the motor neurons and the muscle fibers known as the synapse. Activation involves a change in the balance of electrical charges in the membranous sheath of the muscle fiber, which is depolarized. The depolarization spreads at a speed of approximately one meter per second. It is this change in electrical charge which is measured when electromyography is used to investigate muscle activity. However, the depolarization does not just run along the muscle fibers but penetrates, via tiny channels, into the interior of the cell and leads to a release of calcium which triggers the contraction on a molecular level and permits the binding of the myosin heads to the actin.

Electromyography

Electromyography is a way of measuring the electrical activity in a muscle. An electromyogram (EMG) provides information about whether the part of the muscle being investigated is active or not. Studies using surface electrodes have shown that a muscle is rarely activated as a whole but is subject to activity waves. During cyclic locomotion for instance, different regions of the M. triceps brachii are active at different times. Sometimes, only very small areas of a muscle are activated. During cyclic locomotion, small regions of a muscle containing many type I fibers may be activated while regions containing a large number of type II fibers remain inactive (Fischer 1999, Scholle et al. 2001). Tokuriki (1973a, b, 1974, 1979) and Goslow et al. (1981) investigated the activity patterns of almost all the fore- and hindlimb muscles and some of the back muscles of dogs performing the walk, trot and gallop. The fore- and hindlimb and back and abdominal muscles that are functionally important during locomotion were then reinvestigated by an American research group using more modern EMG techniques (Fife et al. 2001, Ritter et al. 2001, Deban Carrier 2002, Carrier et al. 2006, 2008, Schilling Carrier 2009, Schilling et al. 2009). The animation present the data taken from the literature in such a way that it is clear which muscle or muscle region is activated at which phase in cyclic locomotion.

Muscles as sensors

In addition to its motor properties, muscle is also an important sensory system. Muscle spindles are sensors which are responsible for stabilizing the muscle at the ideal level of initial tension, holding the joints in a particular position and maintaining posture. They form part of a control circuit in which a change in the length of the muscle (reference value) is triggered

by a change in the length of the muscle spindle. Muscle spindles give the muscle and the rest of the body information about the level of contraction of a muscle by reporting changes in muscle length to the spinal cord. Muscle length can actively be adjusted via the gamma motor neurons.
Muscle spindles are modified muscle fibers containing thin muscle fibers whose state of contraction determines the length of the muscle spindle. They are a maximum of three millimeters long and are fixed to the surrounding muscle fibers by a sheath of connective tissue. Their number differs from muscle to muscle, though Buxton Peck (1990) describe some general observations in dogs. The highest muscle spindle density is found in the short muscles of the paw (Mm. interflexorii, Mm. lumbricales), which boast approximately one spindle per square millimeter. The arm flexor muscle M. biceps brachii features approximately one spindle per seven square millimeters, while in the long head of the arm extensor (M. triceps brachii caput longum), one spindle is present per 50 square millimeters. The high density of muscle spindles in the paw muscles is probably connected with their function as "kinesiological monitors". The paws are required to respond to minute changes in uneven substrates and to change their position accordingly. In addition, short muscles which only span one joint have more spindles than long ones which span several joints, which is probably to do with their role in the fine adjustment and monitoring of the position of the joints.
The inbuilt muscle sensor system is an important feedback system for quick corrections and the coordination of the rhythm generators in the spinal cord. Information about the substrate is reflected in changes in the length of the muscle. When a dog steps in a hole, for instance, the muscle is stretched. This information can be used to coordinate and correct the movement of the limbs directly on the level of the spinal cord (Perret Berthoz 1973, Buxton Beck 1990). Muscles inform each other about the state they are in and actively try to prevent overstretching (Hondo et al. 1984, Fischer 2001). Caicoya et al. (1999) describe the muscle information network and

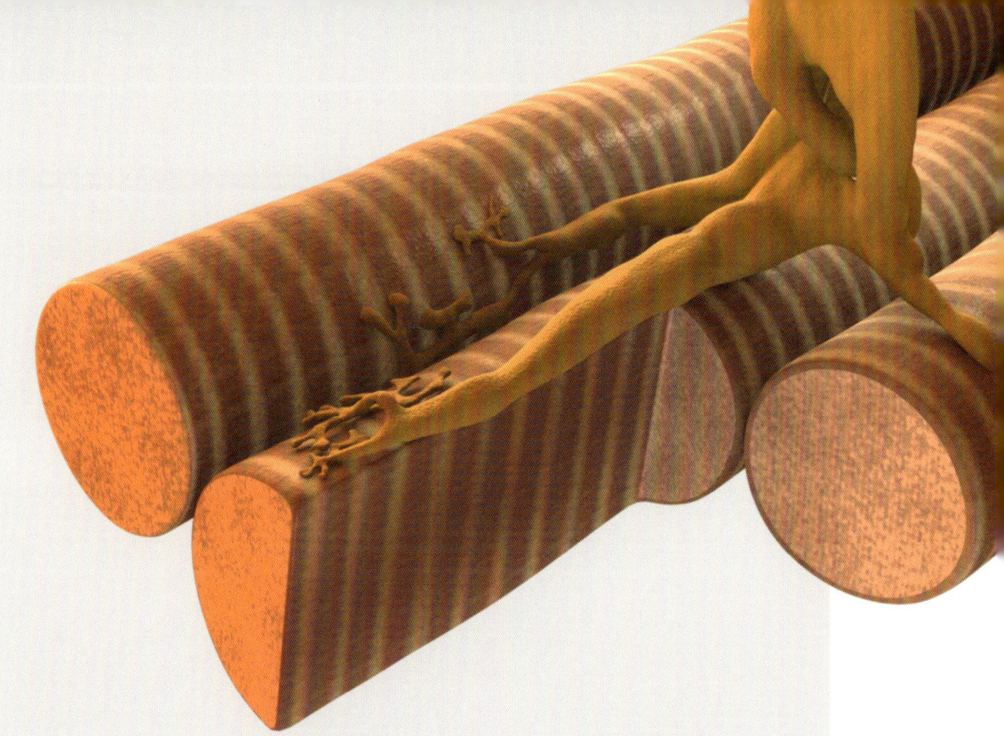

Representation of a motor end plate. The nerve which innervates the muscle fiber branches and dilates at the end until it reaches the motor end plate , where the synapses are located.

Representation of a synapse. The motor end plate of the muscle fiber is shown in yellow. When activation occurs, an electrical signal is transformed into a chemical signal. Tiny bubbles containing the neurotransmitter acetycholin are released via minute channels and reach the receptors via the synaptic divide.

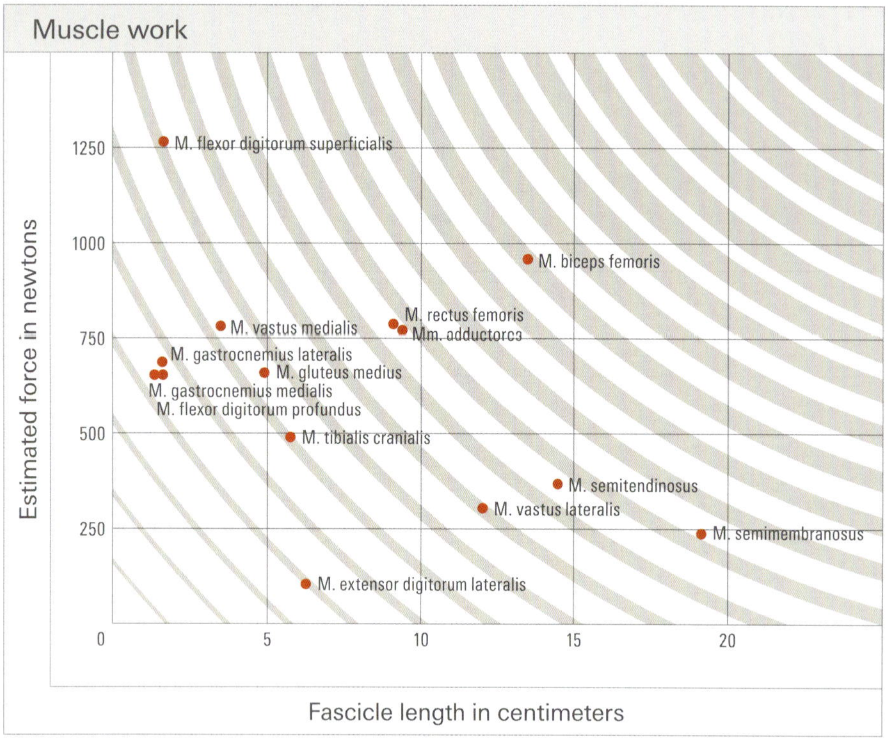

The work of a muscle depends on two factors: the possible force and the length of the muscle fascicles. The muscles which work the hardest are located in the darker region (based on Williams et al. 2008a).

show that the long head of the arm extensor (M. triceps brachii caput longum) plays a central role in the control loop of the forelimbs.

The muscle spindles are not the only sensors in the system. The tendons, too, boast special networks of nerves known as Golgi tendon organs which report to the spinal cord information about the tension of the tendons and thus of the muscles. Golgi tendon organs are capsules of connective tissue which are approximately one millimeter long and a tenth of a millimeter in diameter.

THE GENERATION OF FORCE

Force is generated in the muscle cells by the summation of tiny movements and actin-myosin interaction. It is then transferred via the sheath of the muscle fibers to increasingly taut layers of connective tissue. Via a tendon or other means of insertion, it eventually reaches a skeletal element, which moves. The force of a muscle is thus calculated as the quotient of the muscle volume (the bigger the muscle, the more myofibrils it contains) and the muscle fascicle length (the longer the fascicle, the more sarcomeres it contains. The amount of force a muscle generates corresponds to the number of myosin heads that are activated. Each filament possesses the same number of myosin heads, which mean that muscle force is proportional to the number of filaments in a muscle. The higher the number of filaments, the greater the force generated. The physiological cross-section of a muscle, which is calculated on the basis of the cross-section of the individual muscle fibers, gives a good approximation of the number of serially connected sarcomeres present. The ability of a muscle to generate force depends on the length and position of the muscle fibers. Rapid muscles have long fibers, while powerful muscles boast numerous short fibers which all lie parallel to each other. This is reflected in the "architectural index", the quotient of muscle fiber length and total muscle length. As the muscle fibers in most muscles do not run parallel to each other, the architectural index also takes into account the angle at which the muscle fibers run relative to the length of the muscle, which is also known as the pennation angle. The architectural index is a factor in estimating contraction speed. In the Greyhound, the powerful calf muscles M. gastrocnemius medialis and M. gastrocnemius lateralis have an index of 0.07 to 0.14, while the fast hamstring muscles M. semitendinosus and M. semimembranosus have an index of 0.70 to 0.88 (the fibers in this case are up to ten times longer) (Williams et al. 2008 a). In the forelimbs, too, the rapid muscles with the long fibers are those which connect the limbs to the trunk (M. rhomboideus and M. latissimus dorsi with an index of 0.97), while the powerful muscles are the toe retractors (M. flexor digitorum: 0.06) which have extremely short fibers (0.8 centimeter fiber length to 13.8 centimeters muscle length) (Shahar Milgram 2005).

Force generation is expressed in terms of the relationship between length and force. Maximal force can only be achieved when the sarcomeres are a certain length at the start of a contraction. A sarcomere can only develop the full force of which it is capable from a resting length of between 2.0 and 2.25 micrometers. If a sarcomere is shortened or extended beyond this, its capacity for force generation decreases rapidly. In the first case the actin filaments get in each other's way, while in the second the number of points at which the myosin heads have the opportunity to connect decreases. Although muscle length

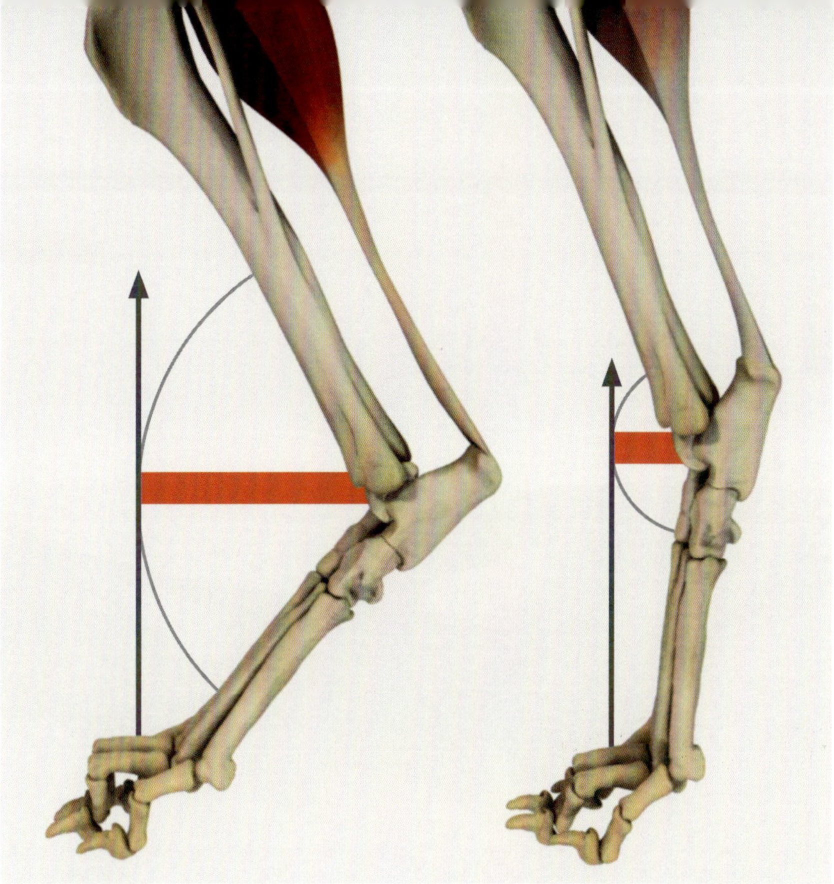

The relationship between the lever arm and the lift arm in the ankle. The extensor muscles need to work much harder at the beginning of the stance phase than in the second half as the relationship between the lever arm, whose length remains almost constant, and the lift arm (red bar) is less favorable (based on Carrier et al. 1998).

can be decreased by almost half, optimal functionality is achieved at between 90 and 120 percent of resting length.

Muscle work

Work is defined as force times displacement. . A muscle can be forceful without being strong. Theoretically then, a muscle would be working at full capacity when all the myosin heads were nodding at the same time, or in other words all the filaments were contracting at the same time. As this is not measurable in practice, however, it is assumed that the maximal work of which a muscle is capable is proportional to its volume. Muscle volume is deemed to be the quotient of muscle mass and muscle density, with muscle density remaining constant at 1.06 g/cm^3 (Mendez Keys 1960). As muscle density thus almost corresponds to the value 1 the only factor that is recorded in practice when it comes to estimating the maximal work of a muscle is muscle weight. The absolute weight of muscles obviously varies in accordance with body weight, and should really be normalized. Unfortunately, however, there are no comprehensive accounts of the weights of individual muscles, which makes it impossible to calculate relative sizes. The most complete set of data pertains to the Greyhound (Williams et al. 2008 a, b). Rauscher (1986) provides the weights of many forelimb muscles in the Dachshund. Williams et al. (2008 a, b) present a summary of the mechanical parameters of the fore- and hindlimb muscles of the Greyhound. A strong muscle, i.e. one that can perform a lot of work, has a specific ratio of muscle volume to muscle fiber length. Williams et al. (2008a) calculated this ratio for a hypothetically very strong muscle and compared the values they obtained with the values obtained for real muscles. In the hindlimb, the M. biceps femoris is the strongest muscle by far. It weighs approximately 500 g and can generate a force of 940 newtons while working at 50 watts. The adductors and the M. rectus femoris are also very strong. It is difficult to transfer these values to other breeds, however, as the M. biceps femoris, for example, is twice the size in the Greyhound as it is in other breeds (Williams et al. 2008 a). Although the M. flexor digitorum superficialis is more powerful than the M. biceps femoris (1.260 as opposed to 940 newtons), it is not stronger (6 as opposed to 50 watts). The strongest forelimb muscle is the long head of the M. triceps brachii, whose values are comparable with those of the M. biceps femoris (1.475 N and 58 W).

Comparisons of muscle mass also highlight the vast differences between the various muscle groups at the level of the different joints. While the ratio of flexors to extensors is 2:10 in the elbow and 3: 10 in the shoulder, there are twice as many flexors as extensors in the wrist. The transmission of force, however, depends not just on the properties of the various muscles but also on the mechanics, and especially the lever arms. The lever arm is the right-angled distance from the point of muscle insertion (the line of action) to the pivot of the movement (e.g. the joint axis). As the ratio between the lift arm and the lever arm changes during a movement, the torque (resulting force x length of the lever arm) is not a constant. 30% of the way through the stance phase the extensor muscles of the ankle joint (M. gastrocnemius medialis and M. gastrocnemius lateralis) need to generate forces that are almost twice as large as those required 70% of the way through the stance phase (Carrier et al. 1998), just because of changes in the position of the leg and the load which the leg has to bear.

Muscle name				
Muscle weight*	Physiological cross-sectional area*	Force*	Work*	Proportion of type I fibers**
Stance phase		Swing phase		
Activity duration walk***				
Activity duration trot***				
Activity duration in the trailing limb in the gallop***				
Activity duration in the leading limb in the gallop****				

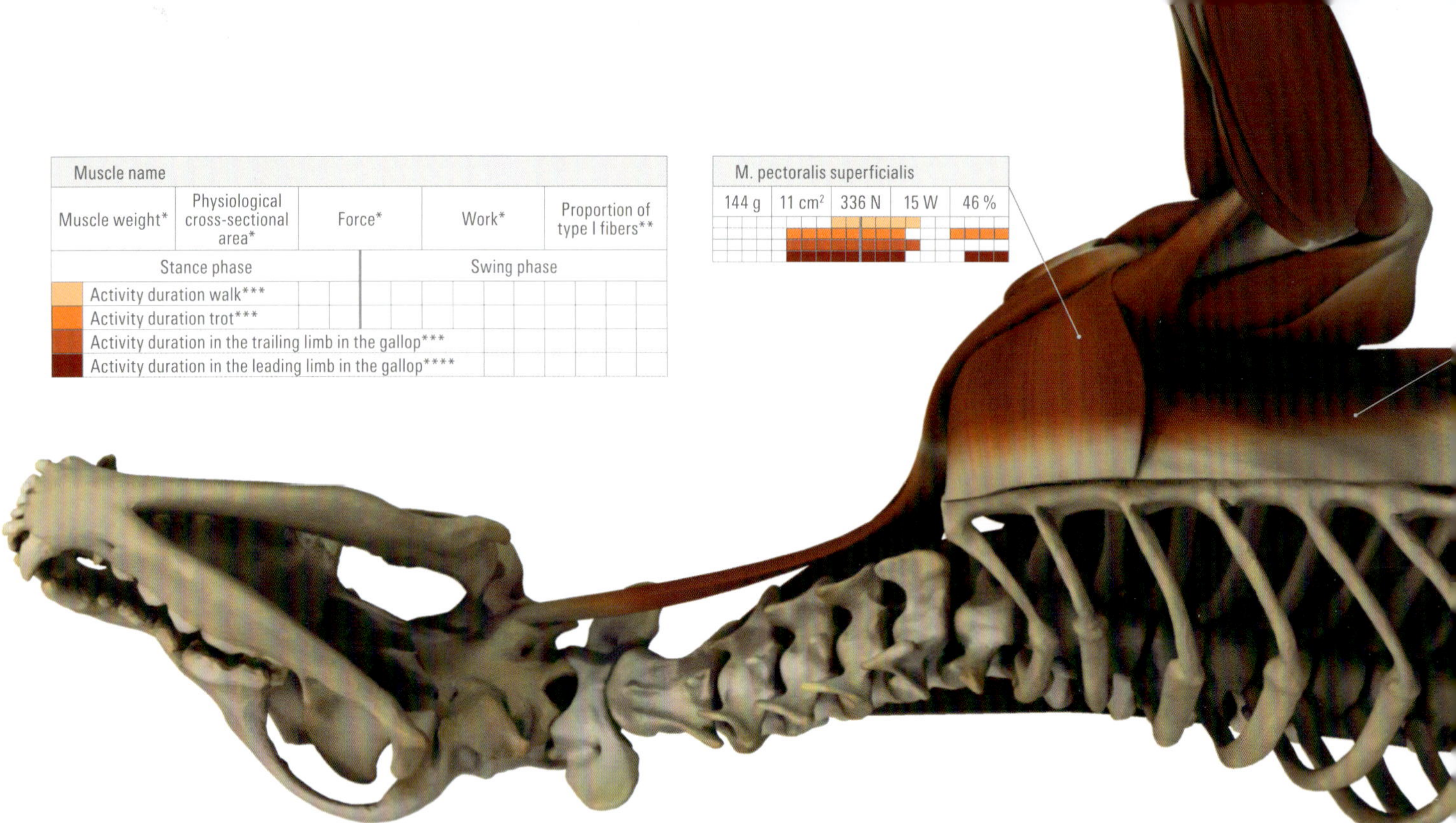

M. pectoralis superficialis				
144 g	11 cm²	336 N	15 W	46 %

THE FUNCTIONAL ANATOMY OF THE MUSCLES OF THE FORE- AND HINDLIMBS

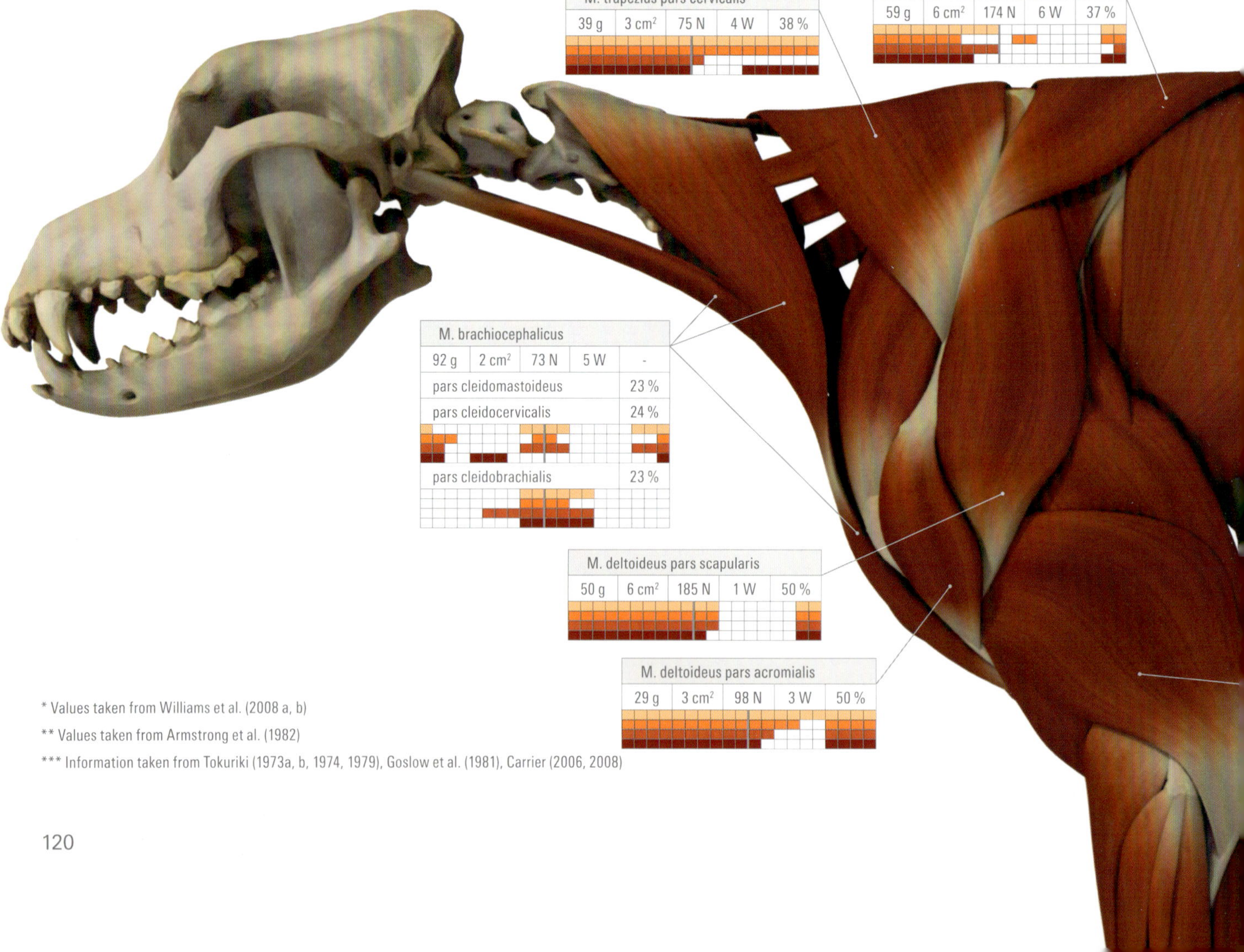

M. trapezius pars cervicalis				
39 g	3 cm²	75 N	4 W	38 %

M. trapezius pars thoracica				
59 g	6 cm²	174 N	6 W	37 %

M. brachiocephalicus				
92 g	2 cm²	73 N	5 W	-
pars cleidomastoideus				23 %
pars cleidocervicalis				24 %
pars cleidobrachialis				23 %

M. deltoideus pars scapularis				
50 g	6 cm²	185 N	1 W	50 %

M. deltoideus pars acromialis				
29 g	3 cm²	98 N	3 W	50 %

* Values taken from Williams et al. (2008 a, b)

** Values taken from Armstrong et al. (1982)

*** Information taken from Tokuriki (1973a, b, 1974, 1979), Goslow et al. (1981), Carrier (2006, 2008)

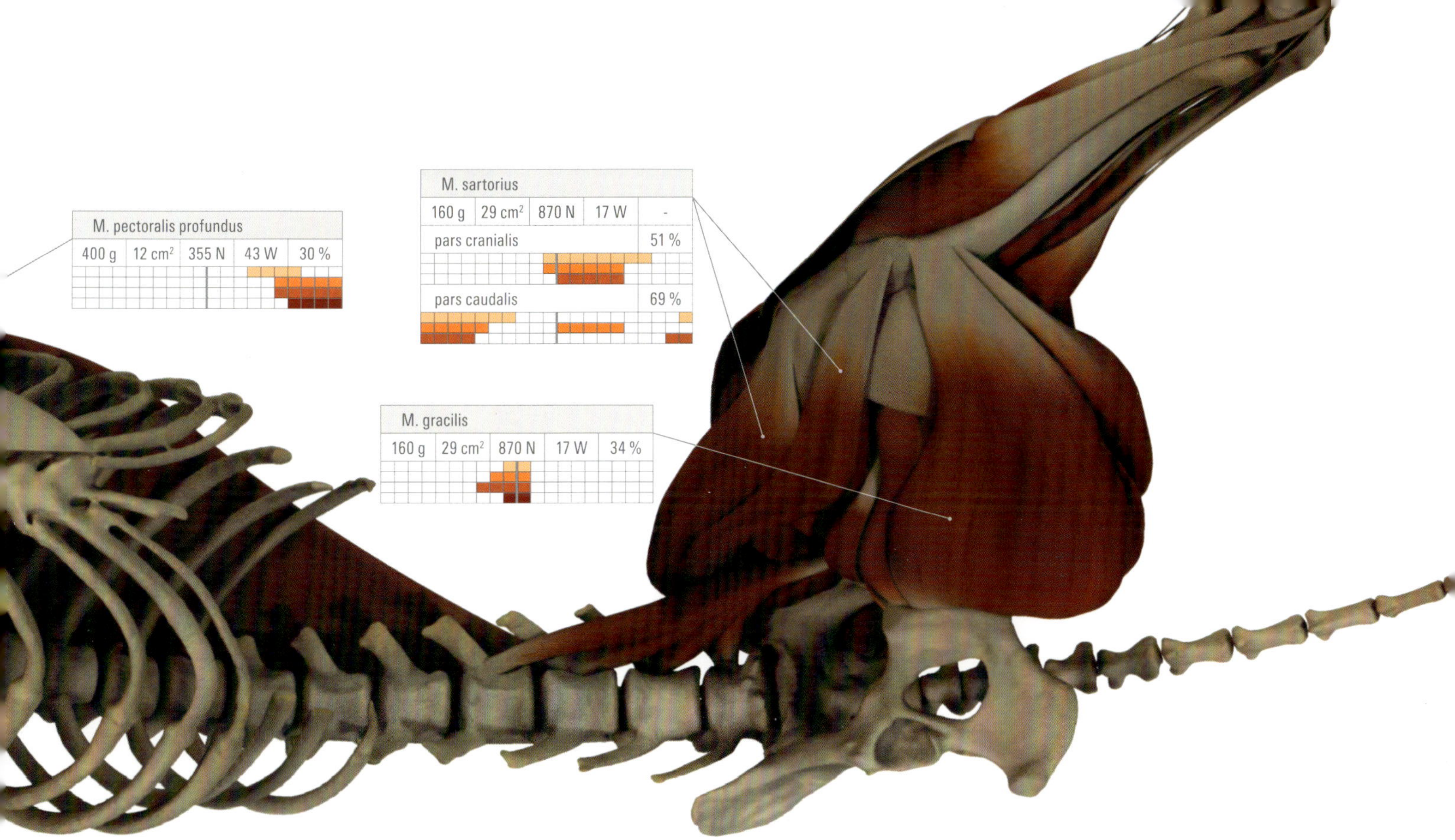
M. pectoralis profundus
400 g | 12 cm² | 355 N | 43 W | 30 %
M. sartorius
160 g | 29 cm² | 870 N | 17 W | -
pars cranialis 51 %
pars caudalis 69 %
M. gracilis
160 g | 29 cm² | 870 N | 17 W | 34 %

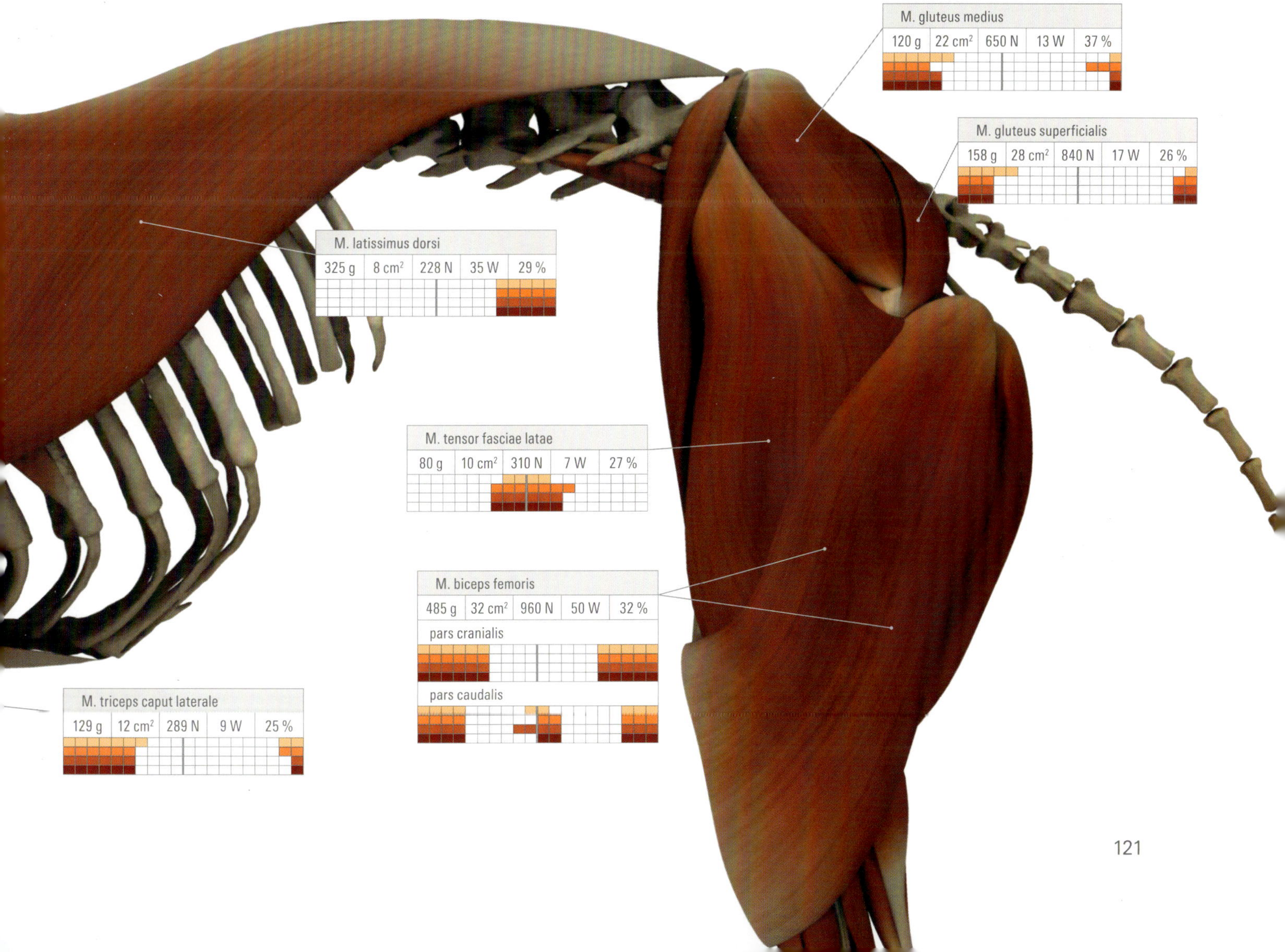
M. gluteus medius
120 g | 22 cm² | 650 N | 13 W | 37 %
M. gluteus superficialis
158 g | 28 cm² | 840 N | 17 W | 26 %
M. latissimus dorsi
325 g | 8 cm² | 228 N | 35 W | 29 %
M. tensor fasciae latae
80 g | 10 cm² | 310 N | 7 W | 27 %
M. biceps femoris
485 g | 32 cm² | 960 N | 50 W | 32 %
pars cranialis
pars caudalis
M. triceps caput laterale
129 g | 12 cm² | 289 N | 9 W | 25 %

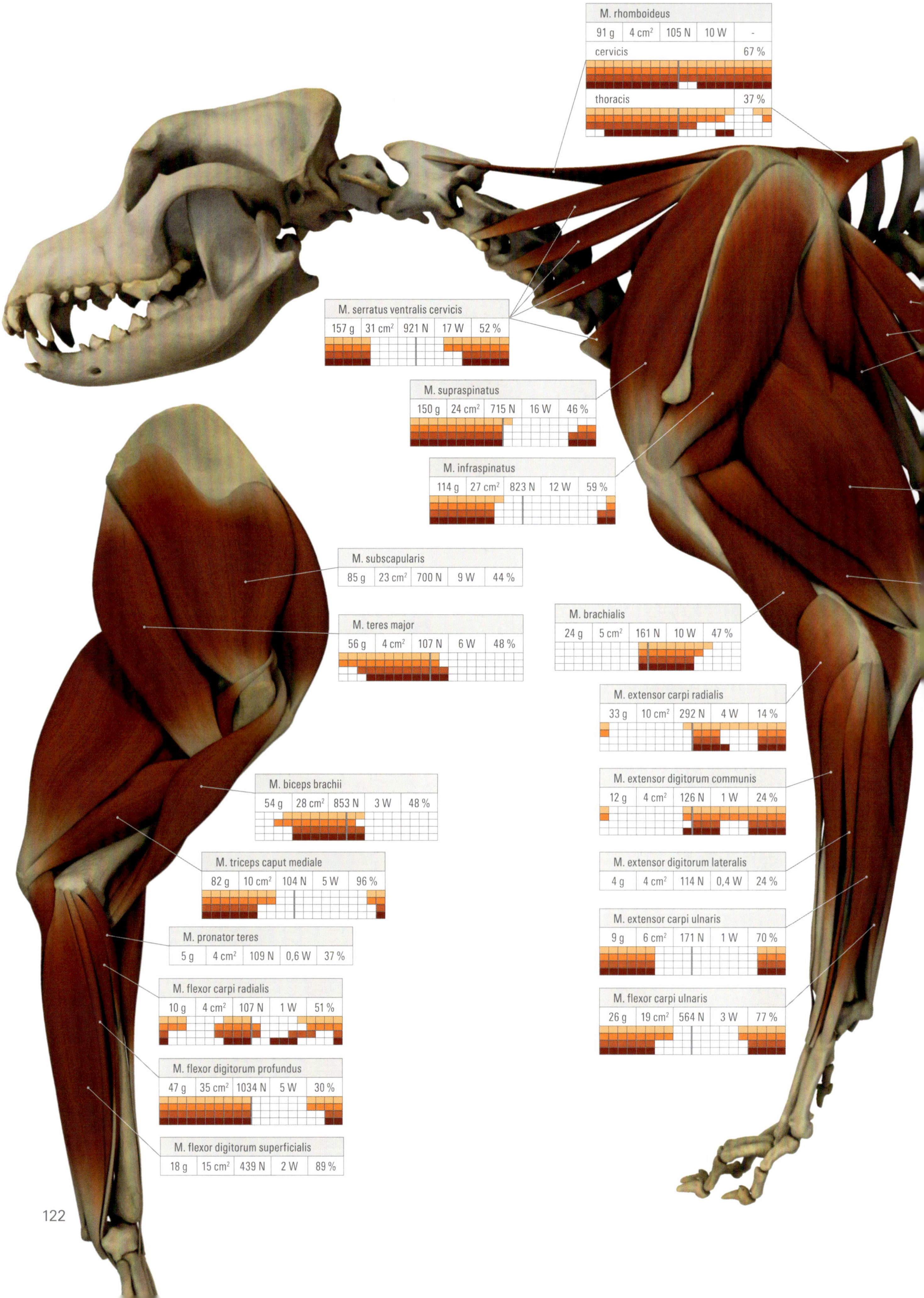
M. rhomboideus
91 g | 4 cm² | 105 N | 10 W | -
cervicis 67 %
thoracis 37 %
M. serratus ventralis cervicis
157 g | 31 cm² | 921 N | 17 W | 52 %
M. supraspinatus
150 g | 24 cm² | 715 N | 16 W | 46 %
M. infraspinatus
114 g | 27 cm² | 823 N | 12 W | 59 %
M. subscapularis
85 g | 23 cm² | 700 N | 9 W | 44 %
M. teres major
56 g | 4 cm² | 107 N | 6 W | 48 %
M. brachialis
24 g | 5 cm² | 161 N | 10 W | 47 %
M. extensor carpi radialis
33 g | 10 cm² | 292 N | 4 W | 14 %
M. biceps brachii
54 g | 28 cm² | 853 N | 3 W | 48 %
M. extensor digitorum communis
12 g | 4 cm² | 126 N | 1 W | 24 %
M. triceps caput mediale
82 g | 10 cm² | 104 N | 5 W | 96 %
M. extensor digitorum lateralis
4 g | 4 cm² | 114 N | 0,4 W | 24 %
M. pronator teres
5 g | 4 cm² | 109 N | 0,6 W | 37 %
M. extensor carpi ulnaris
9 g | 6 cm² | 171 N | 1 W | 70 %
M. flexor carpi radialis
10 g | 4 cm² | 107 N | 1 W | 51 %
M. flexor carpi ulnaris
26 g | 19 cm² | 564 N | 3 W | 77 %
M. flexor digitorum profundus
47 g | 35 cm² | 1034 N | 5 W | 30 %
M. flexor digitorum superficialis
18 g | 15 cm² | 439 N | 2 W | 89 %

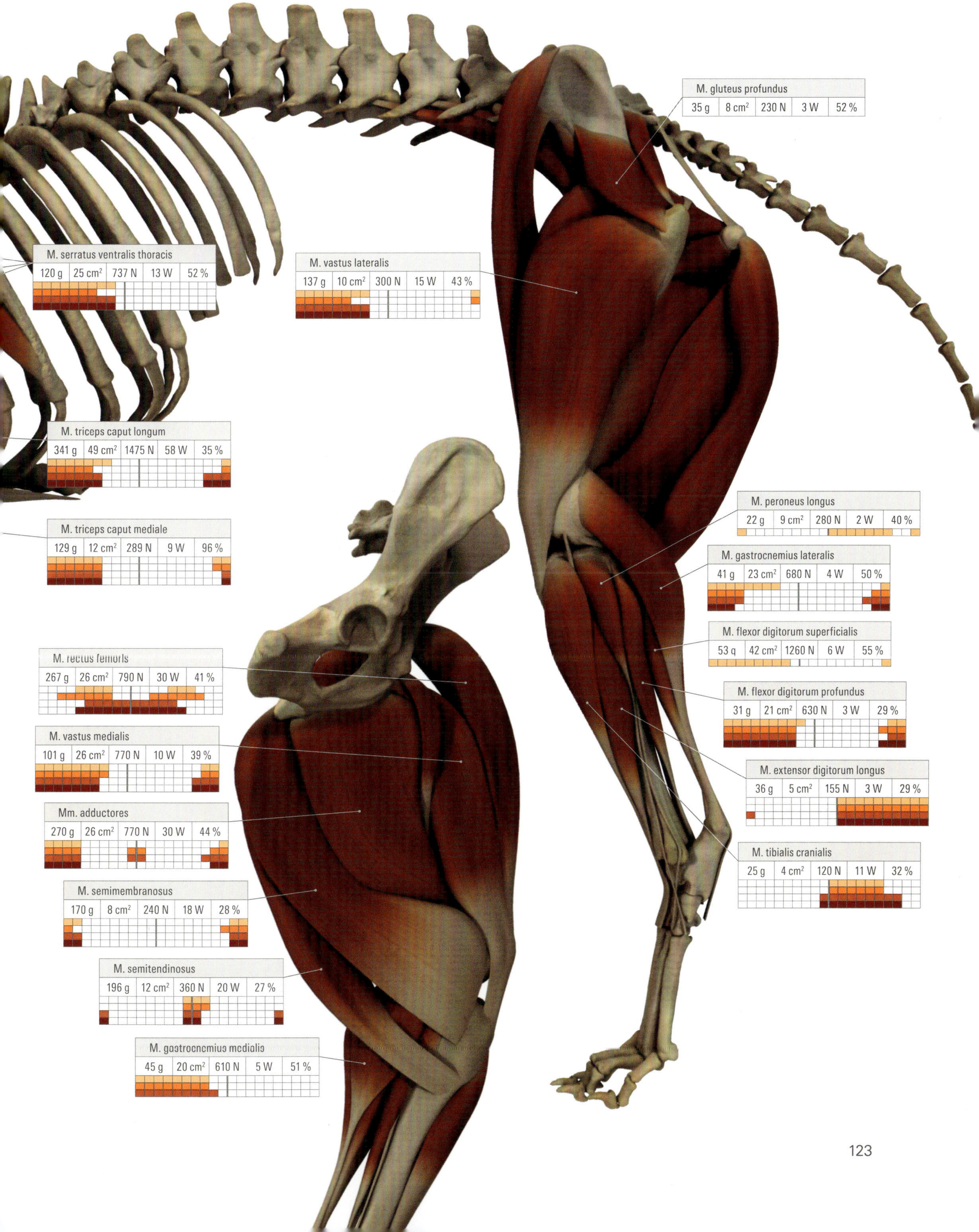
M. gluteus profundus
35 g | 8 cm² | 230 N | 3 W | 52 %
M. serratus ventralis thoracis
120 g | 25 cm² | 737 N | 13 W | 52 %
M. vastus lateralis
137 g | 10 cm² | 300 N | 15 W | 43 %
M. triceps caput longum
341 g | 49 cm² | 1475 N | 58 W | 35 %
M. triceps caput mediale
129 g | 12 cm² | 289 N | 9 W | 96 %
M. peroneus longus
22 g | 9 cm² | 280 N | 2 W | 40 %
M. gastrocnemius lateralis
41 g | 23 cm² | 680 N | 4 W | 50 %
M. flexor digitorum superficialis
53 g | 42 cm² | 1260 N | 6 W | 55 %
M. rectus femoris
267 g | 26 cm² | 790 N | 30 W | 41 %
M. flexor digitorum profundus
31 g | 21 cm² | 630 N | 3 W | 29 %
M. vastus medialis
101 g | 26 cm² | 770 N | 10 W | 39 %
M. extensor digitorum longus
36 g | 5 cm² | 155 N | 3 W | 29 %
Mm. adductores
270 g | 26 cm² | 770 N | 30 W | 44 %
M. tibialis cranialis
25 g | 4 cm² | 120 N | 11 W | 32 %
M. semimembranosus
170 g | 8 cm² | 240 N | 18 W | 28 %
M. semitendinosus
196 g | 12 cm² | 360 N | 20 W | 27 %
M. gastrocnemius medialis
45 g | 20 cm² | 610 N | 5 W | 51 %

MUSCLES AS METABOLIC ORGANS

In recent years it has become increasingly clear that muscle has yet another important function, playing a critical role in metabolic processes and in physical and psychological wellbeing. The same benefit that we humans get from exercise is almost certainly, and keenly, felt by dogs.

Only active muscles make use of the sugar in the blood. When energy is burned up, muscles become receptive to the hormone insulin, which is necessary for the utilization of sugar. When dogs (or humans) do not get enough exercise, instead of sugar being broken down, fat reserves are built up. There is thus a direct connection between movement and diabetes. Dogs that are overweight are almost always so for the same reason - lack of exercise. Being overweight often leads to a serious problem in dogs - they begin to move even less. This leads to muscle atrophy, something which occurs anyway with advancing age, and to the creation of fat deposits in the place of muscle. The lipid catabolism mechanism in the sub-cutaneous connective tissue and the activity of the muscle beneath it has recently been investigated from a biochemical perspective. The locally active signaling molecules that are released through muscle activity are called interleukins. Interleukins are cytokines, or hormone-like proteins. Interleukin-6 has been shown to be responsible not just for fat regulation but to have an anti-inflammatory effect (Petersen Pedersen 2005). With the help of an enzyme produced in the muscle (MIP), it also regulates calcium levels in the body. This implies that muscle aging processes and muscle activity also influence bone density (Romero-Suarez et al. 2010).

Scientists are now aware of almost fifty different messenger substances ("myokins") in muscle, and the biochemical reason behind that "feel-good" sensation experienced by dogs (and their owners) at the end of a long walk is becoming ever clearer.

FUNCTION OF DIFFERENT MUSCLES

The new insights into dog locomotion contained in this book fundamentally change our view of the way in which muscles work, which is why our descriptions of the function of muscles differ from those often found in traditional textbooks. A muscle can serve one of two purposes during locomotion: it either contributes to or prevents movement. Preventing movement becomes particularly important when the limbs are subject to the force of gravity during the stance phase, which can, in turn, induce movement. The role which muscles play in working against gravity is often neglected in traditional descriptions of their function.

In this section, the most important insights which this study provides into the motion of dogs are summarized briefly. The work performed during locomotion is divided unequally between the fore- and the hindlimbs. During regular, cyclic locomotion, propulsion is generated more by the hindlimbs. The forelimbs carry more weight (see p. 50-54). The forelimbs are rotated around a purely force-controlled and movable point (the instantaneous center of rotation) which is located on the upper edge of the shoulder blade (see p. 88-89). The forelimbs are rotated around a point on the upper edge of the shoulder blade (the instantaneous center of rotation) whose precise position changes depending on the forces at play. The hindlimbs are moved from the hip joint, the transmission joint for the whole body, by 40° to 50°. In all gaits, and particularly in the gallop, the movements of the lumbar spine play an important role (see p. 64). The muscles which connect the shoulder blade to the trunk, those which permit the retraction of the hindlimbs and, finally, the muscles of the back all merit particular attention as the work they perform makes the greatest contribution to speed and stride length.

The leg joints actually move very little during locomotion. In some breeds, the joints are held so stiffly during the walk that the functional leg length hardly changes. As a result, the pivot point of the limbs moves up and down with every stride – the limbs act like stiff pendulums, the movement of which conveys the body forwards (see p. 54). During the trot and the gallop the limbs in all breeds are consistently bent by the force of gravity and stretched again as this force decreases. In other words, the limbs act as elastic springs. The body's center of gravity can still be raised and lowered slightly during this process, which permits the transformation of potential energy into kinetic energy (see p. 57). Unlike during the walk, the lowering of the center of gravity occurs as a result of the gravity-induced flexion of the joints. Like springs, the muscles store the elastic energy generated during the flexion of the leg in the tendons or in the titin, the molecular spring of the muscle itself (see p. 113). When, subsequently, the limbs are stretched, the stored energy can be released. The mechanisms involved in dog locomotion are thus essentially geared towards minimizing energy consumption and making "intelligent" use of gravity to save energy.

A muscle can either actively contract or be passively stretched. Muscles can, however, also maintain their length. This too requires force to be exerted and is described as isometric (= equal length) contraction. A good example to illustrate this is that of remaining in a semi-squat over a long period of time. Preventing a joint from moving even when a load is brought to bear on it requires force. Electromyographic investigations (see p. 116) have permitted the initially surprising observation that when a joint is extended, it is not the extensor muscles which are active and when a joint is flexed, it is not the flexor muscles which are active. During locomotion, the job of the extensors is to prevent unintentional flexion in the joints. Rather than living up to their name, the extensors regulate, or "dose", the degree of flexion which occurs in the joint, and they do so via isometric contraction. When all the leg joints are stabilized in this way, the limb becomes a stiff rod. As soon as the "extensors" allow themselves to be passively stretched, however, the joints are flexed and the energy stored during the period of stretching can be released in the next extension, the degree of which is now determined by the "flexors". Terms such as flexor and extensor can, therefore, be confusing. The "dosing" of flexion and extension enables the dog to react quickly to changes in the substrate, which is particularly important on uneven terrain. The compliance of the joints and thus that of the limbs is a precondition for the smooth locomotion of dogs in outdoor spaces. The work that it takes to secure this compliance is divided up between the different parts of the limb. The further they are away from the body, the more the muscles are involved in the storage of elastic energy and in compensating for differences in the level of the substrate. This explains the long, powerful tendons in those muscles which control the wrist and ankle joints. In line with this biomechanical finding, Pit Bull Terriers, which are bred for winning dog fights, are relatively stable as a consequence of that selection. For stability, they have more muscle mass in their forearms and lower legs than Greyhounds, which specialize in racing (Pasi Carrier 2003). Greyhounds have less muscle mass and are better at storing elastic energy, particularly in the Achilles tendon.

Isometric contraction serves another purpose, too. Not all joints are form-fitting hinges or perfect ball and sockets - many are incongruent (see p. 72). This means that muscle work is required to stabilize them. During locomotion, a good many muscles are mainly or exclusively dedicated to this function.

Our knowledge of the work performed by muscles is limited to locomotion. Practically no scientific studies exist into muscle activity during other forms of movement. One exception to this is the study by Tokuriki (1979), who investigated what happens when a dog jumps vertically into the air from standing. It is important to remember that muscles fulfill many other functions beyond locomotion, even if some of them are as "insignificant" as allowing dogs to scratch behind their ears. However, placing locomotion at the forefront of investigations is justified in some respects due to the very particular nature of canine muscles. While the topography of the muscles of dogs – their position and arrangement – does not differ much from that of cats, dogs possess an unusual muscle fiber type, type IIx (see p. 108). This makes the muscles of dogs particularly suited to endurance performance. It is this type of fiber which enables wolves to be able to hound large prey over long distances to the point of exhaustion (see p. 15). As a consequence of their descent from wolves, locomotion is the most important activity in the dog repertoire.

Thanks to Tokuriki (1973 a, b, 1974), it has been known for over thirty years that there is no significant difference between the activity patterns of the limb muscles in the walk and the trot, and only very minor differences in the trot, as the motion sequence which takes place is essentially the same in all three cases. The activity pattern describes the point at which activity occurs and the length of time it lasts for. Differences are observed, however, in the intensity of activity, which provides an indirect indication of the number of active muscle fibers. As in cycling, slow locomotion can be more tiring than rapid locomotion, as during rapid locomotion the energy storage mechanisms built into the muscles can be used more efficiently.

The function of the back muscles

In dog locomotion, the body is propelled forwards not just by the movement of the legs, but by movements which take place in the back as well. The contribution which the motion of the back makes to propulsion depends largely on the gait being used. In the gallop, which is characterized by significant up and down bending in the spine, the contribution of the back to propulsion is fairly significant (see p. 63). Although symmetrical gaits are predominantly movements of the legs, significant lateral displacement of the trunk and rotation around the longitudinal axis of the body also occurs (Jenkins Camazine, 1977), which is visible in the swinging of the hind quarters. Rotation around the longitudinal axis occurs mainly during the phase of three-legged support in the walk and the phase of diagonal support in the trot.

As in other mammals, the back muscles of dogs consist of three main longitudinal systems, the transversospinal system, the longissimus system and the iliocostalis system. While the muscular portions of the latter two systems span several vertebrae, the transverse spinal system consists of both long and short muscles. The deep muscles in particular the Mm. rotatores serve to connect adjacent vertebrae with each other.

The way in which the back muscles function during locomotion was, for a long time, a complete mystery. Even today, EMG recordings only exist for three back muscles in dogs, the M. multifidus lumborum, the M. longissimus thoracis et lumborum and the M. iliocostalis thoracis et lumborum (Tokuriki, 1973 a, b, 1974, Tokuriki Nakata 1991, Ritter et al. 2001, Schilling Carrier 2009, Schilling Carrier 2010). In addition, Wada et al. (1993) investigated the role of the tail muscles in keeping the body balanced.

In the gallop, the activity measured in the muscles is easy to connect to the visible movements of the spine. The muscles on the two sides of the body have one joint activity phase per motion cycle. This begins in the flight phase shortly before the touchdown of the trailing hindlimb and culminates in the stretching of the back. The activity of the back muscles and the stretching of the back end with the liftoff of the trailing hindlimb.

In the walk and the trot, two activity phases per motion cycle were observed for the three back muscles investigated. One occurs in the second half of the stance phase and the other in the second half of the swing phase. The effect of the activation of the muscles differs depending on when it takes place. During the swing phase, the contraction of the muscles leads to the expected lateral bending of the trunk, while during the stance phase this bending does not occur – the muscles stabilize the trunk. In this way, the back muscles create a static basis for the work of those leg muscles which originate in the trunk, and compensate for the torques which are transferred into the trunk (Schilling Carrier, 2009; Schilling Carrier, 2010). By recording the activity of three sections of the back musculature the same authors were able to show that the front sections are activated earlier than the rear sections (2x sectionS = more than three sections!) in the walk and the gallop and that activation takes place successively along the back. In the trot, on the other hand,

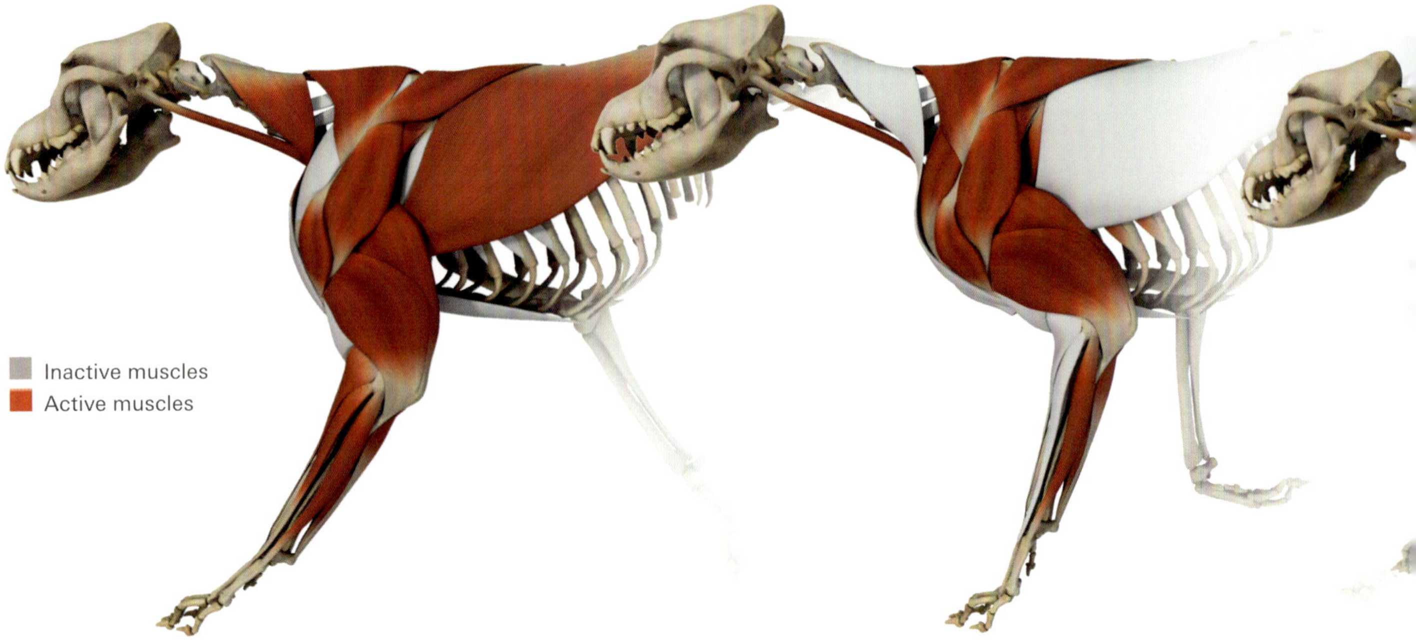

all three sections are activated simultaneously.

The function of the main forelimb muscles

The muscles of the forelimb can be divided roughly into three groups according to their activity pattern during the stance phase (Tokuriki 1973 a, b, 1974, Wentink 1976, Goslow et al. 1981). The first group (M. infraspinatus, M. triceps, M. extensor carpi ulnaris and M. flexor carpi ulnaris) is active from touchdown until the middle of the stance phase, the second group (M. pectoralis superficialis, M. teres major, M. deltoideus pars scapularis and M. flexor carpi radialis) in the second half of the stance phase (fig.). Only a few muscles work throughout the stance phase, and these form the third group (M. serratus ventralis, M. supraspinatus, M. deltoideus pars acromialis, M. biceps brachii, M. flexor digitorum profundus). The muscles of the first group counter the gravity which begins to act on the forelimbs at touchdown (for this reason they are known as "anti-gravity muscles") and stabilize the joints. They thus also play a role in compensating for any unevenness in the terrain. The muscles of the second group support the retraction of the limbs and prepare for liftoff. The muscles of the third group work throughout the stance phase to stabilize the joints for as long as the limb is load-bearing.

The protraction of the limbs during the swing phase is initiated by M. brachiocephalicus, M. brachialis and the flexors of the wrist joint. During the trot and the gallop the activity of the flexors, particularly M. brachialis and M. biceps brachii, is greater than during the walk because the limbs need to be simultaneously brought forward and raised from a trunk position which is usually lower than during the walk. The activity of the muscles which counter the flexion caused by gravity after touchdown is also greater in the trot and the gallop. At the end of the swing phase, M. pectoralis profundus and M. latissimus dorsi cause the forward swing of the forelimbs to slow down.

The function of the scapular muscular sling

The function of the groups of muscles which act on the shoulder is complicated and not well understood, as they serve

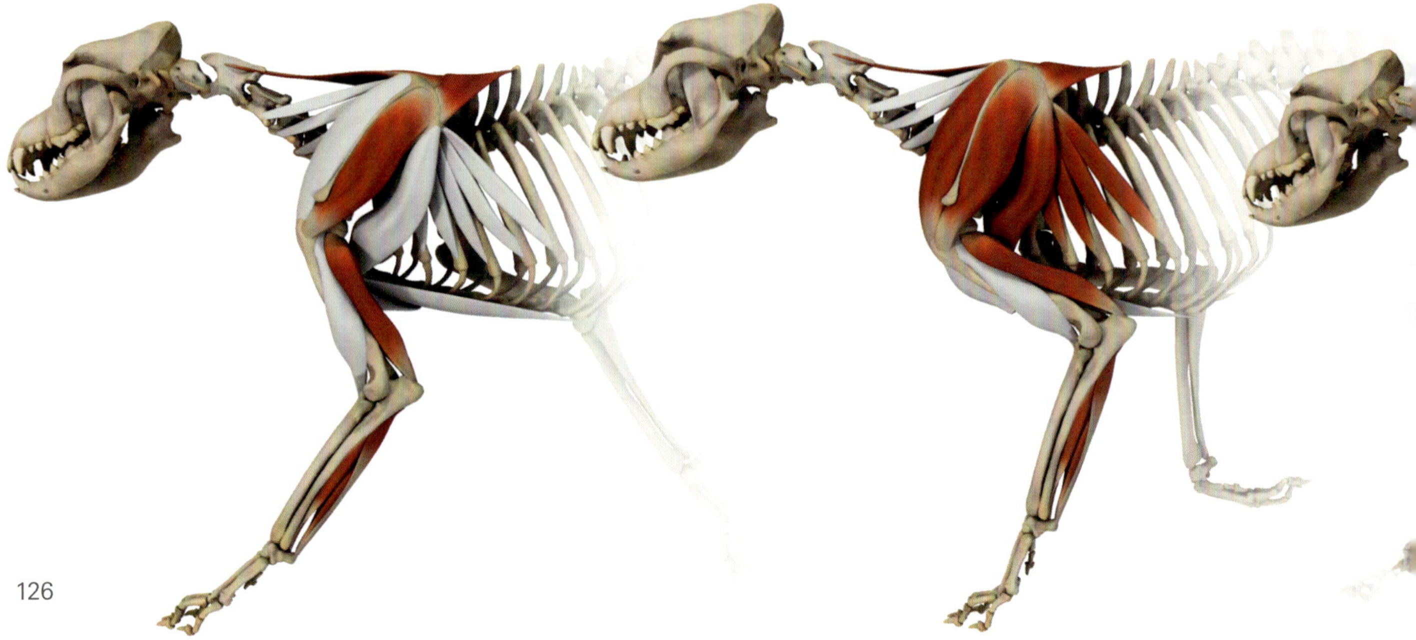

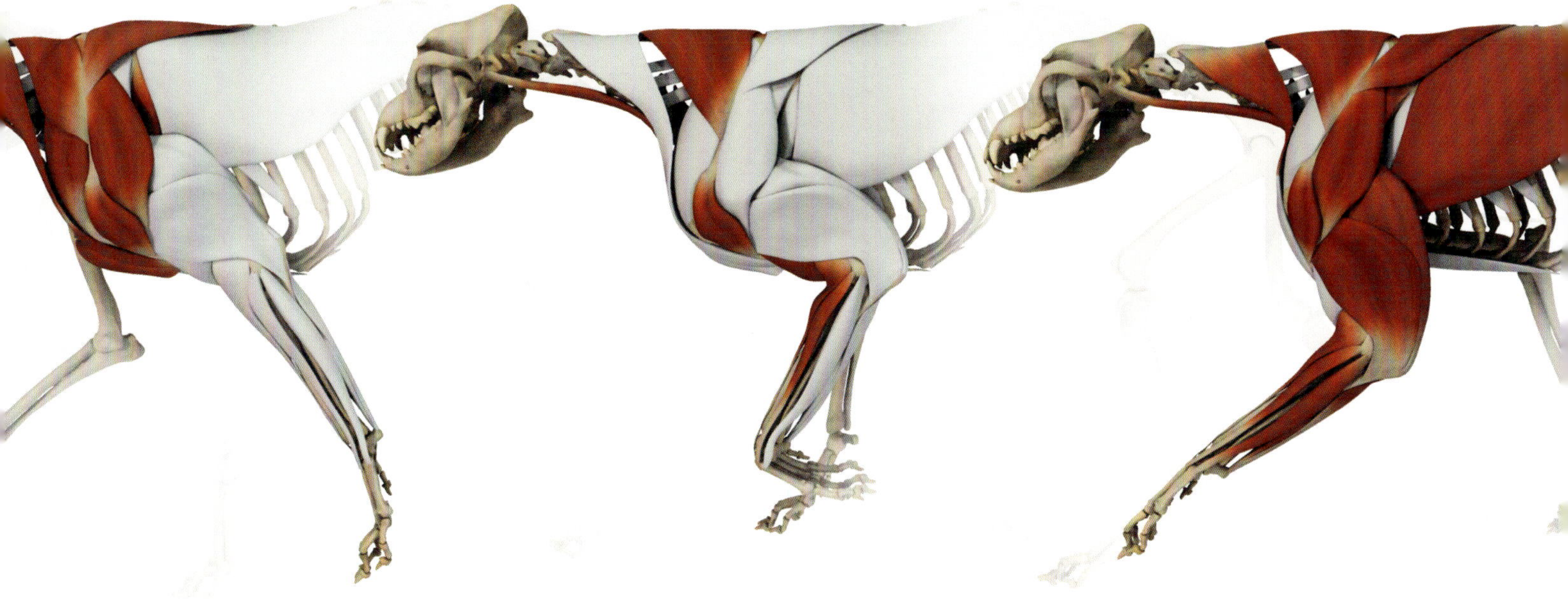

both to fix the extremity to the trunk and to permit its movement independently of the trunk. One of the most complex movements they are involved in is the rotation and translation of the shoulder blade. The same muscles which are responsible for the suspension of the shoulder blade also play a role, at certain times at least, in moving it. Furthermore, the muscular slings are broad at their origin, which means that their activation is likely to be successive. While the cranial portion of the muscle is "fired up", for example, the middle and distal parts can remain "silent". Despite the efforts of various scientists, no complete picture has yet been obtained of the activation pattern of the muscular sling.

The most extensive study to date into the function of these muscles was carried out by Carrier et al. (2006) in trotting dogs. In order to provide a better insight into the work of the individual muscles, the dogs were investigated not just while trotting normally, but also while wearing miniature backpacks containing 8 or 12 percent of their bodyweight to show which muscles exhibited increased activity under additional strain. In another experiment, dogs were made to run up and down an inclined treadmill so that any increase in the activity of individual muscles could be measured. The most important result to come out of the study was the observation that the only muscle involved in the suspension of the trunk is the thoracic portion of the serratus muscle which originates from the ribs, while the other muscles investigated (pectoral muscles, M. rhomboideus) are responsible for the protraction and retraction of the forelimbs and for stabilizing the pivot point of the shoulder blade. The neck portion of the M. trapezius (pars cervicis) has a three-phase activity period: during the first half of the stance phase, at the end of the stance phase and at the beginning of the swing phase (Carrier et al. 2008). The thoracic portion of this muscle is most active in the first third of the swing phase. The main function of the trapezius muscle is to stabilize the pivot point of the shoulder blade. Another surprising observation concerned the low levels of activity which occurred at certain times in the muscles investigated. Low muscle activity indicates that the torques occurring are also only low.

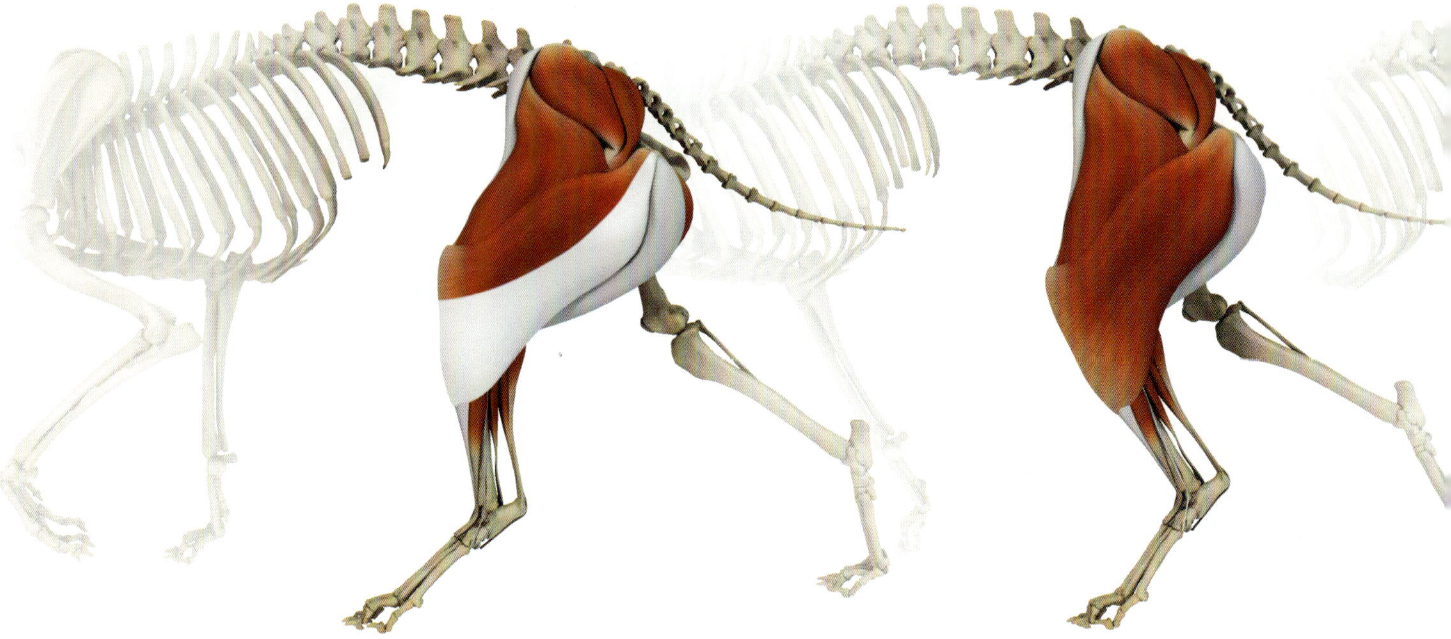

The function of the main hindlimb muscles

Three muscle groups are responsible for the protraction and retraction of the hindlimbs. Hindlimb retraction and thus the propulsion of the body are essentially the domain of the hamstring muscles. This group of muscles comprises the part of the M. biceps femoris which attaches at the femur (the cranial portion), the gluteal muscles, the M. semimembranosus and the M. adductor magnus. The activity of these muscles begins at the end of the swing phase and usually lasts until halfway through the stance phase. The retractors act as levers, pushing the body forwards. When a dog is running uphill, the activity of the retractors can increase enormously – a slope only needs to have a gradient of 10% for retractor activity to increase by a factor of twenty. The second group of muscles is made up of retractors which can, in addition, flex the knee joint. These are the caudal portion of the M. biceps femoris, the M. semitendinosus and the M. gracilis. The activity of this group of muscles begins towards the end of the stance phase and continues into the first part of the swing phase. The muscles are thus responsible for the liftoff of the hindlimbs. If more force is required for

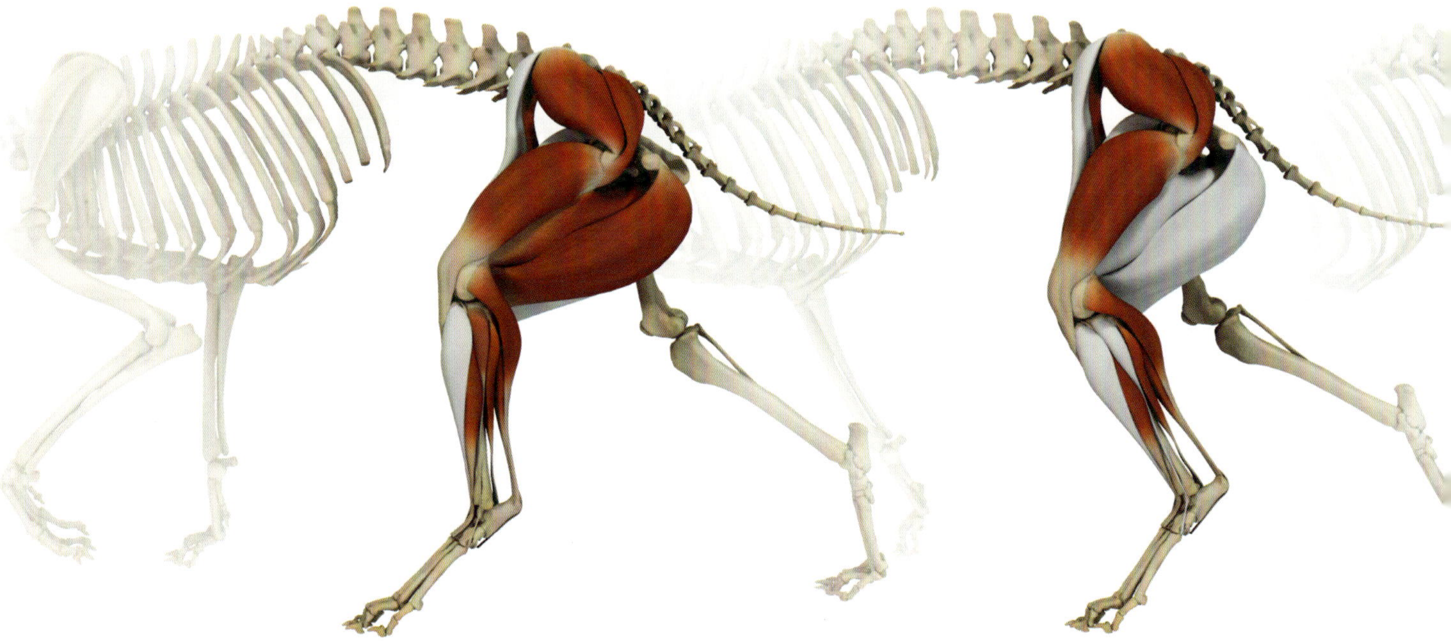

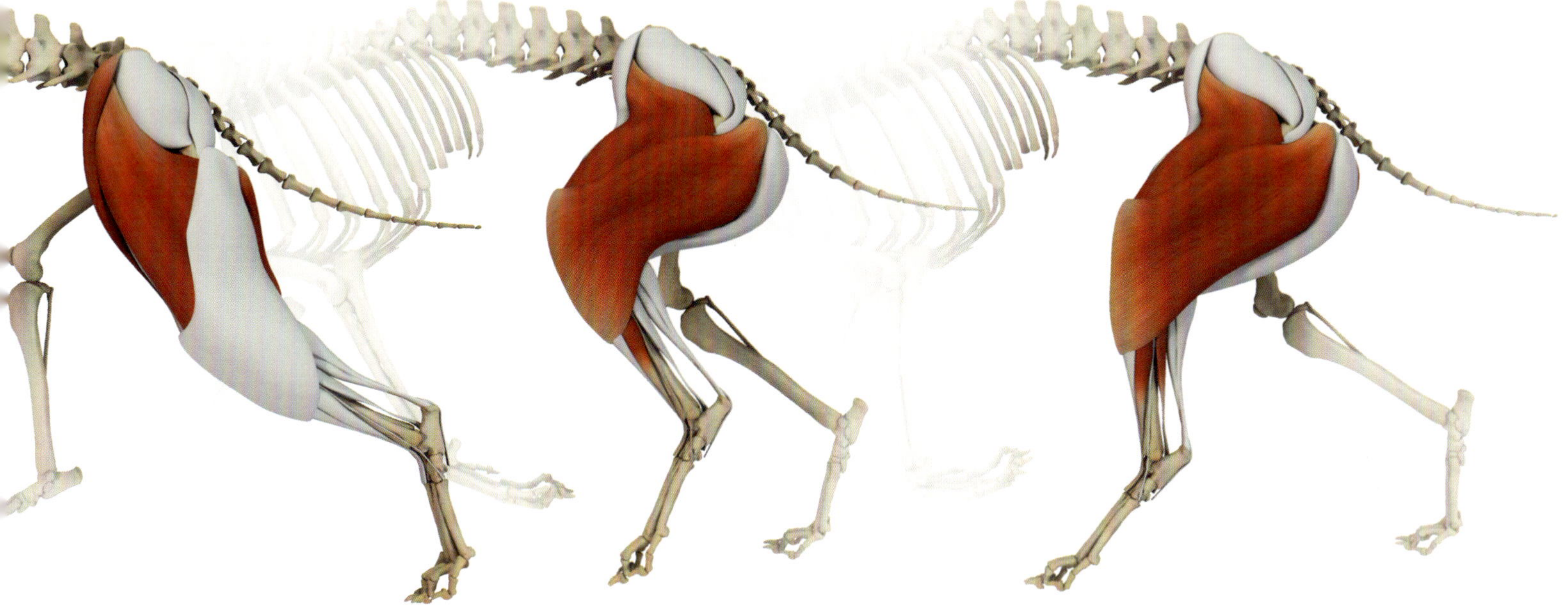

propulsion, however, as is the case when the dog is moving uphill or pulling at the leash, the muscles of the second group are activated early. The lumbar muscle (M. iliopsoas), the M. tensor fasciae latae, the M. sartorius and the M. rectus femoris are the main muscles involved in swinging the limb forwards.
The knee joint, the effective movement of which during the stance phase is minimal is held in position by the vastus muscles, which stabilize it against the flexion induced by gravity. In the ankle, the same role is performed by the two calf muscles (Mm. gastrocnemii), which like the vastus muscles are active at the end of the swing phase and the beginning of the stance phase. They contract almost isometrically, which means that they only become slightly shorter. Neither of these two muscles groups act as extensors when the leg is subjected to gravity, they simply prevent or determine the degree of flexion. The fact that they begin to be active at the end of the swing phase ensures that by the moment of hindlimb touchdown, enough force has been built up in the muscle to ensure that the leg does not buckle upon contact with the ground. The flexion in the ankle joint at liftoff is triggered by the activity of the M. tibialis anterior and the M. extensor digitorum longus.

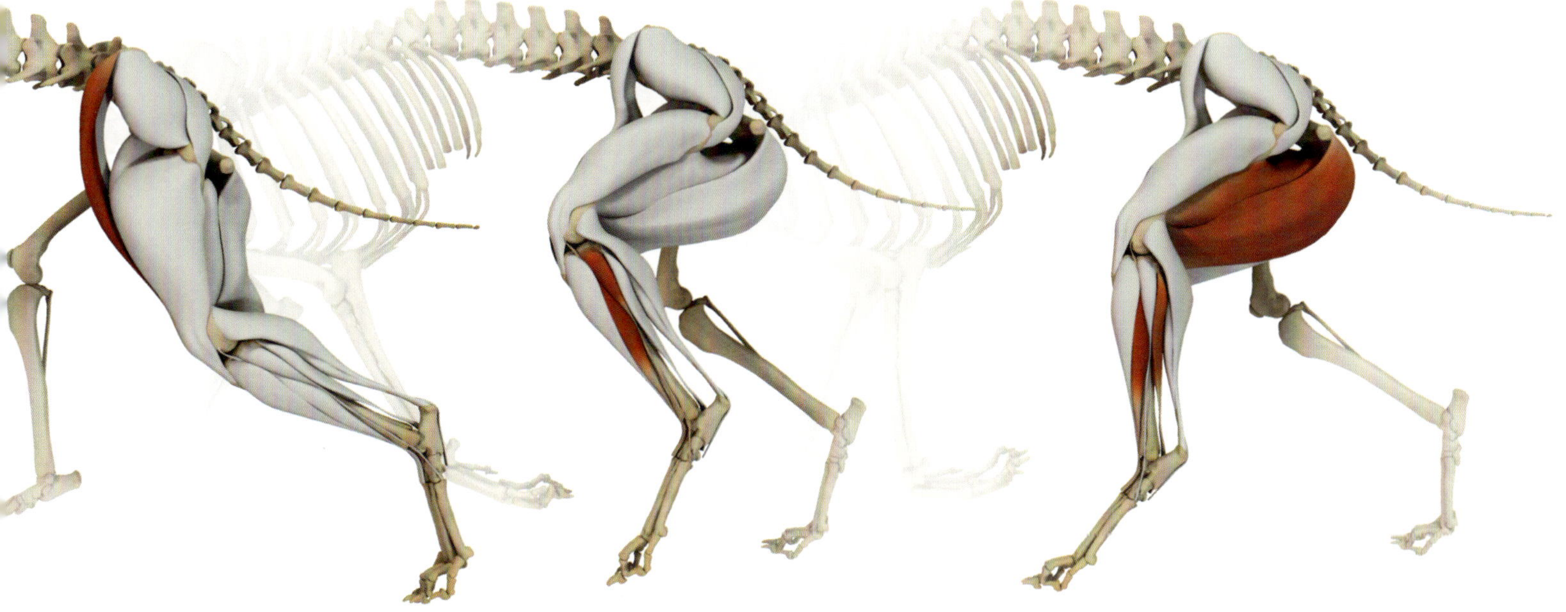

RESULTS OF THE JENA STUDY

Speed is comprised of two parameters, stride length and stride duration. The number of strides a dog makes per second is described as stride frequency. Dogs can thus increase the speed at which they are moving either by increasing stride length, increasing stride frequency, or a combination of the two. As each stride is broken down into a stance phase and a swing phase, there are also a number of different ways in which both stride length and stride frequency can be altered.

INCREASE OF SPEED

All the breeds we investigated increase speed by changing both stride length and stride frequency. The increase in stride frequency associated with higher speeds is effected solely by decreasing the duration of the stance phase. The swing phase is totally unaffected by either speed or gait. The only exception to this rule among the breeds investigated was observed in the Bernese Mountain Dog, in which changes were made to the duration of both the stance phase and the swing phase. Unlike in the stance phase, a limb in the swing phase cannot exchange forces with the ground and thus cannot accelerate the body. As such, for the whole time the limb is swinging forward into the touch down position, normal laws of pendulum motion apply. The duration of the swing phase thus depends on the proportional length of the limb and the distribution of mass along it, neither of which differ much from breed to breed as the following astonishingly uniform values show. All the investigated breeds require between 0.2 and 0.3 seconds to return the limb to the touch down position, with the sole exception of the Dachshund which takes just 0.1 second to swing its forelimbs forward during the walk. The values found in the literature for the German Shepherd (Unkel-Mohrmann 1999) are similar to our results. Other studies put forward slightly higher values for the Greyhound (0.35 and 0.37 seconds, Colborne et al. 2006) and the Doberman (0.45 and 0.37 seconds, Wingfield et al. 1993).

In all breeds except the Tibetan Terrier, changes in stride length are effected by increasing the length of the swing phase. In the Tibetan Terrier, the swing phase of the forelimbs is always the same length.

In most of the breeds investigated, stance phase length is approximately the same in the trot as it is in the walk and only increases during the gallop. Stance phase length increases continuously from gait to gait in the Akita, the work and the performance-line German Shepherd, the English Cocker Spaniel and the White Swiss Shepherd.

CHANGING THE DUTY FACTOR

The duty factor describes whether, during a stride, a limb spends more time on the ground or more time swinging. Generally speaking, within one gait, the duty factor of the fore- and the hindlimb is always the same. There are exceptions, however. During the trot, the ground contact time of the forelimb is always greater than that of the hindlimb in the Pyrenean Shepherd, the Eurasier, the Bedlington Terrier, the Entlebucher Mountain Dog and the show line German Shepherd.

In a cross-breed comparison, the lowest duty factor values during the walk are exhibited by the Airedale Terrier, the Bernese Mountain Dog and the Longhaired Pyrenean Shepherd, the highest by the Akita, the Mastiff and the French Bulldog. During the trot, the lowest values are exhibited by the Bernese Mountain Dog and the Longhaired Pyrenean Shepherd, and the highest by the French Bulldog, the Chihuahua and the Mastiff.

In all breeds, the relative duration of ground contact is shorter during the trot.

RELATIVE EXCURSION

Stance phase length can be increased either by increasing the forward excursion leading up to touch down, the backward excursion leading up to lift off, or both. In most breeds, forward and backward excursion is more or less the same. In the Akita, the Appenzell Mountain Dog, the two Pyrenean Shepherd breeds, the two German Shepherd lines, the English Cocker Spaniel, Entlebucher Mountain Dog, White Swiss Shepherd and Miniature Schnauzer, the forward excursion of the forelimbs is greater in the trot than in the walk. In the Akita, the Longhaired Pyrenean Shepherd, the two German Shepherd lines, the Golden Retriever, Mastiff, Dachshund and Pembroke Welsh Corgi, it is the backward excursion of the forelimbs which increases. The Bernese Mountain Dog is the only breed in which the backward excursion of the forelimbs is smaller in the trot than it is in the walk.

Unlike that of the forelimbs, the forward excursion of the hindlimbs is greater in the walk than it is in the trot. This does not apply to the Akita, German Shepherd, French Bulldog, Shar Pei, White Swiss Shepherd, Pembroke Welsh Corgi or Whippet, however, in which the forward excursion of the hindlimbs is approximately the same in the walk and the trot. The same is true of backward excursion in these breeds. Backward excursion is only smaller in the walk than the trot in the Appenzell Mountain Dog, Chihuahua, show line German Shepherd, Eurasier, Golden Retriever, Scottish Terrier and Whippet. What is striking is that only the Akita, the Longhaired Pyrenean Shepherd and the two German Shepherd lines increase the stance phase in the trot compared to the walk by increasing both the forward and the backward excursion of the forelimbs. In the show line German Shepherd and the Whippet, the forward excursion of the hindlimbs is

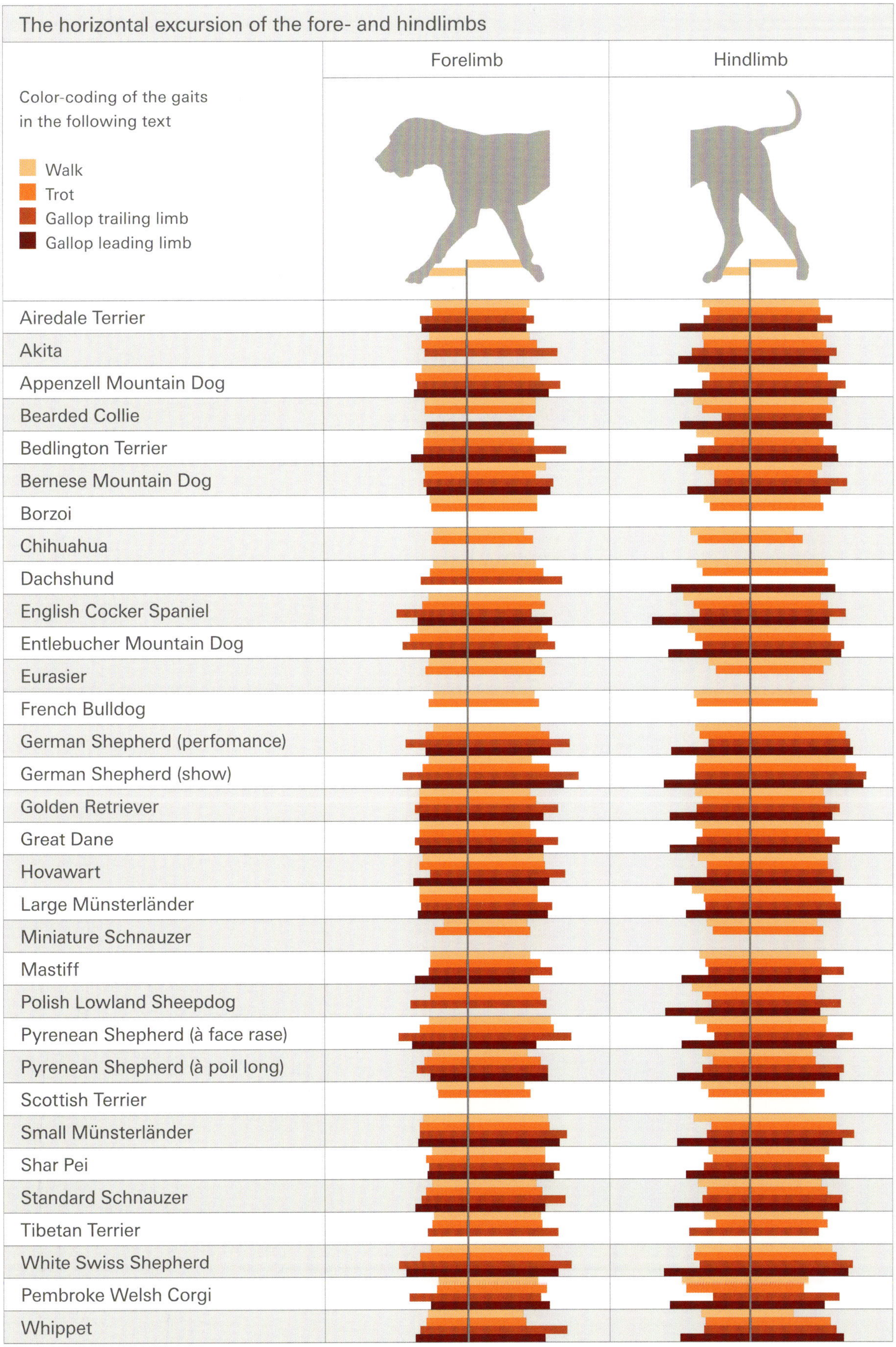
The horizontal excursion of the fore- and hindlimbs
Forelimb
Hindlimb
Color-coding of the gaits
in the following text
Walk
Trot
Gallop trailing limb
Gallop leading limb
Airedale Terrier
Akita
Appenzell Mountain Dog
Bearded Collie
Bedlington Terrier
Bernese Mountain Dog
Borzoi
Chihuahua
Dachshund
English Cocker Spaniel
Entlebucher Mountain Dog
Eurasier
French Bulldog
German Shepherd (perfomance)
German Shepherd (show)
Golden Retriever
Great Dane
Hovawart
Large Münsterländer
Miniature Schnauzer
Mastiff
Polish Lowland Sheepdog
Pyrenean Shepherd (à face rase)
Pyrenean Shepherd (à poil long)
Scottish Terrier
Small Münsterländer
Shar Pei
Standard Schnauzer
Tibetan Terrier
White Swiss Shepherd
Pembroke Welsh Corgi
Whippet

Walk

Forelimb

Touch down – Forwards excursion (ascending)	Lift off – Backwards excursion (ascending)
Miniature Schnauzer	Whippet
Pembroke Welsh Corgi	Chihuahua
Scottish Terrier	Scottish Terrier
PON	Miniature Schnauzer
Tibetan Terrier	Pyrenean Shepherd (à poil long)
Dachshund	Bedlington Terrier
Chihuahua	Great Dane
German Shepherd (perfomance)	Akita
French Bulldog	Mastiff
Shar Pei	Airedale Terrier
Airedale Terrier	Golden Retriever
Pyrenean Shepherd (à poil long)	German Shepherd (show)
Standard Schnauzer	French Bulldog
Mastiff	Appenzell Mountain Dog
Borzoi	Bearded Collie
Akita	Standard Schnauzer
White Swiss Shepherd	Dachshund
English Cocker Spaniel	Large Münsterländer
Pyrenean Shepherd (à face rase)	Pembroke Welsh Corgi
German Shepherd (show)	Borzoi
Eurasier	PON
Whippet	Tibetan Terrier
Golden Retriever	German Shepherd (perfomance)
Bearded Collie	English Cocker Spaniel
Bedlington Terrier	Entlebucher Mountain Dog
Bernese Mountain Dog	Eurasier
Appenzell Mountain Dog	White Swiss Shepherd
Small Münsterländer	Hovawart
Hovawart	Shar Pei
Great Dane	Bernese Mountain Dog
Large Münsterländer	Small Münsterländer
Entlebucher Mountain Dog	Pyrenean Shepherd (à face rase)

Hindlimb

Forwards excursion (ascending)	Backwards excursion (ascending)
Eurasier	Chihuahua
Shar Pei	Whippet
Miniature Schnauzer	Pembroke Welsh Corgi
Borzoi	French Bulldog
Tibetan Terrier	Pyrenean Shepherd (à poil long)
Pyrenean Shepherd (à poil long)	Scottish Terrier
Akita	Golden Retriever
Airedale Terrier	Mastiff
Scottish Terrier	Appenzell Mountain Dog
Whippet	Miniature Schnauzer
Mastiff	PON
Standard Schnauzer	Bedlington Terrier
Appenzell Mountain Dog	Airedale Terrier
Hovawart	Borzoi
Dachshund	English Cocker Spaniel
Bedlington Terrier	Standard Schnauzer
German Shepherd (show)	Bernese Mountain Dog
Bernese Mountain Dog	Great Dane
Pyrenean Shepherd (à face rase)	Tibetan Terrier
Great Dane	Akita
German Shepherd (perfomance)	Dachshund
White Swiss Shepherd	Bearded Collie
Bearded Collie	Pyrenean Shepherd (à face rase)
Small Münsterländer	Hovawart
French Bulldog	Entlebucher Mountain Dog
Large Münsterländer	Shar Pei
PON	Eurasier
Chihuahua	White Swiss Shepherd
Golden Retriever	Large Münsterländer
Entlebucher Mountain Dog	Small Münsterländer
English Cocker Spaniel	German Shepherd (perfomance)
Pembroke Welsh Corgi	German Shepherd (show)

Trot

Forelimb

Touch down – Forwards excursion (ascending)	Lift off – Backwards excursion (ascending)
Scottish Terrier	Whippet
Pembroke Welsh Corgi	Airedale Terrier
Miniature Schnauzer	Scottish Terrier
Airedale Terrier	Miniature Schnauzer
Tibetan Terrier	Chihuahua
Chihuahua	Bedlington Terrier
Borzoi	Great Dane
Dachshund	Bearded Collie
PON	Bernese Mountain Dog
Mastiff	Borzoi
French Bulldog	Akita
Bearded Collie	Large Münsterländer
Golden Retriever	French Bulldog
Standard Schnauzer	Golden Retriever
Shar Pei	Pyrenean Shepherd (à poil long)
Whippet	PON
Eurasier	Mastiff
Bernese Mountain Dog	Appenzell Mountain Dog
German Shepherd (perfomance)	Tibetan Terrier
Pyrenean Shepherd (à poil long)	Standard Schnauzer
Bedlington Terrier	Dachshund
German Shepherd (show)	Hovawart
English Cocker Spaniel	Eurasier
Akita	Shar Pei
Pyrenean Shepherd (à face rase)	English Cocker Spaniel
Hovawart	Pembroke Welsh Corgi
Small Münsterländer	Entlebucher Mountain Dog
Large Münsterländer	Small Münsterländer
White Swiss Shepherd	German Shepherd (perfomance)
Great Dane	White Swiss Shepherd
Appenzell Mountain Dog	German Shepherd (show)
Entlebucher Mountain Dog	Pyrenean Shepherd (à face rase)

Hindlimb

Forwards excursion (ascending)	Backwards excursion (ascending)
Tibetan Terrier	Chihuahua
Eurasier	Pembroke Welsh Corgi
Bernese Mountain Dog	PON
Bedlington Terrier	Pyrenean Shepherd (à poil long)
Shar Pei	Bernese Mountain Dog
Miniature Schnauzer	French Bulldog
Pyrenean Shepherd (à poil long)	Airedale Terrier
Small Münsterländer	Mastiff
Airedale Terrier	Miniature Schnauzer
Appenzell Mountain Dog	Eurasier
Borzoi	Borzoi
Scottish Terrier	Bedlington Terrier
Standard Schnauzer	Great Dane
Pyrenean Shepherd (à face rase)	Scottish Terrier
Hovawart	Shar Pei
Mastiff	Pyrenean Shepherd (à face rase)
Whippet	Akita
Akita	Appenzell Mountain Dog
Large Münsterländer	Tibetan Terrier
Great Dane	Dachshund
Dachshund	English Cocker Spaniel
PON	Standard Schnauzer
Bearded Collie	Hovawart
German Shepherd (perfomance)	Entlebucher Mountain Dog
Golden Retriever	Golden Retriever
Chihuahua	Whippet
French Bulldog	Bearded Collie
Entlebucher Mountain Dog	Large Münsterländer
German Shepherd (show)	Small Münsterländer
White Swiss Shepherd	White Swiss Shepherd
English Cocker Spaniel	German Shepherd (perfomance)
Pembroke Welsh Corgi	German Shepherd (show)

THE HORIZONTAL EXCURSION OF THE FORE- AND HINDLIMBS

The tables compare the degree of relative horizontal excursion of the fore- and hindlimbs in the walk, trot and gallop. The breeds are listed in order of increasing forwards excursion and increasing backwards excursion. In the trot, Scottish Terriers, for instance, display the lowest forward excursion of all breeds and also exhibit a low degree of backwards excursion. Backwards excursion is lower only in Whippets and Airedale Terriers. The degree of forwards excursion displayed by Whippets, however, is fairly middle-of-the range. In German Shepherds, excursion in the gallop is high in both directions in comparison to almost all other breeds. This applies to both the trailing and the leading limb.

Gallop trailing limb

Forelimb

Touch down – Forwards excursion (ascending)	Lift off – Backwards excursion (ascending)
Bearded Collie	Tibetan Terrier
Bernese Mountain Dog	Bearded Collie
Pembroke Welsh Corgi	Airedale Terrier
Golden Retriever	Hovawart
PON	Whippet
Mastiff	Akita
German Shepherd (perfomance)	Bedlington Terrier
Hovawart	Great Dane
Small Münsterländer	Pembroke Welsh Corgi
Large Münsterländer	Golden Retriever
Airedale Terrier	Shar Pei
Shar Pei	PON
Whippet	Large Münsterländer
Appenzell Mountain Dog	Standard Schnauzer
Entlebucher Mountain Dog	Entlebucher Mountain Dog
Standard Schnauzer	Pyrenean Shepherd (à poil long)
Pyrenean Shepherd (à poil long)	Mastiff
White Swiss Shepherd	Appenzell Mountain Dog
English Cocker Spaniel	English Cocker Spaniel
Bedlington Terrier	Bernese Mountain Dog
Great Dane	German Shepherd (perfomance)
German Shepherd (show)	White Swiss Shepherd
Akita	Small Münsterländer
Tibetan Terrier	German Shepherd (show)

Hindlimb

Forwards excursion (ascending)	Backwards excursion (ascending)
Hovawart	English Cocker Spaniel
Mastiff	Airedale Terrier
Shar Pei	Pembroke Welsh Corgi
Tibetan Terrier	PON
Akita	Pyrenean Shepherd (à poil long)
Bedlington Terrier	Mastiff
Bernese Mountain Dog	Large Münsterländer
Dachshund	Bernese Mountain Dog
Standard Schnauzer	Entlebucher Mountain Dog
Large Münsterländer	Akita
Airedale Terrier	Great Dane
Whippet	Tibetan Terrier
Small Münsterländer	Shar Pei
Appenzell Mountain Dog	Appenzell Mountain Dog
Pyrenean Shepherd (à poil long)	Dachshund
Great Dane	Golden Retriever
Golden Retriever	Standard Schnauzer
PON	Hovawart
Pembroke Welsh Corgi	Small Münsterländer
German Shepherd (perfomance)	Whippet
German Shepherd (show)	Bedlington Terrier
Entlebucher Mountain Dog	German Shepherd (perfomance)
White Swiss Shepherd	White Swiss Shepherd
English Cocker Spaniel	German Shepherd (show)

Gallop leading limb

Forelimb

Touch down – Forwards excursion (ascending)	Lift off – Backwards excursion (ascending)
Pembroke Welsh Corgi	Airedale Terrier
Entlebucher Mountain Dog	Mastiff
Pyrenean Shepherd (à poil long)	Bearded Collie
Golden Retriever	Bedlington Terrier
Bearded Collie	Entlebucher Mountain Dog
Shar Pei	Golden Retriever
Bernese Mountain Dog	Great Dane
German Shepherd (perfomance)	Standard Schnauzer
Airedale Terrier	Whippet
German Shepherd (show)	Large Münsterländer
Great Dane	Pyrenean Shepherd (à poil long)
Small Münsterländer	Pembroke Welsh Corgi
English Cocker Spaniel	Appenzell Mountain Dog
Large Münsterländer	Hovawart
Standard Schnauzer	German Shepherd (perfomance)
Mastiff	Bernese Mountain Dog
Whippet	English Cocker Spaniel
Appenzell Mountain Dog	Shar Pei
Hovawart	White Swiss Shepherd
Bedlington Terrier	Small Münsterländer
White Swiss Shepherd	German Shepherd (show)

Hindlimb

Forwards excursion (ascending)	Backwards excursion (ascending)
Bernese Mountain Dog	Airedale Terrier
Shar Pei	PON
Large Münsterländer	Mastiff
Bedlington Terrier	Pembroke Welsh Corgi
Mastiff	English Cocker Spaniel
Airedale Terrier	Akita
Bearded Collie	Bernese Mountain Dog
Whippet	Great Dane
Akita	Bearded Collie
Pyrenean Shepherd (à poil long)	Dachshund
Small Münsterländer	Appenzell Mountain Dog
Appenzell Mountain Dog	Golden Retriever
Hovawart	Bedlington Terrier
Standard Schnauzer	Shar Pei
German Shepherd (perfomance)	Standard Schnauzer
Dachshund	Pyrenean Shepherd (à poil long)
Pembroke Welsh Corgi	Entlebucher Mountain Dog
Great Dane	Large Münsterländer
Entlebucher Mountain Dog	Small Münsterländer
Golden Retriever	Hovawart
PON	Whippet
White Swiss Shepherd	White Swiss Shepherd
German Shepherd (show)	German Shepherd (perfomance)
English Cocker Spaniel	German Shepherd (show)

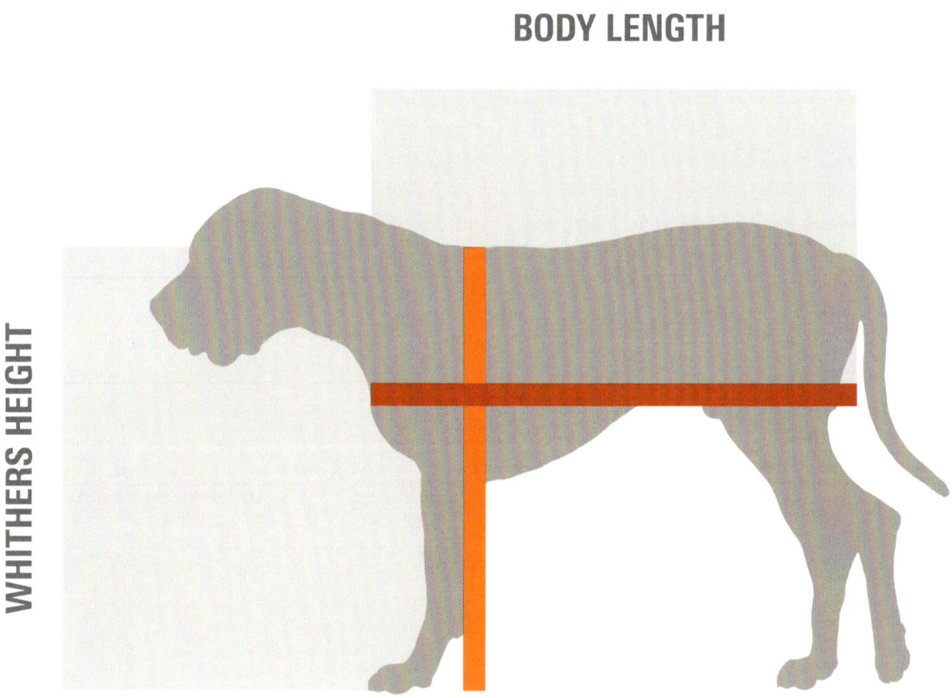

Step parameters in relation to withers height

	Relative stride length								Relative stance lengt								Relative swing lengt							
	Forelimb				Hindlimb				Forelimb				Hindlimb				Forelimb				Hindlimb			
Airedale Terrier	1.3	1.8	3.3	3.2	1.5	2.0	3.9	3.6	0.7	0.7	0.8	0.7	0.8	0.8	0.7	0.9	0.6	1.1	2.4	2.4	0.7	1.3	3.1	2.8
Akita	1.0	1.7	3.0	-	1.4	2.0	3.3	3.5	0.6	0.8	0.9	-	0.9	0.9	0.9	0.9	0.4	0.9	2.1	-	0.5	1.1	2.4	2.5
Appenzell Mountain Dog	1.3	1.9	2.9	2.7	1.4	2.0	3.2	3.2	0.8	0.9	0.9	1.0	0.9	0.8	1.0	1.0	0.5	1.0	2.0	1.7	0.5	1.1	2.2	2.2
Bearded Collie	1.3	1.7	3.1	3.0	1.5	1.9	-	3.6	0.8	0.8	0.7	0.8	0.9	0.9	-	1.0	0.5	0.9	2.4	2.2	0.6	1.0	-	2.6
Bedlington Terrier	1.3	1.8	2.8	2.8	1.6	2.2	3.0	3.2	0.8	0.8	0.9	0.9	0.9	0.8	1.0	1.0	0.5	1.0	1.8	1.9	0.7	1.4	2.0	2.1
Bernese Mountain Dog	1.5	2.0	2.8	2.7	1.7	2.1	3.2	3.1	0.9	0.8	0.9	0.9	0.9	0.8	0.9	0.9	0.7	1.2	1.9	1.8	0.8	1.4	2.3	2.2
Borzoi	1.3	1.6	-	-	1.4	1.9	-	-	0.8	0.8	-	-	0.8	0.8	-	-	0.5	0.9	-	-	0.6	1.1	-	-
Chihuahua	1.0	1.4	-	-	1.2	1.6	-	-	0.6	0.7	-	-	0.7	0.8	-	-	0.4	0.7	-	-	0.5	0.8	-	-
Dachshund	1.3	1.9	-	-	1.6	2.1	2.8	2.7	0.8	0.9	-	-	1.0	0.9	1.1	1.0	0.5	1.0	-	-	0.6	1.2	1.7	1.7
English Cocker Spaniel	1.3	1.9	3.1	2.8	1.6	2.4	3.7	3.4	0.8	0.9	1.0	1.0	1.0	1.0	0.9	1.1	0.5	1.0	2.1	1.8	0.6	1.4	2.7	2.4
Entlebucher Mountain Dog	1.4	2.0	2.9	2.8	1.7	2.3	3.4	3.3	0.9	1.0	0.9	1.0	1.0	0.9	1.1	1.1	0.5	1.0	2.0	1.8	0.7	1.4	2.3	2.2
Eurasier	1.3	1.9	-	-	1.6	2.2	-	-	0.8	0.8	-	-	0.9	0.8	-	-	0.5	1.1	-	-	0.7	1.4	-	-
French Bulldog	1.1	1.5	-	-	1.3	1.8	-	-	0.7	0.8	-	-	0.8	0.9	-	-	0.3	0.7	-	-	0.5	0.9	-	-
German Shepherd (perfomance)	1.3	1.9	3.2	3.2	1.6	2.5	3.5	3.9	0.8	0.9	1.0	1.0	1.0	1.0	1.1	1.2	0.5	1.0	2.2	2.2	0.6	1.5	2.4	2.7
German Shepherd (show)	1.2	2.1	3.6	3.3	1.6	2.7	4.4	4.4	0.8	0.9	1.1	1.1	1.1	1.1	1.2	1.3	0.4	1.2	2.5	2.3	0.6	1.5	3.2	3.1
Golden Retriever	1.2	1.7	2.4	2.4	1.4	2.0	2.8	2.9	0.8	0.8	0.9	0.9	0.9	1.0	1.1	1.1	0.5	0.9	1.6	1.5	0.5	1.1	1.7	1.8
Great Dane	1.2	1.8	1.8	1.8	1.4	2.0	2.1	1.7	0.8	0.7	0.5	0.6	0.9	0.8	0.6	0.5	0.5	1.0	1.3	1.2	0.5	1.2	1.5	1.2
Hovawart	1.4	2.0	2.3	3.0	1.6	2.2	2.8	3.7	0.9	0.9	0.8	1.1	0.9	0.9	1.0	1.2	0.5	1.1	1.5	1.9	0.7	1.4	1.7	2.6
Large Münsterländer	1.4	2.0	3.2	3.1	1.6	2.1	3.7	3.7	0.8	0.8	0.9	0.9	1.0	0.8	0.9	1.0	0.5	1.2	2.3	2.1	0.7	1.3	2.8	2.8
Miniature Schnauzer	1.1	1.7	-	-	1.4	2.0	-	-	0.6	0.7	-	-	0.8	0.8	-	-	0.5	1.0	-	-	0.6	1.2	-	-
Mastiff	1.1	1.6	2.4	1.9	1.3	1.8	1.9	1.9	0.8	0.8	0.8	0.8	0.8	0.8	0.8	0.9	0.4	0.8	1.7	1.1	0.5	0.9	1.1	1.0
Polish Lowland Sheepdog	1.3	1.8	2.3	-	1.6	2.1	2.7	2.7	0.7	0.8	0.8	-	0.9	0.8	1.0	1.0	0.6	1.1	1.5	-	0.7	1.3	1.7	1.7
Pyrenean Shepherd (à face rase)	1.4	2.1	-	-	1.6	2.5	-	-	0.8	0.9	-	-	0.9	0.8	-	-	0.5	1.2	-	-	0.7	1.6	-	-
Pyrenean Shepherd (à poil long)	1.2	2.1	3.1	3.0	1.4	2.6	4.1	3.8	0.7	0.8	0.9	0.9	0.8	0.8	1.0	1.0	0.5	1.3	2.3	2.0	0.6	1.8	3.2	2.8
Scottish Terrier	1.1	1.5	-	-	1.4	1.9	-	-	0.7	0.7	-	-	0.9	0.8	-	-	0.5	0.8	-	-	0.6	1.1	-	-
Small Münsterländer	1.5	2.0	3.1	3.0	1.8	2.3	3.5	3.5	0.9	0.9	1.0	1.0	1.0	0.9	1.0	1.0	0.6	1.1	2.1	2.0	0.8	1.4	2.5	2.4
Shar Pei	1.4	1.8	2.7	2.8	1.6	2.1	3.1	3.2	0.8	0.9	0.9	0.9	0.9	0.8	0.9	1.0	0.6	1.0	1.8	1.9	0.7	1.3	2.1	2.3
Standard Schnauzer	1.2	1.9	2.8	2.7	1.4	2.1	3.0	3.1	0.8	0.8	0.9	1.0	0.9	0.8	1.0	1.0	0.5	1.1	1.9	1.8	0.6	1.3	1.9	2.1
Tibetan Terrier	1.3	1.7	2.5	-	1.5	1.9	2.8	-	0.8	0.8	0.9	-	0.9	0.8	0.9	-	0.5	0.9	1.5	-	0.6	1.1	1.9	-
White Swiss Shepherd	1.3	1.9	3.1	3.0	1.5	2.2	3.7	5.1	0.8	0.9	1.1	1.2	1.0	1.0	1.3	1.2	0.4	0.9	2.0	1.8	0.5	1.2	2.4	2.3
Pembroke Welsh Corgi	1.4	1.8	2.4	2.5	1.6	1.9	2.8	2.9	0.8	0.8	0.8	0.9	1.0	0.9	0.9	1.0	0.5	1.0	1.5	1.6	0.6	1.0	1.9	1.9
Whippet	1.2	1.8	3.4	3.2	1.4	2.0	3.8	3.9	0.7	0.7	0.9	0.9	0.9	0.9	1.0	1.0	0.5	1.1	2.5	2.3	0.6	1.1	2.8	2.8

greater during the walk than the trot, but backward excursion is greater during the trot than during the walk.
Overall, and relatively speaking, forward excursion is greater in the hindlimbs than the forelimbs. This difference is more clearly observable during the walk than during the trot.
During the walk, the forward excursion of the forelimbs is particularly small in the Miniature Schnauzer, Pembroke Welsh Corgi and Scottish Terrier. During the trot, the same applies to the Airedale Terrier. The forward extension of the forelimbs is greatest during the walk in the Swiss Mountain Dog, the Small and Large Münsterländer, the Hovawart and the Great Dane. During the trot, this distinction goes to the Appenzell and the Entlebucher Mountain Dog.
When it comes to the hindlimbs, forward excursion during the walk is smallest in the Borzoi, Eurasier, Shar Pei, Tibetan Terrier and Miniature Schnauzer, and during the trot it is also small in the Bernese Mountain Dog, Bedlington Terrier, Longhaired Pyrenean Shepherd and Small Münsterländer. Forward excursion during the trot is greatest in the Pembroke Welsh Corgi and the English Cocker Spaniel.
During the walk, the backward excursion of the forelimbs is smallest in the Bedlington Terrier, Longhaired Pyrenean Shepherd, Chihuahua, Scottish Terrier, Whippet and Miniature Schnauzer and longest in the Smooth-faced Pyrenean Shepherd, Bernese Mountain Dog, Small Münsterländer and Shar Pei. During the trot, backward excursion is smallest in the Whippet, the Airedale and Scottish Terriers, the Miniature Schnauzer, the Smooth-faced Pyrenean Shepherd, the two German Shepherd lines, the White Swiss Shepherd and the Small Münsterländer.
When it comes to the hindlimbs, backward excursion during the walk is very small in the Chihuahua and the Whippet, and backward excursion during the trot very small in the Chihuahua and the Pembroke Welsh Corgi. Backward excursion is greatest during both the walk and the trot in the two German Shepherd lines.

RELATIVE STRIDE LENGTH

To be able to compare the stride lengths of different dogs it is necessary to eliminate the size factor. In absolute terms, it is obvious that a Tibetan Terrier will make larger strides than a Pembroke Welsh Corgi. The table aside presents the relative stride lengths of the dogs involved in this study, permitting a cross-breed comparison.
During the walk, the shortest forelimb stride (relatively speaking) is seen in the Chihuahua and the Akita and the shortest hindlimb stride in the Chihuahua, the Mastiff and the French Bulldog. The longest strides are made by the Small Münsterländer and the Bernese Mountain Dog.
The picture as a whole delivers some surprises. It is striking that not all small dogs group together, for instance, and that the Smooth-faced Pyrenean Shepherd makes longer strides than the Longhaired Pyrenean Shepherd. Stride length is more or less the same in the Airedale and the Bedlington Terrier, but shorter in the Scottish Terrier. The two German Shepherd lines and the White Swiss Shepherd group together, though representatives of the show line make slightly smaller forelimb strides than the others. The Mastiff makes smaller strides than the Great Dane – the two heaviest breeds in the study.
During the trot, the shortest strides, relatively speaking, are made by some but not all of the small breeds (Chihuahua, Scottish Terrier and French Bulldog), and by the heaviest breed in the study, the Mastiff (though here only in the case of the hindlimbs). The longest strides are made by the two Pyrenean Shepherd breeds and the two German Shepherd lines. As is the case in the walk, not all small breeds group together in the trot, and of the largest breeds, the Great Dane makes larger strides than the Mastiff. Once again, the two German Shepherd lines and the White Swiss Shepherd group together, though here, representatives of the show line make slightly larger forelimb strides than the others.

Stride length at walk

Forelimb		Hindlimb	
1.0	Chihuahua	1.2	Chihuahua
1.0	Akita	1.3	Mastiff
1.1	French Bulldog	1.3	French Bulldog
1.1	Miniature Schnauzer	1.4	Miniature Schnauzer
1.1	Scottish Terrier	1.4	Akita
1.1	Mastiff	1.4	Appenzell Mountain Dog
1.2	Berger des Pyrénées à poil long	1.4	Berger des Pyrénées à poil long
1.2	German Shepherd (show)	1.4	Great Dane
1.2	Golden Retriever	1.4	Borzois
1.2	Whippets	1.4	Scottish Terrier
1.2	Standard Schnauzer	1.4	Whippets
1.2	Great Dane	1.4	Golden Retriever
1.3	Borzois	1.4	Standard Schnauzer
1.3	White Swiss Shepherd	1.5	Airedale Terrier
1.3	German Shepherd (perfomance)	1.5	White Swiss Shepherd
1.3	Tibetan Terrier	1.5	Tibetan Terrier
1.3	Appenzell Mountain Dog	1.5	Bearded Collie
1.3	Airedale Terrier	1.6	PON
1.3	Dachshund	1.6	Eurasier
1.3	Bearded Collie	1.6	Shar Pei
1.3	English Cocker Spaniel	1.6	Pembroke Welsh Corgi
1.3	Bedlington Terrier	1.6	Bedlington Terrier
1.3	PON	1.6	English Cocker Spaniel
1.3	Eurasier	1.6	Hovawart
1.4	Pembroke Welsh Corgi	1.6	Dachshund
1.4	Große Münsterländer	1.6	German Shepherd (perfomance)
1.4	Shar Pei	1.6	Pyrenean Shepherd (à face rase)
1.4	Pyrenean Shepherd (à face rase)	1.6	Große Münsterländer
1.4	Hovawart	1.6	German Shepherd (show)
1.4	Entlebucher Mountain Dog	1.7	Entlebucher Mountain Dog
1.5	Bernese Mountain Dog	1.7	Bernese Mountain Dog
1.5	Kleine Münsterländer	1.8	Kleine Münsterländer

Stride length at trot

Forelimb		Hindlimb	
1.4	Chihuahua	1.6	Chihuahua
1.5	Scottish Terrier	1.8	Mastiff
1.5	French Bulldog	1.8	French Bulldog
1.6	Mastiff	1.9	Borzois
1.6	Borzois	1.9	Scottish Terrier
1.7	Miniature Schnauzer	1.9	Bearded Collie
1.7	Akita	1.9	Pembroke Welsh Corgi
1.7	Tibetan Terrier	1.9	Tibetan Terrier
1.7	Bearded Collie	2.0	Great Dane
1.7	Golden Retriever	2.0	Akita
1.8	Great Dane	2.0	Appenzell Mountain Dog
1.8	Pembroke Welsh Corgi	2.0	Airedale Terrier
1.8	Airedale Terrier	2.0	Miniature Schnauzer
1.8	Whippets	2.0	Whippets
1.8	PON	2.0	Golden Retriever
1.8	Shar Pei	2.1	Shar Pei
1.8	Bedlington Terrier	2.1	Dachshund
1.9	Appenzell Mountain Dog	2.1	Große Münsterländer
1.9	Dachshund	2.1	PON
1.9	White Swiss Shepherd	2.1	Bernese Mountain Dog
1.9	Eurasier	2.1	Standard Schnauzer
1.9	Standard Schnauzer	2.2	White Swiss Shepherd
1.9	English Cocker Spaniel	2.2	Bedlington Terrier
1.9	German Shepherd (perfomance)	2.2	Eurasier
2.0	Hovawart	2.2	Hovawart
2.0	Bernese Mountain Dog	2.3	Entlebucher Mountain Dog
2.0	Kleine Münsterländer	2.3	Kleine Münsterländer
2.0	Entlebucher Mountain Dog	2.4	English Cocker Spaniel
2.0	Große Münsterländer	2.5	Pyrenean Shepherd (à face rase)
2.1	Berger des Pyrénées à poil long	2.5	German Shepherd (perfomance)
2.1	German Shepherd (show)	2.6	Berger des Pyrénées à poil long
2.1	Pyrenean Shepherd (à face rase)	2.7	German Shepherd (show)

Stride length of the trailing limb at gallop

Forelimb		Hinterergliedmaße	
1.8	Great Dane	1.9	Mastiff
2.3	Hovawart	2.1	Great Dane
2.3	PON	2.7	PON
2.4	Pembroke Welsh Corgi	2.8	Hovawart
2.4	Golden Retriever	2.8	Dachshund
2.4	Mastiff	2.8	Tibetan Terrier
2.5	Tibetan Terrier	2.8	Pembroke Welsh Corgi
2.7	Shar Pei	2.8	Golden Retriever
2.8	Bedlington Terrier	3.0	Mittelschnauzer Pf.-Salz
2.8	Bernese Mountain Dog	3.0	Bedlington Terrier
2.8	Mittelschnauzer Pf.-Salz	3.1	Shar Pei
2.9	Entlebucher Mountain Dog	3.2	Appenzell Mountain Dog
2.9	Appenzell Mountain Dog	3.2	Bernese Mountain Dog
3.0	Akita	3.3	Akita
3.1	Kleine Münsterländer	3.4	Entlebucher Mountain Dog
3.1	English Cocker Spaniel	3.5	Kleine Münsterländer
3.1	White Swiss Shepherd	3.5	German Shepherd (perfomance)
3.1	Bearded Collie	3.7	English Cocker Spaniel
3.1	Berger des Pyrénées à poil long	3.7	Große Münsterländer
3.2	Große Münsterländer	3.7	White Swiss Shepherd
3.2	German Shepherd (perfomance)	3.8	Whippets
3.3	Airedale Terrier	3.9	Airedale Terrier
3.4	Whippets	4.1	Berger des Pyrénées à poil long
3.6	German Shepherd (show)	4.4	German Shepherd (show)
-	Borzois	-	Borzois
-	Pyrenean Shepherd (à face rase)	-	Bearded Collie
-	Chihuahua	-	Pyrenean Shepherd (à face rase)
-	Eurasier	-	Chihuahua
-	French Bulldog	-	Eurasier
-	Scottish Terrier	-	French Bulldog
-	Dachshund	-	Scottish Terrier
-	Miniature Schnauzer	-	Miniature Schnauzer

Stride length of the leading limb at gallop

Forelimb		Hindlimb	
3.2	Airedale Terrier	1.7	Great Dane
1.8	Great Dane	1.9	Mastiff
1.9	Mastiff	2.7	PON
2.4	Golden Retriever	2.7	Dachshund
2.5	Pembroke Welsh Corgi	2.9	Golden Retriever
2.7	Bernese Mountain Dog	2.9	Pembroke Welsh Corgi
2.7	Standard Schnauzer	3.1	Bernese Mountain Dog
2.7	Appenzell Mountain Dog	3.1	Standard Schnauzer
2.8	Shar Pei	3.2	Bedlington Terrier
2.8	Bedlington Terrier	3.2	Appenzell Mountain Dog
2.8	English Cocker Spaniel	3.2	Shar Pei
2.8	Entlebucher Mountain Dog	3.3	Entlebucher Mountain Dog
3.0	Berger des Pyrénées à poil long	3.4	English Cocker Spaniel
3.0	White Swiss Shepherd	3.5	Akita
3.0	Hovawart	3.5	Kleine Münsterländer
3.0	Bearded Collie	3.6	Bearded Collie
3.0	Kleine Münsterländer	3.6	Airedale Terrier
3.1	Große Münsterländer	3.7	Hovawart
3.2	German Shepherd (perfomance)	3.7	Große Münsterländer
3.2	Whippets	3.8	Berger des Pyrénées à poil long
3.3	German Shepherd (show)	3.9	Whippets
-	Akita	3.9	German Shepherd (perfomance)
-	Borzois	4.4	German Shepherd (show)
-	Pyrenean Shepherd (à face rase)	5.1	White Swiss Shepherd
-	Chihuahua	-	Borzois
-	Eurasier	-	Pyrenean Shepherd (à face rase)
-	French Bulldog	-	Chihuahua
-	PON	-	Eurasier
-	Scottish Terrier	-	French Bulldog
-	Dachshund	-	Scottish Terrier
-	Tibetan Terrier	-	Tibetan Terrier
-	Miniature Schnauzer	-	Miniature Schnauzer

PROFILE OF THE 32 BREEDS INVOLVED IN THE JENA STUDY

Providing a uniform, comparable analysis of results when faced with data pertaining to over three hundred dogs is no mean feat, especially when space is limited. As a result, certain dog lovers may feel that some of the parameters included here are given too much or too little attention.

The following pages are intended to facilitate comparison between all the breeds involved in the study. Each page is dedicated to one breed. The tables contain the mean values obtained for each measured parameter, while the texts that accompany them discuss the proportions and gait parameters of the breed in question in relation to those of other breeds.

In addition, the brief description given of each breed provides an overview of its age and origins and includes selected excerpts from the breed standard which are of relevance to locomotion.

Discussion of the proportions is limited mainly to the distinguishing features of the breed and to any particular idiosyncrasies displayed by the individuals concerned. Where a breed conforms to the standard in every way, or where segment proportions do not deviate from the standard values calculated in the chapter entitled "Proportions of the 32 breeds involved in the Jena Study", no explicit mention is made of the fact. In most of the breeds, no sex-specific differences in proportions were detected at all, so only those proportions which do actually differ between dogs and bitches are mentioned.

Discussion of the gait parameters is limited to relative values, i.e. those that can be compared between the breeds. The first paragraph of this section explores whereabouts relative stride length, which reflects the ratio between stride length and functional limb length, ranks in a cross-breed comparison. Attention is paid to the way in which this ranking may differ between gaits.

The second paragraph discusses the horizontal excursion of the limbs, i.e. the distance between the point of touchdown or liftoff of a limb and the perpendicular of its uppermost pivot point. It goes without saying that this discussion is based on relative values, i.e. the ratio between excursion and functional leg length, as large dogs would otherwise always be seen to display greater excursion than small ones. This ratio makes the size factor insignificant. We compare how, within the same breed, excursion differs between the various gaits and between the fore and the hindlimbs. The forward excursion of the hindlimbs before touchdown, for instance, is almost always smaller in the trot than in any of the other gaits.

The third paragraph looks at how the relative excursion witnessed in the various gaits compares to that in other breeds. At this point it is possible to say whether the breed in question displays a high degree of forwards excursion ("forward reach") or not.

Finally, where the breed standard makes concrete reference to gait parameters or angulation, we discuss these requirements in the light of our results.

The values provided in the tables, on the other hand, are the mean of the actual values that we obtained in the course of our measurements. The proportions table compares the results obtained for bitches and dogs, and the standard deviation is given alongside the mean value.

The table dealing with gait parameters also provides the mean of the actual values we measured during the study. Here too, the standard deviation (SD) is given alongside the mean value (MV). The kinematics table details the angulation of the joints and the various limb segments as they were measured or calculated during locomotion. The angles given are those observed at the point of touchdown and liftoff and that observed when the paw is directly under the uppermost pivot point of the limb. The difference between the position of touchdown and that of liftoff is given (effective movement), as is the difference between the maximum and minimum excursion to occur within one stride. For an explanation of these values, see the chapter entitled "Motion Sequence". The joint angles provided always refer to the interior angle, the segment angles are given relative to an imagined horizontal line. The values in italics in the kinematics tables indicate that values obtained using the marker system were adjusted on the basis of the X-ray findings.

Proportions

Body	♀		♂	
Number of dogs	10		4	
	MV	SD	MV	SD
Withers height [cm]	57	1.7	60	3.4
Body length / Withers height	1.0	0.0	1.0	0.0
Withers height / croup height	1.0	0.0	1.0	0.0
Legginess index	1.4	0.1	1.4	0.0

Forelimb	♀		♂	
Number of dogs	10		4	
	Proportion of total length			
	%	SD	%	SD
Shoulder blade	26	1.0	26	0.6
Upper arm	27	0.7	26	1.2
Lower arm	32	1.0	32	1.9
Pastern	15	1.7	15	1.3

Hindlimb	♀		♂	
Number of dogs	10		4	
	Proportion of total length			
	%	SD	%	SD
Thigh	38	2.8	38	3.1
Lower leg	37	2.2	37	1.5
Hind pastern	25	2.1	25	2.6

Gait parameters

Forelimb

	Stride length (m)				Stride duration (s)				Stance length (m)				Stance duration (s)			
MV	0.76	1.01	1.78	1.78	0.60	0.45	0.35	0.37	0.40	0.39	0.47	0.45	0.33	0.19	0.10	0.10
SD	0.1	0.1	0.2	0.2	0.1	0.0	0.0	0.0	0.1	0.0	0.0	0.1	0.0	0.0	0.0	0.0
Number of strides					120	162	67	51	All numbers equal							

	Swing length (m)				Swing duration (s)				Dutyfactor				Speed (m/s)			
MV	0.35	0.62	1.32	1.33	0.27	0.27	0.25	0.27	55	41	28	27	1.25	2.23	5.02	4.84
SD	0.1	0.1	0.2	0.1	0.0	0.0	0.0	0.0	4.3	4.3	2.2	3.8	0.2	0.2	0.3	0.4
Number of strides	All numbers equal															

Hindlimb

	Stride length (m)				Stride duration (s)				Stance length (m)				Stance duration (s)			
MV	0.78	1.03	1.94	1.78	0.61	0.45	0.36	0.35	0.42	0.38	0.33	0.43	0.34	0.18	0.07	0.09
SD	0.1	0.1	0.1	0.1	0.1	0.0	0.0	0.0	0.1	0.1	0.0	0.0	0.0	0.0	0.0	0.0
Number of strides					99	150	34	76	All numbers equal							

	Swing length (m)				Swing duration (s)				Dutyfactor				Speed (m/s)			
MV	0.36	0.64	1.61	1.35	0.27	0.27	0.29	0.26	56	40	19	26	1.28	2.26	5.35	5.06
SD	0.1	0.1	0.2	0.1	0.0	0.0	0.0	0.0	4.6	6.3	3.2	1.8	0.2	0.2	0.5	0.3
Number of strides	All numbers equal															

Kinematics

Forelimb

	Shoulder blade				Shoulder joint				Upper arm				Elbow joint			
Touch down	65	65	59	62	147	144	144	141	82	79	85	79	151	147	151	137
Lift off	*98*	*100*	*104*	*97*	*136*	*142*	*141*	*140*	38	42	37	43	151	153	165	158
Mid-stance	-	-	-	-	142	141	138	138	-	-	-	-	153	149	149	138
Effective motion	*33*	*35*	*45*	*35*	*11*	*2*	*3*	*1*	44	37	48	36	1	6	14	21
Stance amplitude	*34*	*37*	*46*	*37*	*9*	*3*	*9*	*6*	43	38	51	41	21	26	30	26
Swing amplitude	*35*	*39*	*57*	*47*	*20*	*23*	*31*	*34*	53	55	74	60	54	63	64	66
Number of strides	120	162	72	49	Values in italics are marker-based values corrected to reflect the X-Ray video and position data.											

	Lower arm				Carpal joint				Pastern			
Touch down	70	68	67	58	197	195	200	190	127	128	134	133
Lift off	112	111	128	115	163	158	176	168	50	48	48	54
Mid-stance	-	-	-	-	197	195	200	190	-	-	-	-
Effective motion	43	43	61	57	34	36	25	22	77	80	86	79
Stance amplitude	50	49	61	56	37	47	46	54	77	80	83	80
Swing amplitude	71	83	104	99	81	96	90	79	101	116	134	125

Hindlimb

	Pelvis				Hip joint				Thigh				Stifle joint			
Touch down	47	40	53	52	108	104	112	107	60	65	59	56	151	153	144	157
Lift off	40	35	43	27	*150*	*146*	144	135	*110*	*111*	102	109	145	146	136	152
Mid-stance	-	-	-	-	107	104	112	107	-	-	-	-	151	153	144	157
Effective motion	7	5	11	25	*42*	*42*	32	28	*50*	*46*	43	53	6	7	8	6
Stance amplitude	15	10	11	23	*43*	*42*	31	35	*51*	*46*	41	51	12	13	23	19
Swing amplitude	10	10	24	31	*47*	*49*	48	40	*53*	*52*	60	68	39	55	57	84
Number of strides	91	150	38	72	Values in italics are marker-based values corrected to reflect the X-Ray video and position data.											

	Lower leg				Ankle joint				Hind pastern			
Touch down	91	88	85	101	146	146	136	148	54	58	50	47
Lift off	43	43	34	43	161	160	136	151	118	117	102	108
Mid-stance	-	-	-	-	146	146	137	148	-	-	-	-
Effective motion	48	45	51	58	15	14	6	10	64	60	52	62
Stance amplitude	48	45	51	58	35	40	42	56	68	61	50	70
Swing amplitude	65	75	95	108	44	58	70	73	71	74	96	102

AIREDALE TERRIER

DESCRIPTION OF BREED

This breed originated in Yorkshire. It received its name in 1880 and was first entered into the Kennel Club stud book in 1882. From the beginning, Airedale Terriers have been bred for versatility, to be used as water dogs, guard dogs and rat catchers, in the hunt for small game and to drive cattle. In the course of their history, they have been crossed with Otterhounds, Bull Terriers, large Old English Terriers, wirehaired Scottish Terriers, shorthaired black and tan Collies and Gordon Setters. According to the breed standard, height at the withers should be 58 to 61 cm in dogs and 56 to 59 cm in bitches. The shoulders should be well laid back, as long as possible, and slope at an angle of 45°. The upper arm should be perpendicular to the shoulder, which according to the standard is only possible when the shoulder blade is appropriately oblique. Deviations from this lead to a restriction of movement and limit the stride length of the forelimbs. Tom Horner and Edwards Brown claim that the Airedale Terrier's steep front is caused in large part by a short upper arm. In the breed compendium of the Club für Terrier e.V., C. von Bardeleben disputes this, however, and our measurements do not indicate that the Airedale Terrier's steep front is the result of a short upper arm either. When it comes to the hindquarters, the breed standard requires that the femur and lower leg be of equal length. Lots of ground should be covered during movement. The hocks should only be raised enough to reveal a glimpse of the pads from behind.

PROPORTIONS

In relation to functional leg length, the stance phase in the walk and trot is relatively short in Airedale Terriers, while the swing phase is long. This means that relative stride length in these gaits is in the middle of the range displayed by the breeds involved in this study. In the gallop, however, the Airedale Terrier's relative stride length is long compared to other breeds, a phenomenon explained by the long swing phase length.

GAIT PARAMETERS

The forelimbs are extended forward by the same amount in the walk and the trot, and backward excursion does not differ either between the gaits. The leading forelimb extends backwards to the same extent in the gallop as it does in the walk and the trot. Both forelimbs, however, reach significantly further forward during the gallop, and the backward excursion of the trailing forelimb is also greater.

The forelimbs are enabled to reach further forward during the gallop by a stronger forwards rotation of the shoulder, a greater opening of the shoulder joint by the upper arm in the trailing limb and a greater closure of the elbow joint by the motion of the forearm in the leading limb.

The backwards excursion of the hindlimbs remains the same across the gaits. When it comes to forward excursion, the leading limb is brought the furthest underneath the body in the gallop, mainly thanks to the movement of the lower leg. This value is lowest in the trot thanks to the low degree of pelvic tilt and the fact that the femur is held closer to the vertical. In comparison with other breeds, horizontal excursion at touchdown and liftoff is in the lower third of the range in the walk and the trot. In the trot, the forelimbs extend both further forwards and further backwards than the hindlimbs. In the gallop, forward excursion in Airedale Terriers lies in the middle of the range, while the degree of backwards excursion is very low compared to other breeds.

The breed standard calls for the position of the shoulder blade to be 45° to the horizontal. According to our measurements, however, this position is never reached, not even at the point of touchdown, when the shoulder blade is rotated particularly far forward. The same applies to the ideal shoulder joint angle of 90°, which was not observed at any point during locomotion.

Proportions

Body	♀		♂	
Number of dogs	6		5	
	MV	SD	MV	SD
Withers height [cm]	60	1.3	66	2.1
Body length / Withers height	1.1	0.0	1.1	0.0
Withers height / croup height	1.0	0.0	1.0	0.0
Legginess index	1.5	0.1	1.4	0.1

Forelimb	♀		♂	
Number of dogs	6		5	
	Proportion of total length			
	%	SD	%	SD
Shoulder blade	26	1.2	27	1.4
Upper arm	27	0.7	27	1.5
Lower arm	32	1.2	32	0.8
Pastern	15	2.0	14	2.1

Hindlimb	♀		♂	
Number of dogs	6		5	
	Proportion of total length			
	%	SD	%	SD
Thigh	37	2.3	38	1.9
Lower leg	36	1.8	37	2.6
Hind pastern	27	1.8	26	2.1

Gait parameters

Forelimb

	Stride length				Stride duration				Stance length				Stance duration			
	m				s				m				s			
MV	0.64	1.00	1.76	-	0.77	0.51	0.35	-	0.42	0.49	0.54	-	0.53	0.26	0.12	-
SD	0.1	0.1	0.1	-	0.1	0.0	0.0	-	0.1	0.1	0.0	-	0.1	0.0	0.0	-
Number of strides					75	127	12	0	All numbers equal							

	Swing length				Swing duration				Dutyfactor				Speed			
	m				s								m/s			
MV	0.21	0.51	1.22	-	0.24	0.25	0.24	-	68	51	33	-	0.84	1.97	4.94	-
SD	0.1	0.1	0.1	-	0.0	0.0	0.0	-	5.0	3.8	1.0	-	0.2	0.3	0.2	-
Number of strides	All numbers equal															

Hindlimb

	Stride length				Stride duration				Stance length				Stance duration			
	m				s				m				s			
MV	0.67	1.00	1.75	1.98	0.77	0.51	0.35	0.42	0.43	0.46	0.49	0.54	0.52	0.24	0.10	0.12
SD	0.1	0.1	0.1	0.0	0.1	0.0	0.0	0.0	0.0	0.1	0.0	0.0	0.1	0.0	0.0	0.0
Number of strides					81	119	10	6	All numbers equal							

	Swing length				Swing duration				Dutyfactor				Speed			
	m				s								m/s			
MV	0.24	0.54	1.26	1.44	0.25	0.26	0.25	0.29	67	48	30	29	0.90	1.96	4.92	4.73
SD	0.1	0.1	0.1	0.0	0.0	0.0	0.0	0.0	4.8	5.0	1.0	0.9	0.2	0.3	0.2	0.2
Number of strides	All numbers equal															

Kinematics

Forelimb

Forelimb	Shoulder blade				Shoulder joint				Upper arm				Elbow joint			
Touch down	67	*62*	*61*	-	148	150	*163*	-	80	83	102	-	148	146	182	-
Lift off	90	99	107	-	127	132	148	-	39	33	41	-	133	138	174	-
Mid-stance	-	-	-	-	132	135	149	-	-	-	-	-	149	139	180	-
Effective motion	23	*37*	*46*	-	21	18	*15*	-	44	50	61	-	15	8	8	-
Stance amplitude	28	*41*	*46*	-	23	23	*46*	-	46	52	68	-	23	28	43	-
Swing amplitude	23	*37*	*50*	-	25	31	*41*	-	45	52	72	-	45	60	73	-
Number of strides	79	127	12	0	Values in italics are marker-based values corrected to reflect the X-Ray video and position data.											

Forelimb	Lower arm				Carpal joint				Pastern			
Touch down	68	63	79	-	190	196	239	-	122	133	159	-
Lift off	94	106	132	-	130	123	176	-	36	17	44	-
Mid-stance	-	-	-	-	190	196	239	-	-	-	-	-
Effective motion	30	43	53	-	64	72	63	-	94	116	115	-
Stance amplitude	52	54	53	-	69	81	70	-	93	114	109	-
Swing amplitude	49	71	89	-	78	101	114	-	102	135	153	-

Hindlimb

Hindlimb	Pelvis				Hip joint				Thigh				Stifle joint			
Touch down	53	40	27	48	113	105	97	96	61	65	70	48	149	153	157	151
Lift off	48	38	24	24	*153*	*152*	*147*	*143*	*105*	*113*	*123*	*119*	132	142	151	149
Mid-stance	-	-	-	-	115	105	96	97	-	-	-	-	149	152	157	149
Effective motion	5	2	3	24	*40*	*47*	*50*	*47*	*44*	*49*	*53*	*71*	18	11	6	2
Stance amplitude	16	12	17	28	*43*	*50*	*55*	*50*	*49*	*51*	*52*	*69*	17	20	20	16
Swing amplitude	8	13	15	28	*41*	*51*	*61*	*60*	*45*	*52*	*58*	*80*	31	54	59	61
Number of strides	77	119	10	6	Values in italics are marker-based values corrected to reflect the X-Ray video and position data.											

Hindlimb	Lower leg				Ankle joint				Hind pastern			
Touch down	89	87	87	102	146	148	137	152	56	60	50	49
Lift off	37	39	38	40	154	166	166	160	117	127	128	120
Mid-stance	-	-	-	-	143	146	137	148	-	-	-	-
Effective motion	52	49	49	62	11	20	29	11	61	71	78	71
Stance amplitude	51	50	50	62	33	54	54	59	70	77	75	81
Swing amplitude	56	73	88	91	32	58	69	56	63	80	99	101

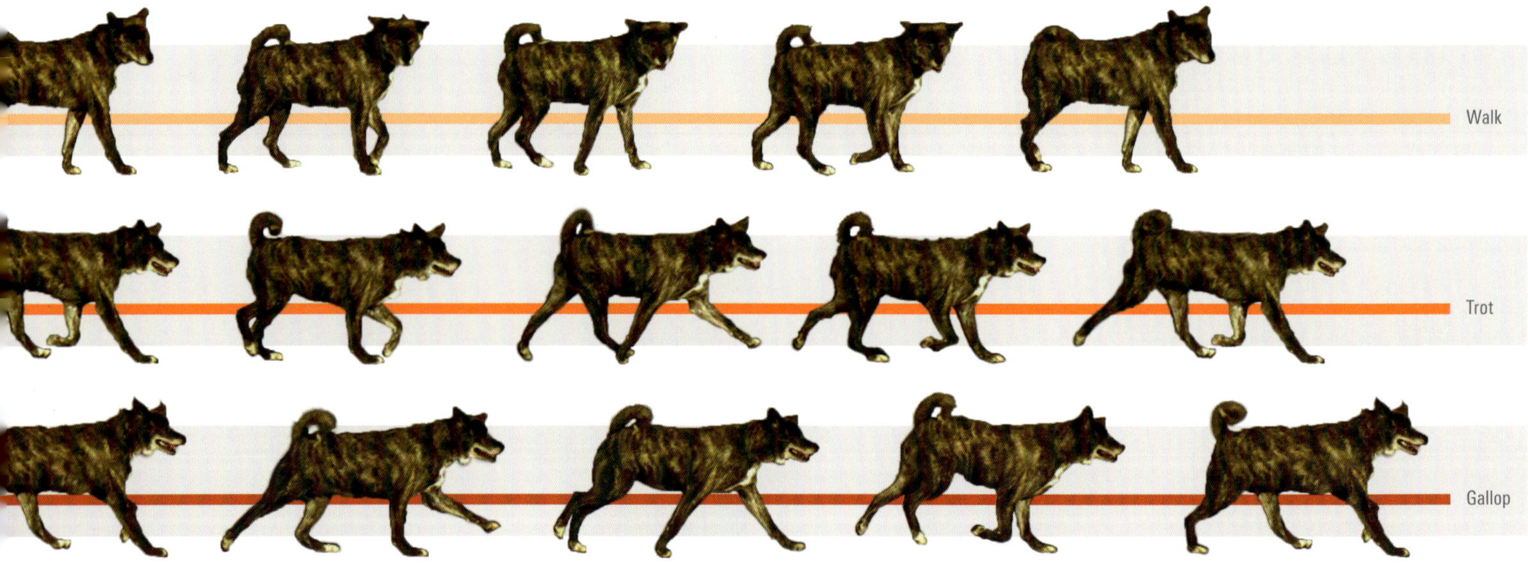

PROPORTIONS

In Akita bitches, the proportional length of the upper arm is more or less constant. In dogs, this applies to the forearm but not the upper arm. The shoulder blade and upper arm are the same length, as are the femur and lower leg. The relatively high legginess index of this breed indicates that the chest is not particularly deep in relation to leg length.

GAIT PARAMETERS

The stance and swing phases are short in the walk and slightly longer in the trot. In the gallop, these phases are long (forelimbs), which puts the stride length into the top third of the range observed here. The hindlimbs have a short stance but a much longer swing phase, putting their stride length into the middle of the range. The forward excursion of the forelimbs is greatest during the gallop and smallest during the walk. The same applies to backwards excursion. The further increase in forward excursion in the gallop is mainly down to the movement of the upper arm and front pastern. The increase in backwards excursion in the trot is also brought about by an increase in the movement of all the limb segments, and the further increase observed in the gallop is caused mainly by the movements of the shoulder blade and forearm. The situation is less uniform in the hindlimb, where the smallest degree of forward excursion is observed in the trailing limb in the gallop and the greatest degree of forward excursion in the leading limb in the same gait. The low degree of forward extension in the trailing limb results from the position of the pelvis and femur, and the long forward reach of the leading limb from the motion of the femur and the lower leg. In the walk and the trot the hindlimbs reach out to the same extent. Backwards excursion is greatest in the gallop (trailing limb). The degree of backwards excursion in the leading limb does not vary between the walk, trot and gallop, but is not as high as that seen in the trailing limb. In the walk, the forward excursion of the forelimbs lies in the middle of the range for the breeds investigated here, while their backwards excursion lies in the bottom quarter of the range. The situation is reversed in the hindlimbs. In the trot, the degree of forward excursion of the forelimbs is high, putting this breed in the top third of the range. The forelimbs are extended the furthest forward in the gallop, while the backward excursion of the forelimbs and the forward excursion of the hindlimbs lies in the bottom third of the range. The degree of backwards excursion of the hindlimbs, on the other hand, is high, with the value situated in the top third of the range. Our measurements do not indicate that the angulation of the hindlimbs is any more moderate in Akitas than in other breeds of a similar size.

AKITA

DESCRIPTION OF BREED

The Akita is a variety of Japanese Spitz, dogs which Räber (1993) claims are closely related to the Scandinavian Spitzes. The breed originated in the province Akita. The name therefore reflects the dog's geographic origins. No definitive information is available on the ancestry and history of this breed. Genetic studies (Parker et al. 2004) show that the Akita is a very old breed, though it is not mentioned in the first Japanese stud book, which dates from around 500 years ago. The first written references to the breed are from approximately one hundred years ago and are found in books about bear and deer hunting. When dog fights became popular in Japan around 1900, the native breeds were crossed with large, heavy Japanese and European breeds. Fans of the breed searched the mountain regions almost in vain for purebred Akitas, and eventually the old breed was reconstructed by crossing the few remaining purebreds with a closely related hunting dog (Matagi). In 1927, the Akita Hozonkai was established, and in 1931 the Akita received the status of national cultural asset. To ensure that the best breeding stock remained in Japan, a law was passed forbidding the export of the dogs. The Japanese Kennel Club was set up in 1948. In the course of the re-establisment after World War II, two different lines, a light one (Ichinoseki) and a heavier one (Dewa), emerged at the hands of different breeders. The first Akitas came to America as early as around 1900, but the breed was not registered in the AKC stud book until 1964. The dogs that were imported were from the heavy Dewa line and the Akitas bred in the USA today differ in body shape, coat type and markings from the Japanese Akitas.
The FCI standard requires a height at the withers of 64 – 70 centimeters for dogs and 58 – 64 centimeters for bitches. The ribcage should be low, the hind legs moderately angled and movement elastic and powerful.

Proportions

Body	♀		♂	
Number of dogs	8		3	
	MV	SD	MV	SD
Withers height [cm]	50	2.0	56	0.6
Body length / Withers height	1.2	0.0	1.1	0.0
Withers height / croup height	1.0	0.0	1.0	0.0
Legginess index	1.1	0.1	1.1	0.1

Forelimb	♀		♂	
Number of dogs	8		3	
	Proportion of total length			
	%	SD	%	SD
Shoulder blade	28	0.8	31	3.1
Upper arm	27	1.5	26	2.8
Lower arm	29	0.8	28	1.0
Pastern	16	1.7	15	1.3

Hindlimb	♀		♂	
Number of dogs	8		3	
	Proportion of total length			
	%	SD	%	SD
Thigh	37	2.3	38	3.3
Lower leg	36	1.5	36	3.1
Hind pastern	27	1.7	26	0.1

Gait parameters

Forelimb

	Stride length				Stride duration				Stance length				Stance duration				Swing length				Swing duration				Dutyfactor				Speed			
	m				s				m				s				m				s								m/s			
MV	0.69	0.97	1.43	1.39	0.65	0.47	0.37	0.37	0.43	0.46	0.47	0.51	0.42	0.23	0.13	0.14	0.26	0.51	0.96	0.87	0.23	0.24	0.24	0.23	65	50	35	39	1.05	2.05	3.83	3.72
SD	0.1	0.1	0.1	0.1	0.0	0.0	0.0	0.0	0.0	0.0	0.0	0.0	0.1	0.0	0.0	0.0	0.1	0.1	0.1	0.1	0.0	0.0	0.0	0.0	5.8	4.6	3.4	3.3	0.2	0.2	0.4	0.3
Number of strides					83	120	145	102	All numbers equal																							

Hindlimb

	Stride length				Stride duration				Stance length				Stance duration				Swing length				Swing duration				Dutyfactor				Speed			
	m				s				m				s				m				s								m/s			
MV	0.63	0.90	1.49	1.48	0.66	0.46	0.37	0.37	0.40	0.39	0.47	0.46	0.44	0.21	0.13	0.12	0.23	0.52	1.02	1.02	0.22	0.25	0.24	0.25	66	45	34	33	0.95	1.96	4.01	3.96
SD	0.1	0.1	0.2	0.1	0.0	0.0	0.0	0.0	0.1	0.0	0.1	0.0	0.1	0.0	0.0	0.0	0.1	0.1	0.2	0.1	0.0	0.0	0.0	0.0	5.9	3.6	5.8	3.7	0.2	0.2	0.6	0.4
Number of strides					90	124	145	152	All numbers equal																							

Kinematics

Forelimb	Shoulder blade				Shoulder joint				Upper arm				Elbow joint				Lower arm				Carpal joint				Pastern			
Touch down	65	60	59	62	149	145	142	142	84	85	83	79	145	146	145	133	62	61	62	53	199	201	202	200	138	141	141	148
Lift off	*98*	*105*	109	98	*130*	*137*	137	132	32	32	28	34	140	148	157	158	109	116	129	124	142	146	157	153	34	30	27	29
Mid-stance	-	-	-	-	136	129	132	127	-	-	-	-	146	141	144	134	-	-	-	-	199	201	202	200	-	-	-	-
Effective motion	*33*	*45*	50	35	*19*	*8*	5	10	52	53	55	45	5	2	12	25	47	55	67	70	57	55	45	47	104	111	114	119
Stance amplitude	*38*	*48*	52	44	*20*	*12*	28	22	54	54	61	57	26	30	34	41	65	65	70	74	64	73	63	72	102	107	111	118
Swing amplitude	*33*	*45*	53	41	*24*	*23*	33	35	55	58	69	60	48	60	62	58	66	85	96	95	97	115	106	121	126	155	155	168
Number of strides	83	120	148	102	Values in italics are marker-based values corrected to reflect the X-Ray video and position data.																							

Hindlimb	Pelvis				Hip joint				Thigh				Stifle joint				Lower leg				Ankle joint				Hind pastern			
Touch down	40	33	34	46	99	99	100	95	59	66	65	51	155	157	161	157	96	91	95	105	148	149	153	155	51	58	57	48
Lift off	37	28	27	21	*144*	*142*	*155*	*143*	*107*	*117*	*128*	*124*	127	141	148	150	34	39	35	42	148	163	170	169	113	123	135	127
Mid-stance	-	-	-	-	100	99	104	99	-	-	-	-	154	157	157	156	-	-	-	-	146	149	147	154	-	-	-	-
Effective motion	3	5	8	26	*45*	*43*	*55*	*48*	*48*	*50*	*53*	*73*	29	16	12	6	62	52	60	64	5	14	21	18	62	65	78	79
Stance amplitude	19	13	19	28	*50*	*44*	*55*	*58*	*54*	*51*	*52*	*72*	27	19	32	25	61	50	63	64	36	51	69	77	77	71	86	95
Swing amplitude	6	11	18	31	*47*	*49*	*64*	*59*	*49*	*54*	*58*	*83*	43	67	63	74	69	83	102	108	31	55	61	65	64	75	94	101
Number of strides	90	124	146	153	Values in italics are marker-based values corrected to reflect the X-Ray video and position data.																							

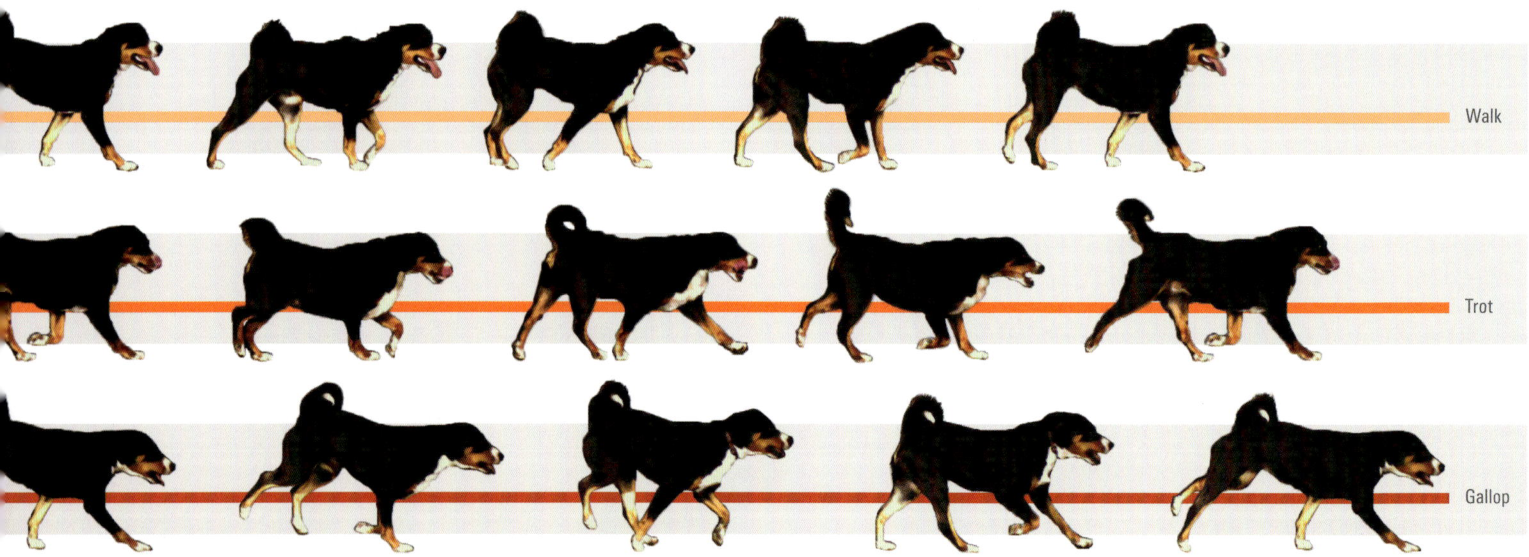

PROPORTIONS

Unlike in the other breeds in the study, the variability of the length of the upper arm in the Appenzell individuals we investigated was greater than that of the shoulder blade and forearm. The chest descends to about half the height at the withers, as the legginess index shows. Three of the bitches we investigated were slightly smaller than the breed standard requires, though no differences were observed between their limb segment proportions and those of other dogs of the same breed.

GAIT PARAMETERS

Stride length is middle-of-the-range in all three gaits and is made up of a long stance phase and a shorter swing phase, though both phases are, in relation to functional limb length, longer in the forelimbs than the hindlimbs. There is no difference between the gaits when it comes to the forward excursion of the forelimbs, and no difference between the walk and the trot when it comes to their backwards excursion. In the gallop, backwards excursion in the leading forelimb is slightly greater thanks to the movement of the forearm, and even greater again in the trailing forelimb as the result of a greater backwards movement in all segments. The degree of forward excursion of the hindlimb is smallest in the trot and highest in the gallop (leading limb). In the trailing hindlimb, no difference is observed between the walk and the gallop. The high degree of forward excursion in the walk is caused by the position of the pelvis and femur, and the high degree of forward excursion observed in the trailing hindlimb in the gallop is caused mainly by the motion of the lower leg. The backwards excursion of the hindlimbs increases from the walk to the trot to the gallop, and here from the leading to trailing limb. This is mainly due to the motion of the proximal segments of the limb (pelvis and femur). In comparative terms, the degree of forward excursion of the forelimbs is always high in Appenzell Mountain Dogs, with the exception of the trailing forelimb in the gallop. The degree of backwards excursion lies in the middle of the range for the walk and the trot, and the top third of the range for the gallop. The forward excursion of the hindlimbs is middle-of-the range in all gaits except the trot, where it is lower down the range. The backward excursion of the hindlimbs is in the bottom third of the range in the walk, the middle of the range in the trot and the top third of the range in the gallop, which means that this breed extend their hindlimbs further backwards than other breeds when moving at high speeds. Our data do not show that Appenzell Mountain Dogs exhibit particularly "steep" hindquarters during locomotion.

APPENZELL MOUNTAIN DOG

DESCRIPTION OF BREED

This breed probably originated in the 15th or 16th century in the foothills of the Alps in connection with a particular form of cattle farming practiced at the time. Herds of over a hundred cows would be walked for long distances into the mountains, which, of course, required a good cattle dog. As well as driving the cows and keeping the herd together, the dog would be expected to guard the possessions of the cowherd. In the 19th century, this form of farming was made redundant by the boom in the number of cheese dairies in the valleys. The cattle dog lost its significance and would have disappeared altogether if certain cynologists had not fought to preserve it. Around 1900, there were distinct local varieties of cattle dog in the various regions, but no actual breeds. This was to change at the initiative of the cynologist Max Siber, and the Appenzell Mountain Dog was one of the first targets of his campaign. The Swiss Cynological Society formulated a breed standard, and farmers who kept dogs which conformed to this standard and/or bred such dogs were rewarded financially. The Appenzell Mountain Dog Club, which kept a stud book and issued pedigrees, was set up in 1906. The dogs used for breeding today basically all descend from just eleven individuals.

Appenzell Mountain Dog have a withers height of 52 to 56 centimeters; females are required to be two centimeters smaller. Body length should be slightly greater than height (withers height to length of body = 9:10), and the chest should be broad and deep. The shoulder blade should be the same length as or slightly longer than the upper arm, the hind legs straight and parallel. The upper thigh is slightly longer than or the same length as the lower thigh.

Proportions

Body	♀		♂	
Number of dogs	5		5	
	MV	SD	MV	SD
Withers height [cm]	71	1.7	80	3.2
Body length / Withers height	1.0	0.1	1.0	0.0
Withers height / croup height	1.0	0.0	1.0	0.0
Legginess index	1.5	0.1	1.5	0.1

Forelimb	♀		♂	
Number of dogs	5		5	
	Proportion of total length			
	%	SD	%	SD
Shoulder blade	25	1.0	24	0.1
Upper arm	28	1.6	28	0.4
Lower arm	32	1.0	33	2.1
Pastern	15	1.2	15	2.5

Hindlimb	♀		♂	
Number of dogs	5		5	
	Proportion of total length			
	%	SD	%	SD
Thigh	37	2.4	36	1.4
Lower leg	40	1.3	39	1.9
Hind pastern	23	1.3	25	1.1

Gait parameters

Forelimb

	Stride length				Stride duration				Stance length				Stance duration				Swing length				Swing duration				Dutyfactor				Speed			
	m				s				m				s				m				s								m/s			
MV	0.93	1.26	-	-	0.89	0.59	-	-	0.57	0.57	-	-	0.56	0.28	-	-	0.36	0.69	-	-	0.33	0.31	-	-	63	48	-	-	1.05	2.15	-	-
SD	0.1	0.1	-	-	0.1	0.0	-	-	0.1	0.1	-	-	0.1	0.0	-	-	0.1	0.1	-	-	0.0	0.0	-	-	2.4	5.6	-	-	0.2	0.3	-	-
Number of strides					78	131	-	-	All numbers equal																							

Hindlimb

	Stride length				Stride duration				Stance length				Stance duration				Swing length				Swing duration				Dutyfactor				Speed			
	m				s				m				s				m				s								m/s			
MV	0.92	1.29	-	-	0.89	0.58	-	-	0.53	0.53	-	-	0.53	0.25	-	-	0.39	0.76	-	-	0.37	0.33	-	-	59	43	-	-	1.03	2.21	-	-
SD	0.1	0.2	-	-	0.1	0.0	-	-	0.1	0.1	-	-	0.1	0.0	-	-	0.1	0.2	-	-	0.0	0.0	-	-	2.7	4.8	-	-	0.2	0.4	-	-
Number of strides					71	93	-	-	All numbers equal																							

Kinematics

Forelimb	Shoulder blade				Shoulder joint				Upper arm				Elbow joint				Lower arm				Carpal joint				Pastern			
Touch down	*68*	*69*	-	-	*156*	*156*	-	-	88	88	-	-	150	151	-	-	63	63	-	-	189	189	-	-	128	127	-	-
Lift off	*106*	*108*	-	-	*142*	*145*	-	-	36	37	-	-	140	141	-	-	105	105	-	-	146	142	-	-	42	38	-	-
Mid-stance	-	-	-	-	158	159	-	-	-	-	-	-	147	145	-	-	-	-	-	-	189	189	-	-	-	-	-	-
Effective motion	*38*	*39*	-	-	*14*	*11*	-	-	52	51	-	-	10	10	-	-	42	41	-	-	44	47	-	-	86	89	-	-
Stance amplitude	*40*	*41*	-	-	*14*	*14*	-	-	51	51	-	-	19	21	-	-	53	48	-	-	46	57	-	-	82	86	-	-
Swing amplitude	*38*	*41*	-	-	*24*	*30*	-	-	54	57	-	-	51	63	-	-	64	70	-	-	85	106	-	-	107	128	-	-
Number of strides	78	131	0	0	Values in italics are marker-based values corrected to reflect the X-Ray video and position data.																							

Hindlimb	Pelvis				Hip joint				Thigh				Stifle joint				Lower leg				Ankle joint				Hind pastern			
Touch down	62	51	-	-	119	113	-	-	58	61	-	-	143	146	-	-	86	85	-	-	145	144	-	-	58	58	-	-
Lift off	52	48	-	-	*162*	*163*	-	-	*111*	*115*	-	-	134	140	-	-	38	40	-	-	157	160	-	-	119	120	-	-
Mid-stance	-	-	-	-	118	113	-	-	-	-	-	-	143	146	-	-	-	-	-	-	145	144	-	-	-	-	-	-
Effective motion	10	3	-	-	*43*	*50*	-	-	*53*	*54*	-	-	9	6	-	-	48	45	-	-	12	17	-	-	61	62	-	-
Stance amplitude	13	6	-	-	*47*	*51*	-	-	*56*	*54*	-	-	13	20	-	-	46	45	-	-	25	40	-	-	62	63	-	-
Swing amplitude	11	8	-	-	*51*	*55*	-	-	*59*	*61*	-	-	32	51	-	-	55	62	-	-	33	52	-	-	64	69	-	-
Number of strides	71	93	0	0	Values in italics are marker-based values corrected to reflect the X-Ray video and position data.																							

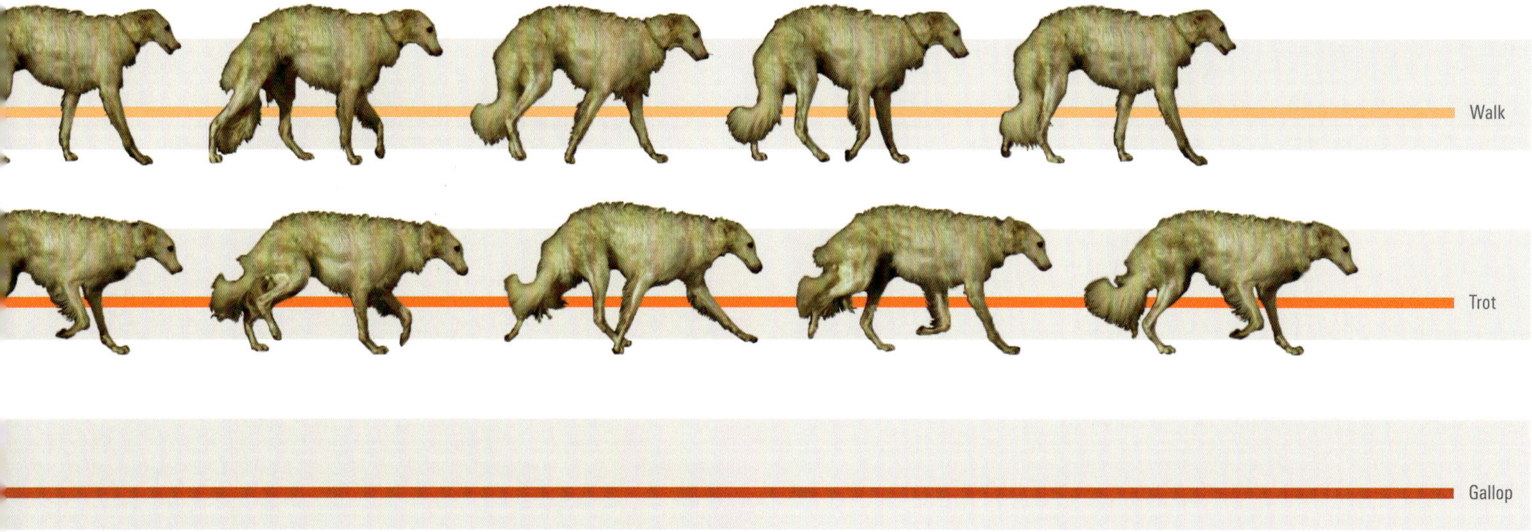

PROPORTIONS

All the individuals we investigated were the size required by the standard. This breed is square-built, meaning that trunk length corresponds to height at the withers. The legginess index is relatively high, indicating that the animals have a lot of legroom under the chest, which in turn reaches to less than half of height at the withers. The proportional length of the upper arm is shorter than that of the shoulder blade. The shoulder blade in Borzois is slightly shorter than that of most other breeds, but the forearm is longer. The length of the upper arm varies much more between bitches than between dogs. The lower leg is slightly longer than average for all breeds we investigated and the femur and hind pastern slightly shorter. Apart from the difference in variability, we did not observe any sex-specific differences.

GAIT PARAMETERS

In the walk, relative stride length is middle-of-the-range for the breeds investigated here. In the forelimbs, both the stance and the swing phase are of medium length, while in the hindlimbs the stance phase is short and the swing phase of medium length. In the trot, both the stance and the swing phase are short, and so too, therefore, is the relative stride length. The horizontal excursion of the forelimbs is identical in the walk and the trot, as can be seen from the uniform values in the kinematics table. In the hindlimbs, only forwards excursion differs between the gaits, with the limb extended slightly further forward in the walk than in the trot, mainly as the result of a greater pelvic tilt.

From a comparative perspective, the backwards and forwards excursion of the forelimbs lies in the middle of the range in the walk, as does the backward excursion of the hindlimbs. Two thirds of the breeds in this study extend their hindlimbs further forward than the Borzoi before touchdown. The forwards excursion of the forelimbs is fairly short in the trot, in the bottom quarter of the range for this study, while backwards excursion lies in the middle of the range. Both the forwards excursion and the backwards excursion of the hindlimbs lie in the bottom third of the range.

BORZOI

DESCRIPTION OF BREED

The Borzoi is undisputedly a hunting dog with Russian origins which was kept for hare, fox and wolf coursing. What is not quite as clear is its prehistoric ancestry. It is conjectured that Asiatic peoples came to Russia bringing their sighthounds with them and, once there, began to breed a Russian sighthound purely for performance (Räber, 1995). The oldest reference to a Borzoi-like dog is an illustration in a 16th century missal, and the first description is found in the "Book of Hunting Rules" from 1635. Until 1861, many Russian landlords kept packs of hundreds of hunting dogs. The literature makes reference to one such pack of 500 Borzois. Until the beginning of the 20th century, Borzois were bred exclusively for performance. Because they were crossed on numerous occasions with other sighthounds, individuals tended to vary greatly in appearance. In 1873, the "Imperial Society for the Breeding of Game and Working Animals for Proper Hunting" was established in Germany, out of which the Association of Borzoi Breeders developed. In 1888, the first breed standard was written. The first Borzois were shown in Germany in 1890.

According to the breed standard, height at the withers should be between 75 and 85 centimeters for dogs and between 68 and 78 centimeters for bitches. Body length should be slightly greater than withers height. Withers and croup should be approximately on a level, and the chest should be deep, reaching to approximately half of withers height. The shoulder blade and upper arm should be of similar length.

Proportions

Body	♀		♂	
Number of dogs	5		5	
	MV	SD	MV	SD
Withers height [cm]	52	2.1	55	1.0
Body length / Withers height	1.2	0.0	1.1	0.0
Withers height / croup height	1.0	0.0	1.0	0.0
Legginess index	1.4	0.2	1.4	0.1

Forelimb	♀		♂	
Number of dogs	5		5	
	Proportion of total length			
	%	SD	%	SD
Shoulder blade	27	1.7	26	1.1
Upper arm	26	1.3	27	0.4
Lower arm	30	2.0	29	0.6
Pastern	17	2.7	18	1.0

Hindlimb	♀		♂	
Number of dogs	5		5	
	Proportion of total length			
	%	SD	%	SD
Thigh	36	1.5	35	1.0
Lower leg	37	2.0	37	1.3
Hind pastern	27	1.1	28	1.5

Gait parameters

Forelimb

	Stride length				Stride duration				Stance length				Stance duration				Swing length				Swing duration				Dutyfactor				Speed			
	m				s				m				s				m				s								m/s			
MV	0.69	0.91	1.65	1.58	0.71	0.45	0.33	0.32	0.42	0.42	0.37	0.41	0.45	0.22	0.08	0.09	0.27	0.49	1.28	1.17	0.26	0.23	0.25	0.23	63	49	25	28	0.97	2.04	4.91	4.89
SD	0.1	0.2	0.2	0.1	0.1	0.0	0.0	0.0	0.0	0.0	0.0	0.0	0.1	0.0	0.0	0.0	0.1	0.2	0.2	0.1	0.0	0.0	0.0	0.0	2.8	5.8	3.5	2.3	0.1	0.5	0.1	0.1
Number of strides					111	212	5	21	All numbers equal																							

Hindlimb

	Stride length				Stride duration				Stance length				Stance duration				Swing length				Swing duration				Dutyfactor				Speed			
	m				s				m				s				m				s								m/s			
MV	0.67	0.85	-	1.56	0.71	0.45	-	0.32	0.41	0.41	-	0.43	0.45	0.23	-	0.10	0.26	0.44	-	1.13	0.26	0.22	-	0.22	63	51	-	30	0.95	1.92	-	4.84
SD	0.0	0.2	-	0.0	0.1	0.0	-	0.0	0.0	0.0	-	0.0	0.1	0.0	-	0.0	0.0	0.2	-	0.1	0.0	0.0	-	0.0	4.5	7.5	-	1.8	0.1	0.5	-	0.1
Number of strides					125	208	-	30	All numbers equal																							

Kinematics

Forelimb	Shoulder blade				Shoulder joint				Upper arm				Elbow joint				Lower arm				Carpal joint				Pastern			
Touch down	*63*	*63*	*61*	*57*	156	156	123	133	87	86	63	68	151	148	124	120	63	63	62	52	193	196	181	183	130	135	119	132
Lift off	*101*	*100*	*100*	*98*	132	136	117	111	36	35	31	33	147	144	160	158	112	108	129	125	157	136	180	182	45	28	51	57
Mid-stance	-	-	-	-	145	145	116	116	-	-	-	-	149	142	134	128	-	-	-	-	193	196	181	183	-	-	-	-
Effective motion	*38*	*35*	*39*	*41*	24	20	6	22	52	50	31	35	4	5	36	38	48	45	67	73	37	60	1	1	85	107	68	75
Stance amplitude	*39*	*38*	*35*	*50*	24	22	10	22	51	51	33	39	22	25	33	37	59	55	64	70	34	70	28	23	80	104	61	70
Swing amplitude	*38*	*36*	*50*	*56*	30	30	36	47	54	53	60	56	47	58	62	65	71	78	105	105	90	114	96	111	119	146	151	162
Number of strides	111	212	5	21	Values in italics are marker-based values corrected to reflect the X-Ray video and position data.																							

Hindlimb	Pelvis				Hip joint				Thigh				Stifle joint				Lower leg				Ankle joint				Hind pastern			
Touch down	54	48	-	54	112	109	-	97	58	61	-	43	150	151	-	137	92	89	-	95	141	141	-	129	48	51	-	33
Lift off	42	36	-	28	*152*	*148*	-	*136*	*109*	*112*	-	*108*	135	134	-	138	36	32	-	30	157	156	-	143	121	124	-	113
Mid-stance	-	-	-	-	113	109	-	97	-	-	-	-	149	150	-	138	-	-	-	-	141	140	-	129	-	-	-	-
Effective motion	12	11	-	26	*40*	*40*	-	*39*	*52*	*51*	-	66	15	17	-	1	56	57	-	65	16	15	-	15	73	73	-	80
Stance amplitude	17	13	-	25	*43*	*42*	-	*41*	*54*	*52*	-	63	17	19	-	19	56	56	-	68	37	51	-	57	78	76	-	87
Swing amplitude	12	15	-	38	*43*	*42*	-	*53*	*53*	*53*	-	83	41	56	-	55	66	77	-	81	43	59	-	62	75	81	-	103
Number of strides	125	208	0	30	Values in italics are marker-based values corrected to reflect the X-Ray video and position data.																							

PROPORTIONS

The proportional length of the limb segments varies slightly more between bitches than dogs, except in the case of the hind pastern. As the legginess index indicates, the chest is not significantly deeper in Bearded Collies than in other races of a similar body size. One of the bitches we investigated was, at 49 cm at the withers, slightly smaller than required by the standard. In this individual, the shoulder blade was proportionally larger than in the other Bearded Collie bitches.

GAIT PARAMETERS

In the gallop, relative stride length is fairly long and is made up of a long swing phase and a short stance phase. In the trot, both the stance phase and the swing phase are short. Step length in the walk is middle-of-the-range, though here the stance phase is, relatively speaking, longer than the swing phase in all limbs. Horizontal excursion prior to touchdown is smallest in the gallop in the trailing forelimb, and largest in the leading forelimb. This is because the upper arm is not extended as far forward in the trailing limb, and the motion of the shoulder blade and forearm are greater in the leading limb. No difference in forwards excursion is observed between the walk and the trot. Backwards excursion is greatest in the trailing forelimb in the gallop, other than that it does not differ between the gaits. The relatively high degree of excursion shown by the trailing limb is achieved by the positioning of the upper arm and forearm. In the hindlimbs, forward extension is greater in the walk than the trot. This is the result of an accumulation of small movements in every segment. Forward excursion is even greater in the gallop (leading limb), mainly as the result of increased movement in the femur. When it comes to backwards excursion, no differences are observed between the gaits. In a cross-breed comparison, the forward excursion of the forelimbs is situated towards the top of the range in the walk, the middle of the range in the trot and the bottom of the range in the gallop. Backwards excursion of the forelimbs is middle-of-the-range in the walk and lower down the range in the trot and gallop. The same applies to both the forwards and backwards excursion of the hindlimbs.
One striking feature of Bearded Collie locomotion is the fact that the knee joint of the leading limb is more angulated than in most other breeds, though this is not the case in the other gaits. The gait which the longest forward excursion is the walk.

BEARDED COLLIE

DESCRIPTION OF BREED

This breed originates in the Scottish Highlands, where the harsh climate necessitated a sheepdog with a thick coat. The dogs' primary purpose was to drive the sheep from the hills with lots of barking, but they were also expected to protect the flock against predators and poachers. Bearded Collies would be left alone for long periods of time during their work, and independence (sometimes bordering on stubbornness) is still a recognizable personality trait to this day. For a long time, the Bearded Collie, as the name suggests, was considered to belong to the Collie family. However, at the beginning of the twentieth century, certain Old English Sheepdogs looked very similar to today's Bearded Collies (Strebel 1905). According to some sources (e.g. www.cfbrh.de), Bearded Collies were to be found in Scotland as early as the 15th century. The breed almost died out in the mid-twentieth century, and today's lines are the result of the efforts of one G.O. Willison, who in the 1940s sought out the last Bearded Collies in Britain and began breeding them. The breed was officially recognized by the English Kennel Club in 1944.
Desired height at withers is 53 – 56 centimeters for dogs and 51 – 53 centimeters for bitches. The standard's other specifications are fairly general: the chest should be deep, the forelegs straight and vertical, the shoulders should slope well back, the stifles should be well bent and the hocks low. The hind pastern should be at right angles to the ground, and movement should be smooth and long-reaching.

Proportions

Body	♀		♂	
Number of dogs	8		2	
	MV	SD	MV	SD
Withers height [cm]	41	0.9	45	1.1
Body length / Withers height	1.0	0.0	1.0	0.0
Withers height / croup height	1.0	0.0	1.0	0.0
Legginess index	1.3	0.1	1.4	0.0

Forelimb	♀		♂	
Number of dogs	8		2	
	Proportion of total length			
	%	SD	%	SD
Shoulder blade	25	0.6	26	1.1
Upper arm	26	0.9	28	1.0
Lower arm	32	1.2	30	3.2
Pastern	17	1.8	16	3.1

Hindlimb	♀		♂	
Number of dogs	8		2	
	Proportion of total length			
	%	SD	%	SD
Thigh	36	2.5	36	0.2
Lower leg	38	2.3	39	0.8
Hind pastern	26	1.7	25	0.6

Gait parameters

Forelimb

	Stride length				Stride duration				Stance length				Stance duration				Swing length				Swing duration				Dutyfactor				Speed			
	m				s				m				s				m				s								m/s			
MV	0.55	0.77	1.19	1.16	0.61	0.41	0.37	0.37	0.32	0.33	0.40	0.39	0.37	0.18	0.13	0.13	0.23	0.45	0.79	0.76	0.24	0.23	0.24	0.24	61	45	35	36	0.89	1.89	3.22	3.12
SD	0.1	0.1	0.1	0.1	0.0	0.0	0.0	0.0	0.0	0.0	0.0	0.0	0.0	0.0	0.0	0.0	0.0	0.1	0.1	0.1	0.0	0.0	0.0	0.0	3.8	3.5	1.4	3.0	0.1	0.3	0.1	0.2
Number of strides					133	214	27	10	All numbers equal																							

Hindlimb

	Stride length				Stride duration				Stance length				Stance duration				Swing length				Swing duration				Dutyfactor				Speed			
	m				s				m				s				m				s								m/s			
MV	0.56	0.78	1.13	1.20	0.61	0.40	0.36	0.37	0.31	0.27	0.38	0.36	0.35	0.15	0.13	0.12	0.26	0.51	0.75	0.83	0.27	0.25	0.24	0.25	56	37	35	33	0.91	1.91	3.07	3.21
SD	0.1	0.1	0.1	0.0	0.0	0.0	0.0	0.0	0.0	0.0	0.0	0.0	0.0	0.0	0.0	0.0	0.0	0.1	0.1	0.0	0.0	0.0	0.0	0.0	4.2	4.4	1.8	1.3	0.1	0.3	0.2	0.1
Number of strides					134	207	34	23	All numbers equal																							

Kinematics

Forelimb	Shoulder blade				Shoulder joint				Upper arm				Elbow joint				Lower arm				Carpal joint				Pastern			
Touch down	*63*	*61*	*58*	*60*	150	151	151	154	82	85	89	89	145	147	147	142	63	62	58	53	191	188	188	183	128	127	130	130
Lift off	*98*	*99*	*104*	*93*	*129*	*135*	*131*	*136*	31	36	27	43	130	140	144	145	99	105	118	102	133	143	148	129	34	38	31	28
Mid-stance	-	-	-	-	144	144	135	136	-	-	-	-	142	143	146	145	-	-	-	-	191	188	188	183	-	-	-	-
Effective motion	*35*	*38*	*46*	*33*	*31*	*26*	*31*	*28*	51	49	62	46	15	7	2	3	36	43	59	49	57	45	40	53	94	89	100	111
Stance amplitude	*38*	*39*	*46*	*41*	*31*	*27*	*34*	*27*	52	50	63	50	22	23	24	27	54	53	63	54	59	69	61	104	93	94	97	120
Swing amplitude	*37*	*40*	*55*	*41*	*38*	*38*	*41*	*46*	56	56	74	57	50	64	64	73	60	81	91	107	98	116	118	137	120	141	160	165
Number of strides	133	214	27	10	Values in italics are marker-based values corrected to reflect the X-Ray video and position data.																							

Hindlimb	Pelvis				Hip joint				Thigh				Stifle joint				Lower leg				Ankle joint				Hind pastern			
Touch down	61	51	56	61	118	114	119	110	58	64	63	49	149	146	151	141	92	83	88	93	152	147	147	142	61	64	59	48
Lift off	52	46	41	32	145	144	150	138	93	97	109	107	135	139	141	147	42	41	32	40	167	164	169	168	124	122	138	127
Mid-stance	-	-	-	-	118	114	124	124	-	-	-	-	149	146	144	137	-	-	-	-	152	146	136	107	-	-	-	-
Effective motion	9	4	15	29	27	30	31	29	35	34	46	58	14	8	10	6	49	42	56	52	14	17	22	26	64	58	79	79
Stance amplitude	15	8	21	29	30	29	34	33	37	33	46	57	14	16	26	20	48	41	54	55	30	42	60	75	67	61	78	95
Swing amplitude	12	11	20	36	36	38	41	42	41	42	49	73	39	58	52	50	62	71	84	88	35	57	52	66	72	78	87	113
Number of strides	134	207	34	23	Values in italics are marker-based values corrected to reflect the X-Ray video and position data.																							

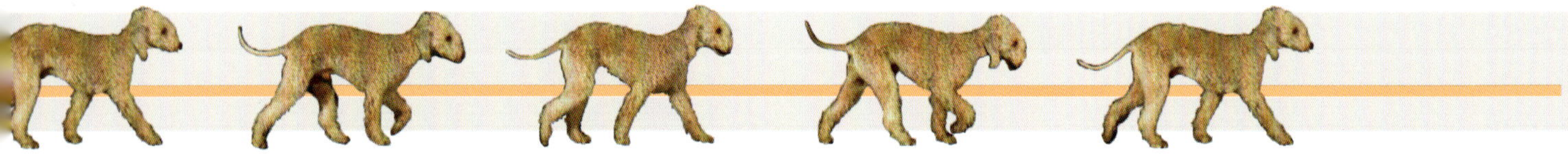
Walk

Trot

Gallop

BEDLINGTON TERRIER

DESCRIPTION OF BREED

The first dogs of this type are reported to have emerged around 1780 in the English-Scottish borders. Among their ancestors, without doubt, is a group of working terriers which lived in the region around the same time. The original Bedlington Terriers were the result of cross-breeding between these working terriers and Otterhounds, Greyhounds and Bull Terriers. Subsequently, large and medium-sized dogs were separated from short-legged ones. In rural areas, Bedlington Terriers were popular rat catchers and travelled around with itinerant tinkers. Organized breeding began in 1820, which makes Bedlingtons the oldest recognized breed of terrier. In the early days, the dogs were bred predominantly for hunting, which they were particularly suited to thanks to their speed and silence. The first Bedlington Terriers arrived in America in 1890, and by the end of the 1920s they were also present in Germany.
Bedlington Terriers are long-legged dogs but the body should ideally be slightly greater in length than height. The back has a natural arch with the highest point over the loin. The hindlegs should have the appearance of being longer than the forelegs, the chest should be deep and reach to the elbow. The shoulders should be flat and sloping. Height at the withers should be around 41 centimeters, with slight variation below allowed for bitches and above allowed for dogs.

PROPORTIONS

In seven of the Bedlington Terriers involved in the study, trunk length slightly exceeded height at the withers. The reverse was true in two individuals, and in one dog the two measurements were the same. There does not appear to be a connection between body proportions and the relative length of the limb segments, however. The variability of the individual segments of the forelimb is higher in dogs than in bitches, while the opposite is true of the hindlimb. One of the bitches we investigated was, at 43 centimeters, slightly taller than required by the standard, but her limb segment proportions did not differ from those of other bitches.

GAIT PARAMETERS

Stride length is middle-of-the-range in most gaits. In the trot, the stance phase is strikingly short, but this is compensated by a long swing phase. Forward excursion before touchdown is the same in the walk and the trot but greater in the gallop, with no difference detected between the trailing and leading limb. The difference between synchronous and symmetrical gaits is caused by the movements of the upper and forearm. Backwards excursion is lowest in the walk, as a result of the position of the upper arm and forearm. It is greatest in the trailing forelimb in the gallop, a phenomenon to which all the limb segments contribute. There is no difference between the backwards excursion displayed in trot and that displayed by the leading forelimb in the gallop. The forward excursion of the trailing hindlimb is greater in the gallop than the trot. This can be put down to the position of the pelvis and forearm. The forward excursion of the leading hindlimb is lower in the walk than the gallop. These differences can be traced back to differences in the movement of all the limb segments. Backwards excursion does not differ between the walk and the trot. It is greater in the gallop, though less so in the leading limb than the trailing limb. In a cross-breed comparison, the forward excursion of the forelimbs comes close to the top of the range in Bedlington Terriers in the walk and the gallop, but is middle-of-the-range in the trot. Backwards excursion is low in all gaits. The forwards excursion of the hindlimbs is very low in the trot and the gallop and middle-of-the-range in the walk. Backwards excursion is also middle-of-the range in the walk and the gallop (leading limb), but very high in the trailing limb in the gallop. The sloping shoulder required by the breed standard is not noticeable during locomotion: the position of the shoulder blade at touchdown and liftoff corresponds to the values observed in other breeds of a similar size.

Proportions

Body	♀		♂	
Number of dogs	3		1	
	MV	SD	MV	SD
Withers height [cm]	44	1.3	49	-
Body length / Withers height	1.1	0.0	1.0	-
Withers height / croup height	1.0	0.0	1.0	-
Legginess index	1.6	0.4	1.3	-

Forelimb	♀		♂	
Number of dogs	3		1	
	Proportion of total length			
	%	SD	%	SD
Shoulder blade	25	1.1	25	-
Upper arm	28	1.0	28	-
Lower arm	30	0.7	30	-
Pastern	17	0.5	17	-

Hindlimb	♀		♂	
Number of dogs	3		1	
	Proportion of total length			
	%	SD	%	SD
Thigh	38	0.5	38	-
Lower leg	36	1.8	38	-
Hind pastern	26	2.3	24	-

Gait parameters

Forelimb

	Stride length				Stride duration				Stance length				Stance duration				Swing length				Swing duration				Dutyfactor				Speed			
	m				s				m				s				m				s								m/s			
MV	0.62	0.95	-	-	0.59	0.42	-	-	0.37	0.41	-	-	0.37	0.19	-	-	0.24	0.54	-	-	0.22	0.23	-	-	63	46	-	-	1.03	2.24	-	-
SD	0.1	0.1	-	-	0.0	0.0	-	-	0.0	0.0	-	-	0.0	0.0	-	-	0.0	0.1	-	-	0.0	0.0	-	-	2.9	5.5	-	-	0.1	0.3	-	-
Number of strides					61	89	-	-	All numbers equal																							

Hindlimb

	Stride length				Stride duration				Stance length				Stance duration				Swing length				Swing duration				Dutyfactor				Speed			
	m				s				m				s				m				s								m/s			
MV	0.64	0.98	-	-	0.59	0.42	-	-	0.36	0.33	-	-	0.35	0.15	-	-	0.28	0.65	-	-	0.25	0.27	-	-	59	37	-	-	1.07	2.31	-	-
SD	0.1	0.1	-	-	0.0	0.0	-	-	0.0	0.0	-	-	0.0	0.0	-	-	0.0	0.1	-	-	0.0	0.0	-	-	3.0	4.8	-	-	0.1	0.3	-	-
Number of strides					64	91	-	-	All numbers equal																							

Kinematics

Forelimb

	Shoulder blade				Shoulder joint				Upper arm				Elbow joint				Lower arm				Carpal joint				Pastern			
Touch down	*59*	*62*	-	-	*142*	*152*	-	-	78	85	-	-	138	141	-	-	61	56	-	-	186	189	-	-	125	133	-	-
Lift off	*97*	*106*	-	-	*133*	*136*	-	-	36	30	-	-	150	148	-	-	114	117	-	-	148	159	-	-	35	41	-	-
Mid-stance	-	-	-	-	*147*	*157*	-	-	-	-	-	-	138	141	-	-	-	-	-	-	186	189	-	-	-	-	-	-
Effective motion	*38*	*44*	-	-	*9*	*16*	-	-	42	54	-	-	11	6	-	-	53	61	-	-	39	31	-	-	91	93	-	-
Stance amplitude	*36*	*42*	-	-	*11*	*24*	-	-	43	56	-	-	33	36	-	-	66	65	-	-	39	38	-	-	88	85	-	-
Swing amplitude	*34*	*43*	-	-	*17*	*30*	-	-	47	60	-	-	53	68	-	-	74	91	-	-	78	100	-	-	106	137	-	-
Number of strides	61	89	0	0	Values in italics are marker-based values corrected to reflect the X-Ray video and position data.																							

Hindlimb

	Pelvis				Hip joint				Thigh				Stifle joint				Lower leg				Ankle joint				Hind pastern			
Touch down	62	55	-	-	111	112	-	-	50	57	-	-	135	139	-	-	85	83	-	-	141	142	-	-	55	59	-	-
Lift off	51	45	-	-	*162*	*157*	-	-	*111*	*112*	-	-	131	133	-	-	35	36	-	-	157	156	-	-	122	120	-	-
Mid-stance	-	-	-	-				-	-	-	-	-	135	139	-	-	-	-	-	-	141	142	-	-	-	-	-	-
Effective motion	10	10	-	-	*51*	*46*	-	-	*61*	*56*	-	-	4	6	-	-	50	47	-	-	17	14	-	-	67	61	-	-
Stance amplitude	19	11	-	-	*51*	*48*	-	-	*61*	*55*	-	-	14	20	-	-	50	48	-	-	37	46	-	-	71	64	-	-
Swing amplitude	12	17	-	-	*53*	*56*	-	-	*63*	*66*	-	-	40	60	-	-	61	75	-	-	43	64	-	-	72	76	-	-
Number of strides	64	91	0	0	Values in italics are marker-based values corrected to reflect the X-Ray video and position data.																							

PROPORTIONS

No conclusions can be drawn about sex-specific differences in this breed as only one male took part in the study. The shoulder blade is slightly shorter than in other breeds. Smooth-faced bitches are, on average, larger than rough-faced bitches, have a less deep chest and very slightly longer forearms and lower legs. The low level of variability in the proportional length of the front pastern is striking.

GAIT PARAMETERS

The stance phase and the swing phase of the forelimbs are long in all gaits, while in the hindlimbs the stance phases tend to be short and the swing phases and stride lengths long. Forward excursion of the forelimbs is lower in the walk than in the trot, a difference which can be put down to the movement of the upper and forearm. No difference is observed between the walk and the trot when it comes to the backwards excursion of the forelimbs, however. In the hindlimbs, forward excursion is lower in the trot than in the walk due to differences in the position of the pelvis and femur. When it comes to backwards excursion, however, no difference is seen between the gaits. In a cross-breed comparison, the forwards excursion of the forelimbs is situated in the middle of the range in the walk and towards the top of the range in the trot. Backwards excursion is fairly high in all gaits. The forwards excursion of the hindlimbs is middle-of-the-range in the walk and the trot, while backwards excursion is high in the walk and moderate in the trot.

The breed standard requires steeper limb angulation in this variety than in the Rough-faced Pyrenean Shepherd, but according to our data, this is not always the case. However, the opening of the shoulder joint at touchdown, the hip joint at liftoff and the knee joint at both touchdown and liftoff is slightly greater in the smooth-faced variety.

PYRENEAN SHEPHERD (Smooth-Faced)

DESCRIPTION OF BREED

Bred from the beginning for performance, the job of these dogs was to herd sheep, cows, horses and goats, though they were not put to work in the high mountains. The breed standard was recognized in 1926, as was that of the Long-Haired Pyrenean Shepherd.

The Smooth-Faced differs visibly from the other coat types in the shortness of the hair on its face. The rest of the coat can be longer, but is never tangled. Height at the withers is the same for the Smooth-Faced and the Long-Haired Shepherd (dogs between 40 and 48 centimeters, bitches between 38 and 46 centimeters), but smooth-faced dogs should be slightly shorter in the trunk and the angulation should be steeper than in the Long-Haired Pyrenean Shepherd.

Proportions

Body	♀		♂	
Number of dogs	7		4	
	MV	SD	MV	SD
Withers height [cm]	41	1.4	42	0.6
Body length / Withers height	1.1	0.1	1.1	0.0
Withers height / croup height	1.0	0.0	1.0	0.1
Legginess index	1.3	0.0	1.5	0.1

Forelimb	♀		♂	
Number of dogs	7		4	
	Proportion of total length			
	%	SD	%	SD
Shoulder blade	26	1.0	26	0.4
Upper arm	27	0.8	27	0.8
Lower arm	32	1.3	32	1.7
Pastern	15	1.9	15	0.9

Hindlimb	♀		♂	
Number of dogs	7		4	
	Proportion of total length			
	%	SD	%	SD
Thigh	37	2.2	36	3.1
Lower leg	38	1.3	39	2.0
Hind pastern	25	2.6	25	1.2

Gait parameters

Forelimb

	Stride length				Stride duration				Stance length				Stance duration			
	m				s				m				s			
MV	0.52	0.83	1.22	1.25	0.56	0.39	0.33	0.33	0.29	0.32	0.34	0.37	0.32	0.16	0.10	0.10
SD	0.1	0.1	0.1	0.1	0.0	0.0	0.0	0.0	0.1	0.0	0.0	0.1	0.0	0.0	0.0	0.0
Number of strides					71	89	45	21	All numbers equal							

	Swing length				Swing duration				Dutyfactor				Speed			
	m				s								m/s			
MV	0.23	0.51	0.89	0.87	0.24	0.23	0.23	0.22	57	40	30	32	0.93	2.15	3.70	3.75
SD	0.1	0.1	0.1	0.1	0.0	0.0	0.0	0.0	3.2	4.0	2.5	4.3	0.2	0.3	0.3	0.3

Hindlimb

	Stride length				Stride duration				Stance length				Stance duration			
	m				s				m				s			
MV	0.54	0.91	1.34	1.38	0.56	0.39	0.33	0.33	0.26	0.27	0.32	0.32	0.29	0.12	0.09	0.08
SD	0.1	0.1	0.1	0.1	0.1	0.0	0.0	0.0	0.1	0.0	0.0	0.1	0.1	0.0	0.0	0.0
Number of strides					82	87	6	57	All numbers equal							

	Swing length				Swing duration				Dutyfactor				Speed			
	m				s								m/s			
MV	0.28	0.51	1.02	1.06	0.27	0.27	0.24	0.25	52	32	27	26	0.94	2.33	4.02	4.13
SD	0.1	0.3	0.1	0.1	0.1	0.0	0.0	0.0	10.9	3.9	2.5	4.9	0.2	0.3	0.4	0.2

Kinematics

Forelimb

Forelimb	Shoulder blade				Shoulder joint				Upper arm				Elbow joint			
Touch down	*59*	*61*	*60*	*67*	141	144	148	161	77	77	77	80	145	140	140	137
Lift off	*93*	*102*	116	108	*129*	*135*	141	137	36	33	25	29	149	151	153	155
Mid-stance	-	-	-	-	139	143	144	155	-	-	-	-	145	139	137	134
Effective motion	*34*	*41*	*56*	*41*	*32*	*28*	*7*	25	41	44	53	51	4	11	13	18
Stance amplitude	*32*	*37*	46	32	*34*	*32*	*21*	33	42	45	55	54	24	30	26	44
Swing amplitude	*32*	*42*	61	38	*40*	*42*	*34*	52	48	55	77	70	50	70	58	74
Number of strides	71	89	45	21	Values in italics are marker-based values corrected to reflect the X-Ray video and position data.											

Forelimb	Lower arm				Carpal joint				Pastern			
Touch down	68	62	63	57	195	194	207	214	127	132	145	157
Lift off	112	117	129	127	160	172	173	189	48	55	44	62
Mid-stance	-	-	-	-	195	194	207	214	-	-	-	-
Effective motion	45	55	66	70	35	22	34	25	79	77	101	95
Stance amplitude	53	57	66	70	29	29	44	41	73	71	92	88
Swing amplitude	73	93	100	110	78	99	87	126	104	130	139	165

Hindlimb

Hindlimb	Pelvis				Hip joint				Thigh				Stifle joint			
Touch down	55	45	53	51	114	103	97	98	56	58	45	49	144	142	132	140
Lift off	46	37	32	29	*144*	*142*	131	140	*98*	*104*	99	106	128	131	130	135
Mid-stance	-	-	-	-	114	104	97	98	-	-	-	-	144	142	133	140
Effective motion	9	8	20	23	*31*	*38*	33	42	*42*	*46*	54	57	16	11	2	5
Stance amplitude	13	10	20	25	*32*	*37*	36	46	*42*	*44*	51	55	17	19	25	21
Swing amplitude	13	14	32	33	*39*	*50*	48	51	*52*	*58*	63	74	50	59	63	74
Number of strides	82	87	6	57	Values in italics are marker-based values corrected to reflect the X-Ray video and position data.											

Hindlimb	Lower leg				Ankle joint				Hind pastern			
Touch down	88	84	87	91	143	134	134	131	54	49	47	38
Lift off	40	38	31	30	153	150	153	136	113	113	122	106
Mid-stance	-	-	-	-	143	134	135	132	-	-	-	-
Effective motion	48	47	56	62	15	18	19	15	59	63	75	68
Stance amplitude	47	46	56	62	27	39	52	50	60	62	72	67
Swing amplitude	68	77	103	115	45	61	72	63	78	87	102	108

PROPORTIONS

The only sex-specific difference we observed was in the legginess index: bitches have a deeper chest than dogs. As in the other Pyrenean Shepherd variety, the shoulder blade is slightly shorter than the cross-breed average, but the forearm slightly longer. There is a striking lack of variability in shoulder blade length in males.

GAIT PARAMETERS

In comparative terms, stride length is very long in the trot and the gallop but short in the walk. This ties in with the requirements of the standard.
The forward excursion of the forelimbs does not differ between the walk and the gallop (leading limb). It is smaller in these two gaits than in the trot, while the greatest degree of forward excursion is exhibited by the trailing limb in the gallop. Backward excursion is lowest in the walk and greatest in the gallop (where it is greater in the trailing limb than the leading limb). All the limb segments are involved in increasing backward excursion. Forward excursion of the hindlimbs is lowest in the trot and slightly greater in the walk and the gallop (trailing limb). The greatest forward excursion is displayed by the leading hindlimb in the gallop. The proximal segments pelvis and femur take most of the responsibility here, though the movement of the lower leg plays a role in the gallop. There is no difference between the walk and the trot when it comes to the backwards excursion of the hindlimbs, and backwards excursion is greatest in the gallop (leading hindlimb), once again thanks to the motion of the proximal segments. The forward excursion of the forelimbs lies in the middle of the range in all gaits. The only exception is provided by the leading limb in the gallop, which displays very low forward extension. The lowest degree of backward excursion is seen in the walk; in all the other gaits the values are middle-of-the-range. The forward excursion of the hindlimbs is low in the walk, trot and gallop (leading limb) and middle-of-the-range in the trailing limb in the gallop. Backwards excursion is moderate in the walk and the gallop (leading limb) and rather low in the trot and the trailing limb in the gallop.

PYRENEAN SHEPHERD (Long-Haired)

DESCRIPTION OF BREED

These dogs were bred for performance and put to work herding sheep, cows, horses and goats. As they were used in the high mountains, they also needed to be agile. On the basis of their ability and agility, they were used in the First World War as messenger dogs, which increased awareness of the breed dramatically. Groups of fans soon formed, and the first standard was formulated as early as 1921, though not recognized until five years later.
There are two varieties of Long-Haired Pyrenean Shepherd, one with a thick undercoat which is matted from the mid-point of the back and protects against wind and rain, and one with medium-long hair which does not mat. According to the standard, height at the withers in both varieties should be between 40 and 48 centimeters in dogs and between 38 and 46 centimeters in bitches. Body length should be greater than withers height. The shoulder blade and upper arm should be fairly long, the upper thigh rather short and the lower thigh very long. The limbs should be well-angulated. In the walk, strides should be short.

Proportions

Body	♀		♂	
Number of dogs	9		3	
	MV	SD	MV	SD
Withers height [cm]	63	2.0	68	0.5
Body length / Withers height	1.1	0.0	1.1	0.0
Withers height / croup height	1.0	0.0	1.0	0.0
Legginess index	-	-	-	-

Forelimb	♀		♂	
Number of dogs	9		3	
	Proportion of total length			
	%	SD	%	SD
Shoulder blade	30	2.8	30	1.5
Upper arm	27	2.7	29	1.1
Lower arm	30	1.7	28	2.1
Pastern	13	2.1	13	2.8

Hindlimb	♀		♂	
Number of dogs	9		3	
	Proportion of total length			
	%	SD	%	SD
Thigh	41	2.9	40	4.1
Lower leg	37	2.5	37	2.6
Hind pastern	22	1.3	22	2.1

Gait parameters

Forelimb

	Stride length				Stride duration				Stance length				Stance duration				Swing length				Swing duration				Dutyfactor				Speed			
	m				s				m				s				m				s								m/s			
MV	0.96	1.26	1.69	1.70	0.76	0.49	0.40	0.38	0.53	0.50	0.54	0.56	0.43	0.20	0.13	0.14	0.43	0.76	1.15	1.14	0.33	0.29	0.26	0.25	56	41	34	36	1.26	2.54	4.23	4.37
SD	0.0	0.1	0.2	0.1	0.1	0.0	0.0	0.0	0.0	0.0	0.1	0.0	0.1	0.0	0.0	0.0	0.1	0.1	0.2	0.1	0.0	0.0	0.0	0.0	8.4	4.2	3.4	2.6	0.1	0.2	0.5	0.5
Number of strides					59	45	19	18	All numbers equal																							

Hindlimb

	Stride length				Stride duration				Stance length				Stance duration				Swing length				Swing duration				Dutyfactor				Speed			
	m				s				m				s				m				s								m/s			
MV	0.96	1.22	1.71	1.79	0.76	0.51	0.39	0.40	0.50	0.43	0.49	0.51	0.41	0.19	0.12	0.12	0.46	0.79	1.22	1.27	0.35	0.32	0.27	0.27	54	37	31	31	1.27	2.38	4.42	4.50
SD	0.1	0.1	0.2	0.2	0.0	0.0	0.0	0.0	0.1	0.0	0.1	0.1	0.0	0.0	0.0	0.0	0.1	0.1	0.2	0.2	0.0	0.0	0.0	0.0	4.7	4.2	2.4	5.4	0.1	0.2	0.8	0.5
Number of strides					64	89	21	21	All numbers equal																							

Kinematics

Forelimb	Shoulder blade				Shoulder joint				Upper arm				Elbow joint				Lower arm				Carpal joint				Pastern			
Touch down	61	61	57	72	145	139	143	155	84	79	85	82	148	145	165	149	65	67	79	66	205	205	195	216	140	139	112	150
Lift off	97	101	112	109	124	136	160	139	27	35	48	29	149	150	139	150	122	115	92	121	176	176	171	172	54	60	80	52
Mid-stance	-	-	-	-	136	125	132	132	-	-	-	-	141	130	150	138	-	-	-	-	205	205	194	216	-	-	-	-
Effective motion	36	40	59	37	21	3	-17	16	57	44	73	53	1	5	25	1	57	49	46	54	28	29	56	44	85	78	98	98
Stance amplitude	38	44	61	41	25	28	56	38	56	48	83	65	31	35	57	37	62	51	46	56	37	48	73	61	83	79	101	96
Swing amplitude	36	42	70	56	38	38	77	61	65	59	86	73	64	73	86	74	86	84	85	96	102	113	102	130	127	136	134	153
Number of strides	54	32	19	18	Values in italics are marker-based values corrected to reflect the X-Ray video and position data.																							

Hindlimb	Pelvis				Hip joint				Thigh				Stifle joint				Lower leg				Ankle joint				Hind pastern			
Touch down	53	40	42	47	111	106	102	98	58	64	59	51	155	150	138	146	97	86	79	95	148	145	138	126	50	59	57	30
Lift off	47	35	38	38	*156*	*142*	*160*	*149*	*109*	*107*	*122*	111	138	136	137	147	39	40	25	36	156	157	152	149	117	117	127	113
Mid-stance	-	-	-	-	111	106	101	101	-	-	-	-	155	150	139	143	-	-	-	-	146	146	141	119	-	-	-	-
Effective motion	5	5	4	9	*45*	*37*	*59*	*51*	*51*	*43*	*63*	60	17	13	1	1	58	46	55	59	13	15	23	26	67	58	69	83
Stance amplitude	27	9	18	22	*47*	*38*	*65*	*57*	*51*	*43*	*62*	59	20	18	25	27	57	46	55	60	35	51	67	66	74	67	77	90
Swing amplitude	14	20	27	34	*56*	*49*	*81*	*68*	*54*	*52*	*69*	74	48	63	48	64	74	78	81	93	53	67	72	68	77	83	110	104
Number of strides	58	77	20	21	Values in italics are marker-based values corrected to reflect the X-Ray video and position data.																							

BERNESE MOUNTAIN DOG

DESCRIPTION OF BREED

The Bernese Mountain Dog has been being bred since around 1900 and is the direct descendent of a variety of farm dog found in the canton of Bern whose origin and age is unclear. The dogs in question were three-colored animals which occurred in a small, isolated and fairly inaccessible region. Referred to as "Dürrbächler" after a local hamlet, their main purpose was to pull sledges and act as guard dogs. In 1904, four "Dürrbächler" were presented at a dog show in Bern and subsequently entered into the Swiss Stud Book. In the course of this show, the decision was taken to create a breed which was based on these old farm dogs. The Dürrbach Club was set up in 1907, and the process of establishing the characteristics of the breed began. It wasn't until 1913 that the Club's members gave in to five years of pressure from the Swiss Cynological Society and renamed their breed the Bernese Mountain Dog. It took the breeders until well into the 1920s to obtain today's Bernese Mountain Dogs using the initially very different-looking Dürrbächler.

The breed standard requires a height to length ratio of 9:10, making the dogs compact rather than long. The chest should be deep, reaching at least to the elbows, and the shoulders and upper arm, upper thighs and shank should be long. Movement should be expansive with good reach in front.

PROPORTIONS

No sex-specific differences in segment proportions were detected apart from the fact that the upper arm is very slightly longer in dogs and the forearm very slightly longer in bitches. As a result, the proportional lengths of the shoulder blade and upper arm are the same in dogs, while in bitches the shoulder blade is slightly longer. In both sexes, the femur is slightly longer than the lower leg. Relatively speaking, the shoulder blade in Bernese Mountain Dogs is slightly longer than the crossbreed average, the femur is longer than average and the hind pastern is shorter. The variability of upper arm length is relatively high in bitches, while femur length is particularly variable in the males of the breed.

GAIT PARAMETERS

Stride length in this breed is high in the walk and the trot, but moderate in the gallop. The long stance phases which occur in the walk and the trot are caused by the high degree of forward excursion displayed by the forelimbs in these gaits. Forwards excursion is lower in the gallop, mainly as the result of increased shoulder blade motion. Backwards excursion, on the other hand, is lower in the trot than in the walk or the gallop, and in the gallop it is highest of all in the trailing forelimb. All the limb segments contribute to these differences. When it comes to the hindlimbs, excursion in both directions is lowest in the trot. The greatest degree of forward excursion is displayed by the leading hindlimb in the gallop. Forward excursion in the trailing limb is lower in the gallop than in the walk. The low level of excursion observed in the trot can be put down to the position of the femur. Backwards excursion is greater in the gallop than in the walk, a state of affairs to which all the limb segments contribute. Comparatively speaking, the excursion of the forelimbs in both directions is high in the walk and middle-of-the-range in the trot. In the gallop, forwards excursion is low while backwards excursion is high. When it comes to the hindlimbs, excursion in both directions is moderate in the walk and fairly low in all the other gaits. In the walk, the forelimbs provide a good example of the "sound and balanced movement...covering a lot of ground" required by the breed standard.

Proportions

Body	♀		♂	
Number of dogs	11		1	
	MV	SD	MV	SD
Withers height [cm]	21	1.0	24	-
Body length / Withers height	1.3	0.1	1.1	-
Withers height / croup height	1.0	0.0	1.0	-
Legginess index	0.7	0.1	0.8	-

Forelimb	♀		♂	
Number of dogs	11		1	
	Proportion of total length			
	%	SD	%	SD
Shoulder blade	30	1.7	28	-
Upper arm	27	1.6	26	-
Lower arm	25	1.8	30	-
Pastern	18	1.9	16	-

Hindlimb	♀		♂	
Number of dogs	11		1	
	Proportion of total length			
	%	SD	%	SD
Thigh	39	2.5	38	-
Lower leg	34	2.9	33	-
Hind pastern	27	2.6	29	-

Gait parameters

Forelimb	Stride length				Stride duration				Stance length				Stance duration				Swing length				Swing duration				Dutyfactor				Speed			
	m				s				m				s				m				s								m/s			
MV	0.21	0.31	-	-	0.59	0.36	-	-	0.14	0.16	-	-	0.40	0.20	-	-	0.08	0.15	-	-	0.19	0.16	-	-	68	56	-	-	0.38	0.85	-	-
SD	0.0	0.1	-	-	0.1	0.0	-	-	0.0	0.0	-	-	0.1	0.0	-	-	0.0	0.1	-	-	0.0	0.0	-	-	4.1	5.7	-	-	0.1	0.2	-	-
Number of strides					126	205	-	-	All numbers equal																							

Hindlimb	Stride length				Stride duration				Stance length				Stance duration				Swing length				Swing duration				Dutyfactor				Speed			
	m				s				m				s				m				s								m/s			
MV	0.21	0.29	-	-	0.59	0.37	-	-	0.13	0.14	-	-	0.40	0.19	-	-	0.08	0.15	-	-	0.20	0.18	-	-	66	51	-	-	0.37	0.79	-	-
SD	0.0	0.1	-	-	0.1	0.0	-	-	0.0	0.0	-	-	0.1	0.0	-	-	0.0	0.1	-	-	0.0	0.0	-	-	7.0	6.8	-	-	0.1	0.2	-	-
Number of strides					134	222	-	-	All numbers equal																							

Kinematics

Forelimb	Shoulder blade				Shoulder joint				Upper arm				Elbow joint				Lower arm				Carpal joint				Pastern			
Touch down	*69*	*65*	-	-	*148*	*143*	-	-	79	78	-	-	136	134	-	-	57	56	-	-	185	188	-	-	128	132	-	-
Lift off	102	106	-	-	140	138	-	-	38	32	-	-	125	129	-	-	87	97	-	-	131	125	-	-	45	30	-	-
Mid-stance	-	-	-	-	154	149	-	-	-	-	-	-	137	130	-	-	-	-	-	-	185	188	-	-	-	-	-	-
Effective motion	*33*	*41*	-	-	*8*	*6*	-	-	4	4	-	-	11	5	-	-	30	41	-	-	54	63	-	-	-2	-1	-	-
Stance amplitude	*41*	*49*	-	-	*15*	*17*	-	-	45	48	-	-	27	31	-	-	55	60	-	-	68	73	-	-	81	96	-	-
Swing amplitude	*37*	*47*	-	-	*18*	*22*	-	-	48	56	-	-	42	49	-	-	52	71	-	-	72	95	-	-	91	122	-	-
Number of strides	126	205	0	0	Values in italics are marker-based values corrected to reflect the X-Ray video and position data.																							

Hindlimb	Pelvis				Hip joint				Thigh				Stifle joint				Lower leg				Ankle joint				Hind pastern			
Touch down	57	53	-	-	109	108	-	-	51	55	-	-	140	137	-	-	88	82	-	-	142	134	-	-	53	52	-	-
Lift off	51	45	-	-	*142*	*141*	-	-	*91*	*97*	-	-	107	110	-	-	36	33	-	-	146	147	-	-	110	114	-	-
Mid-stance	-	-	-	-	110	110	-	-	-	-	-	-	135	135	-	-	-	-	-	-	135	127	-	-	-	-	-	-
Effective motion	7	8	-	-	*47*	*50*	-	-	*40*	*42*	-	-	32	27	-	-	53	50	-	-	-4	-12	-	-	-1	-3	-	-
Stance amplitude	11	11	-	-	*52*	*54*	-	-	*41*	*40*	-	-	30	27	-	-	51	48	-	-	34	48	-	-	62	67	-	-
Swing amplitude	8	12	-	-	*53*	*59*	-	-	*41*	*47*	-	-	45	55	-	-	62	71	-	-	34	55	-	-	64	77	-	-
Number of strides	134	221	0	0	Values in italics are marker-based values corrected to reflect the X-Ray video and position data.																							

PROPORTIONS

In all the dogs we measured, trunk length exceeded height at the withers. However, we only managed to obtain data for one male dog. The legginess index is relatively low, indicating that the chest is deep in this breed. The proportional lengths of the segments of the forelimbs are less variable than those of the hindlimbs. Chihuahua bitches have a relatively long shoulder blade and a short forearm compared to the cross-breed average. Both dogs and bitches(?) have a shorter-than-average lower leg.

GAIT PARAMETERS

Comparatively speaking, stride length is relatively short in Chihuahuas and is made up of a short stance phase and a short swing phase. This is the case both in the walk and the trot. The extent of the forward excursion of the forelimbs before touchdown does not differ between the walk and the trot, but backward excursion before liftoff is greater in the trot than the walk due to the larger movements which occur in the shoulder blade, upper arm and forearm. In the hindlimbs, forward excursion before touchdown is greater in the walk than the trot, while backward excursion before liftoff is greater in the trot than the walk. In both cases, all the hindlimb segments contribute to these differences.

A cross-breed comparison shows that the excursion of the forelimbs in both directions is very low in both the walk and the trot. In the hindlimbs, forward excursion is high in the walk and moderate in the trot. Backward extension prior to liftoff is very low in both the walk and the trot.

The angulation of the shoulder and ankle joint in Chihuahuas does not deviate to any significant extent from that of other breeds, but the knee joint opens slightly more than in other breeds at both touchdown and liftoff.

CHIHUAHUA

DESCRIPTION OF BREED

The Chihuahua's true region of origin is unclear. Stories and drawings point to the breed having originated in Mexico, China or even on the European island of Malta, but none of these are backed up by any concrete evidence. It is said, that the smallest dog of the world was used in the sacred rituals of pre-Columbian Indian nations to accompagny the souls to the afterworld. The direct ancestors of the Chihuahua came from Mexico, but the breed itself was established in the USA, which means that today's Chihuahuas can actually be said to be a North American breed (Räber 1995). The first Chihuahuas were registered in the Kennel Club's stud book in 1904. At the end of the 19th and beginning of the 20th centuries, a few Chihuahuas were found in England, but breeding came to a complete halt during the Second World War. The first Chihuahuas bred in England after the end of the war were not purebred, but crossed with Papillons, Pomeranians and Tibet Spaniels. Chihuahuas were not introduced to Germany until the middle of the 20th century. They are allowed to be short- and long-haired.

The standard requires a square body, though length can exceed height slightly, particularly in bitches. The chest should be deep, the forelimbs straight and a good length, the angle between the shoulder blade and the upper arm moderate, the stifle and hock joints well-angulated and movement characterized by long strides.

Proportions

Body	♀		♂	
Number of dogs	5		5	
	MV	SD	MV	SD
Withers height [cm]	79	2.8	86	6.6
Body length / Withers height	1.1	0.1	1.0	0.0
Withers height / croup height	1.0	0.0	1.0	0.0
Legginess index	1.3	0.0	1.3	0.2

Forelimb	♀		♂	
Number of dogs	5		5	
	Proportion of total length			
	%	SD	%	SD
Shoulder blade	26	1.2	26	1.4
Upper arm	26	1.2	27	0.5
Lower arm	31	0.8	31	1.4
Pastern	16	1.2	16	2.0

Hindlimb	♀		♂	
Number of dogs	5		5	
	Proportion of total length			
	%	SD	%	SD
Thigh	36	2.8	35	3.5
Lower leg	38	3.4	37	1.6
Hind pastern	26	2.2	28	2.1

Gait parameters

Forelimb

	Stride length				Stride duration				Stance length				Stance duration				Swing length				Swing duration				Dutyfactor				Speed			
	m				s				m				s				m				s								m/s			
MV	1.03	1.59	2.41	2.30	0.93	0.58	0.48	0.48	0.64	0.68	0.72	0.74	0.60	0.27	0.16	0.16	0.39	0.91	1.68	1.56	0.33	0.31	0.33	0.31	64	45	32	34	1.12	2.87	4.95	4.82
SD	0.1	0.2	0.3	0.2	0.1	0.1	0.0	0.0	0.1	0.1	0.1	0.1	0.1	0.1	0.0	0.0	0.1	0.2	0.3	0.2	0.0	0.0	0.0	0.0	3.6	7.2	3.1	3.1	0.2	0.8	0.6	0.5
Number of strides					95	143	106	99	All numbers equal																							

Hindlimb

	Stride length				Stride duration				Stance length				Stance duration				Swing length				Swing duration				Dutyfactor				Speed			
	m				s				m				s				m				s								m/s			
MV	1.01	1.56	2.39	2.36	0.93	0.58	0.48	0.47	0.63	0.60	0.68	0.72	0.60	0.25	0.15	0.15	0.38	0.96	1.71	1.64	0.33	0.33	0.33	0.32	64	42	31	33	1.09	2.83	4.95	4.98
SD	0.1	0.3	0.2	0.2	0.1	0.1	0.0	0.0	0.1	0.1	0.1	0.1	0.1	0.1	0.0	0.0	0.1	0.3	0.2	0.2	0.0	0.0	0.0	0.0	3.5	9.0	3.8	3.2	0.2	0.8	0.4	0.5
Number of strides					94	137	103	87	All numbers equal																							

Kinematics

Forelimb	Shoulder blade				Shoulder joint				Upper arm				Elbow joint				Lower arm				Carpal joint				Pastern			
Touch down	59	58	55	65	144	141	145	147	84	84	90	82	143	145	151	140	60	62	62	59	187	193	195	190	127	131	135	133
Lift off	*97*	*102*	106	91	*129*	*132*	135	128	32	31	29	37	136	144	155	159	105	113	126	122	145	152	163	165	41	39	37	44
Mid-stance	-	-	-	-	*128*	*127*	131	134	-	-	-	-	146	140	141	141	-	-	-	-	187	193	195	190	-	-	-	-
Effective motion	*37*	*44*	50	26	*15*	*9*	10	18	52	53	60	44	7	2	3	19	45	51	63	63	42	41	32	25	86	92	97	89
Stance amplitude	*40*	*45*	50	32	*14*	*12*	23	23	52	53	64	50	21	24	28	26	58	55	62	62	43	52	54	53	83	89	96	89
Swing amplitude	*37*	*44*	53	37	*18*	*26*	36	46	53	59	74	54	41	60	61	70	61	84	95	105	78	101	84	98	106	132	132	138
Number of strides	95	143	104	107	Values in italics are marker-based values corrected to reflect the X-Ray video and position data.																							

Hindlimb	Pelvis				Hip joint				Thigh				Stifle joint				Lower leg				Ankle joint				Hind pastern			
Touch down	50	44	53	55	104	104	110	101	54	60	56	46	149	152	152	152	95	92	96	106	158	158	162	157	62	65	66	49
Lift off	44	39	41	27	*147*	*147*	*157*	*142*	*104*	*109*	*117*	*115*	133	144	144	150	39	45	37	45	166	172	174	175	127	127	136	130
Mid-stance	-	-	-	-	104	104	109	101	-	-	-	-	149	152	152	151	-	-	-	-	158	157	162	156	-	-	-	-
Effective motion	6	5	13	28	*44*	*44*	*48*	*41*	*50*	*49*	*60*	*69*	16	9	8	2	56	48	58	61	9	15	15	19	65	62	71	80
Stance amplitude	14	8	17	28	*46*	*43*	*47*	*42*	*52*	*48*	*59*	*68*	15	21	27	26	55	48	59	63	31	48	56	70	69	65	74	89
Swing amplitude	7	13	22	33	*46*	*47*	*58*	*52*	*51*	*54*	*69*	*80*	35	56	52	65	61	71	90	91	36	57	59	57	65	75	89	100
Number of strides	95	136	104	86	Values in italics are marker-based values corrected to reflect the X-Ray video and position data.																							

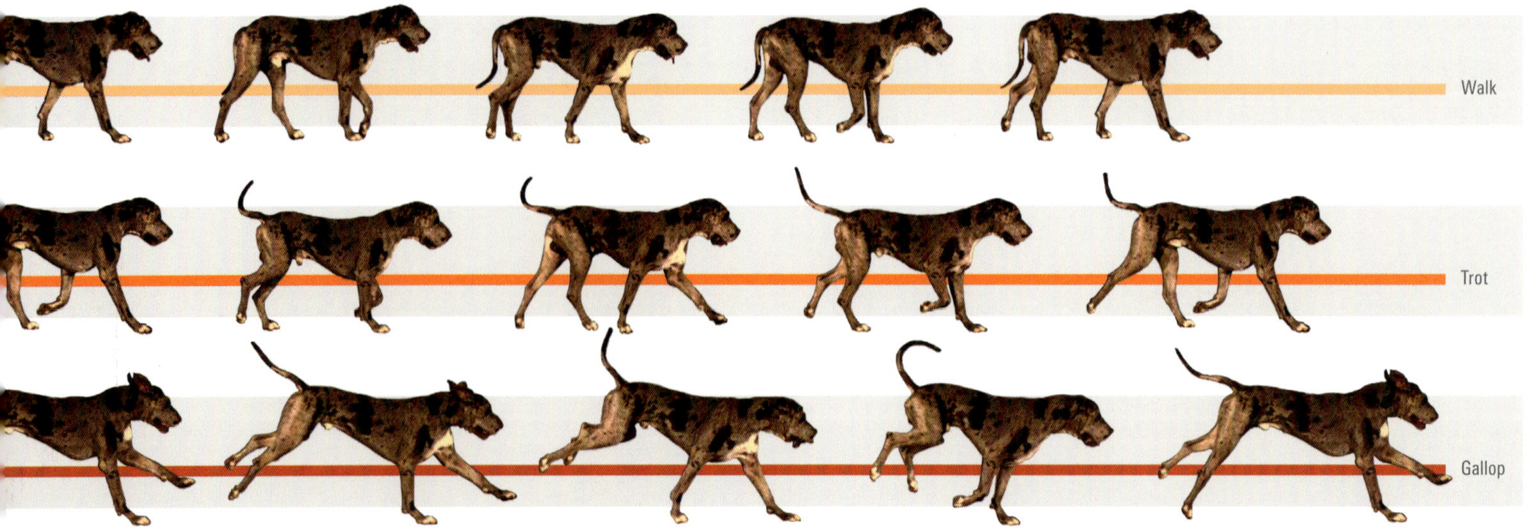

PROPORTIONS

In two of the dogs we investigated, trunk length slightly exceeded that required by the breed standard, leading to small deviations in the proportional lengths of the limb segments. Generally, the upper arm and the shoulder blade are the same length in this breed. The lower leg is, on average, slightly longer than the femur, though this is not reflected in any noticeable differences in the kinematic values. Overall, the relative lengths of the forelimb segments are less variable than those of the hindlimb segments. The hind pastern is slightly shorter in bitches than in dogs, and while the length of the upper arm is variable in bitches, the length of the forearm is very stable.

GAIT PARAMETERS

Compared with other breeds, the relative stride length of the forelimbs in Great Danes is middle-of-the-range in the walk and the trot and very short in the gallop. The same applies to the stride length of the hindlimbs in the gallop, and even in the walk and the trot it is short rather than medium length.

Essentially, forward excursion of the forelimbs before touchdown does not differ between the gaits. The only exception is the gallop, in which the forward excursion of the trailing forelimb is slightly greater due to the motion of the proximal limb segments (shoulder blade and upper arm). Backward excursion increases from the walk to the trot to the gallop, and is greater in the trailing limb than the leading limb. All the limb segments play a role in this increase. Forward excursion in the hindlimbs is also essentially constant between the walk, trot and gallop (trailing limb), but is much greater in the leading hindlimb in the gallop as the result of the position in which the femur and lower leg are held. Backwards excursion prior to liftoff does not differ between the walk and the trot but is greater in the gallop, with higher values displayed by the trailing hindlimb than the leading hindlimb. Differences in the angulation of the hip and knee joint between the walk and the trot cancel each other out so that excursion remains the same. Essentially, the forward excursion of the forelimbs is top-of-the-range in all gaits. In the gallop, however, the excursion of the leading limb is middle-of-the-range. Backward excursion is situated towards the bottom of the range in the walk and the trot and the middle of the range in the gallop. In the hindlimbs, excursion in both directions is middle-of-the-range. The angle of 100 - 110° [between the shoulder blade and upper arm] required by the standard is not detectable in our kinematic data.

GREAT DANE

DESCRIPTION OF BREED

Originally, the term mastiff referred to large, strong dogs of no particular breed which were used for fighting, herding and as guard dogs. Later, mastiffs were put to work in the hunt for large game, crossed for this purpose with large sighthounds. The various varieties were given different names (e.g. Dänische, Englische and Große Dogge, Saupacker, Hatzrüde, Ulmer Dogge), and in 1878 it was decided in Berlin to term all these varieties Great Dane (Great Dane). The first standard for the Great Dane was laid down in 1880 and has not changed significantly since. The Great Dane Club, founded in 1888, was the first cynological breed club in Germany.
According to the current standard, height at the withers should be at least 80 centimeters in dogs and at least 72 centimeters in bitches, while the body should appear to be almost square. Length should not exceed height at the withers by more than five percent in dogs or more than ten percent in bitches. Shoulder blades should be long and sloping, meeting the upper arm at an angle of 100 – 110 degrees. The upper arm should be slightly longer than the shoulder blade. The thigh and stifle should also be long, and approximately the same length.

Proportions

Body	♀		♂	
Number of dogs	11		3	
	MV	SD	MV	SD
Withers height [cm]	61	1.8	66	1.8
Body length / Withers height	1.1	0.0	1.1	0.1
Withers height / croup height	1.1	0.0	1.1	0.0
Legginess index	1.2	0.1	1.1	0.0

Forelimb	♀		♂	
Number of dogs	11		3	
	Proportion of total length			
	%	SD	%	SD
Shoulder blade	26	1.0	26	1.6
Upper arm	27	0.7	27	1.5
Lower arm	33	1.0	32	0.9
Pastern	15	1.0	14	3.6

Hindlimb	♀		♂	
Number of dogs	11		3	
	Proportion of total length			
	%	SD	%	SD
Thigh	36	1.9	35	2.2
Lower leg	38	1.3	39	1.7
Hind pastern	26	1.2	26	1.3

Gait parameters

Forelimb

	Stride length				Stride duration				Stance length				Stance duration			
	m				s				m				s			
MV	0.75	1.32	2.26	2.06	0.73	0.51	0.41	0.38	0.48	0.57	0.69	0.67	0.52	0.23	0.14	0.13
SD	0.1	0.2	0.2	0.1	0.2	0.1	0.0	0.0	0.1	0.1	0.1	0.0	0.1	0.0	0.0	0.0
Number of strides					94	138	26	29	All numbers equal							

	Swing length				Swing duration				Dutyfactor				Speed			
	m				s								m/s			
MV	0.27	0.75	1.58	1.39	0.26	0.27	0.27	0.25	66	46	33	34	0.97	2.63	5.52	5.36
SD	0.1	0.2	0.2	0.1	0.0	0.0	0.0	0.0	3.9	4.7	4.9	2.3	0.2	0.6	0.6	0.4
Number of strides																

Hindlimb

	Stride length				Stride duration				Stance length				Stance duration			
	m				s				m				s			
MV	0.81	1.33	2.23	2.20	0.77	0.51	0.40	0.40	0.52	0.57	0.62	0.62	0.51	0.23	0.12	0.12
SD	0.1	0.2	0.2	0.2	0.1	0.0	0.0	0.0	0.0	0.1	0.1	0.0	0.0	0.0	0.0	0.0
Number of strides					95	150	47	59	All numbers equal							

	Swing length				Swing duration				Dutyfactor				Speed			
	m				s								m/s			
MV	0.29	0.76	1.61	1.59	0.25	0.27	0.28	0.28	67	45	30	30	1.05	2.64	5.56	5.53
SD	0.1	0.2	0.2	0.2	0.0	0.0	0.0	0.0	3.1	4.6	2.4	2.2	0.1	0.7	0.7	0.6
Number of strides																

Kinematics

Forelimb

Forelimb	Shoulder blade				Shoulder joint				Upper arm				Elbow joint			
Touch down	*62*	*60*	*67*	*62*	143	138	170	132	76	77	93	70	135	135	153	126
Lift off	*93*	*98*	*116*	*107*	124	123	164	132	31	25	28	25	126	140	157	151
Mid-stance	-	-	-	-	134	128	162	127	-	-	-	-	138	124	152	126
Effective motion	*31*	*38*	*49*	*45*	19	15	6	0	45	52	65	45	9	5	4	25
Stance amplitude	*32*	*42*	*62*	*52*	22	22	38	17	48	53	72	56	30	35	37	38
Swing amplitude	*31*	*41*	*64*	*48*	30	38	59	52	49	61	84	66	47	71	77	86
Number of strides	94	138	26	29	Values in italics are marker-based values corrected to reflect the X-Ray video and position data.											

Forelimb	Lower arm				Carpal joint				Pastern			
Touch down	60	58	61	57	191	195	213	196	132	139	155	141
Lift off	95	115	129	126	114	137	150	152	19	22	20	26
Mid-stance	-	-	-	-	191	195	213	196	-	-	-	-
Effective motion	35	57	69	70	77	58	64	44	113	116	135	115
Stance amplitude	64	65	71	71	83	68	86	64	114	111	131	116
Swing amplitude	60	95	107	113	97	123	140	116	124	158	187	171
Number of strides												

Hindlimb

Hindlimb	Pelvis				Hip joint				Thigh				Stifle joint			
Touch down	68	68	60	67	116	123	120	115	48	56	59	48	121	134	148	149
Lift off	59	55	52	51	*167*	*175*	*177*	175	*108*	*120*	*127*	124	115	132	138	147
Mid-stance	-	-	-	-	116	123	119	118	-	-	-	-	122	134	148	147
Effective motion	9	13	8	16	*51*	*52*	*45*	60	*60*	*64*	*68*	76	6	2	10	1
Stance amplitude	17	14	16	17	*61*	*56*	*58*	61	*69*	*65*	*66*	75	16	21	33	30
Swing amplitude	10	24	18	22	*56*	*60*	*67*	74	*62*	*74*	*75*	86	36	58	60	66
Number of strides	95	150	47	45	Values in italics are marker-based values corrected to reflect the X-Ray video and position data.											

Hindlimb	Lower leg				Ankle joint				Hind pastern			
Touch down	74	78	90	101	106	108	125	132	32	29	35	30
Lift off	18	22	21	23	136	150	149	147	118	128	128	124
Mid-stance	-	-	-	-	107	108	125	126	-	-	-	-
Effective motion	56	56	68	77	29	42	26	15	86	98	93	94
Stance amplitude	56	58	69	78	56	72	69	71	96	100	94	102
Swing amplitude	66	80	99	108	46	74	67	69	86	102	107	117
Number of strides												

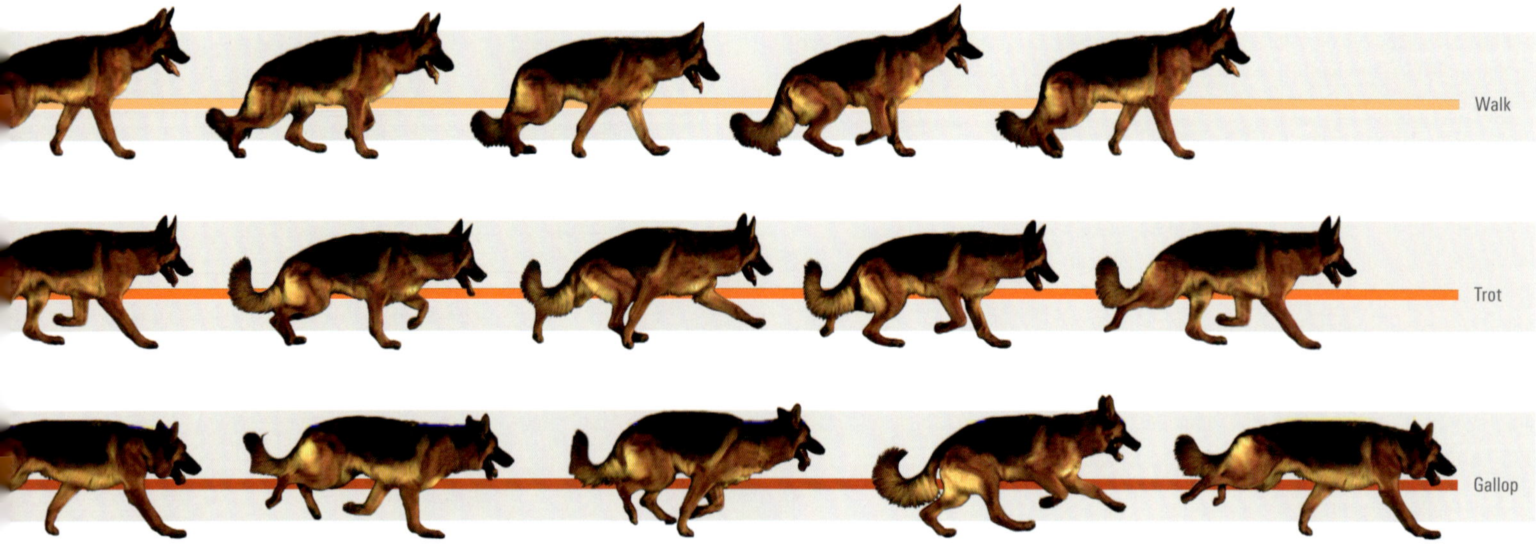

PROPORTIONS

The segment proportions in show line German Shepherds are very similar to those of the working line, except that show line animals have a slightly shorter femur and a slightly longer lower leg. The proportional lengths of the limb segments vary slightly more in dogs than in bitches, with the exception of the femur and lower leg, which vary more in bitches.

GAIT PARAMETERS

Forelimb stride length in this breed line is shorter in the walk than it is in most other breeds, but in all other gaits it is very long, as is that of the hindlimbs.

Excursion is more or less the same in this line of the breed as it is in the working line, except for the fact that the forward excursion of the leading and trailing forelimbs in the gallop is not equal but longer in the trailing limb. This is mainly down to differences in the excursion of the upper arm. The forwards excursion of the hindlimbs does not differ from that observed in the working line. The backwards excursion of the hindlimbs is greater in the trot than the walk, and greater still in the gallop. The difference between the degree of excursion in the walk and the trot is mainly down to the position of the pelvis and femur. Overall, excursion is almost always greater in show line animals than in working line animals. Comparatively speaking, forwards excursion is always moderate in the forelimbs, except in the gallop where it is high in the trailing limb. Backwards excursion is always very high in the forelimbs, except in the walk where it is very low. With the exception of the walk, the forwards excursion of the hindlimbs is also very high. The backwards excursion of the hindlimbs is at the very top of the range in all gaits.

As mentioned for the working line, the 90° shoulder joint ankle and 120° knee joint angle are not observed in the locomotion of this line either.

GERMAN SHEPHERD (Show)

DESCRIPTION OF BREED

The German Shepherd was bred as a herding dog, but was also expected to guard and defend the property of the shepherd. The breed has its ancestral roots in herding dog lines from Thuringia and Württemberg. Although German Shepherds began to be bred in the second half of the 19th century, methodological breeding did not start until the 1890s. The first standard was formulated in 1890. In 1899, the German Shepherd Club was set up under Max von Stephanitz, and at the first meeting of its members the description of the breed contained in the standard was amended to a version which is still valid today. A stud book was started in the Club's first year, as much emphasis was placed on lineage, training and the marks achieved at shows. Unlike other breed clubs, the German Shepherd Club noted in the stud book not only the winners of a show, but all of their offspring, which led to a very broad breeding base. The decline of sheep-keeping at the beginning of the 20th century resulted in a decline in the number of sheepdogs required, so new areas of activity were sought, and the animals found to be suitable as police dogs. During World War I, German Shepherd numbers decreased, and during the Second World War the club and its program to improve the breed almost disappeared entirely.

Proportions

Body	♀		♂	
Number of dogs	6		4	
	MV	SD	MV	SD
Withers height [cm]	58	1.2	63	1.4
Body length / Withers height	1.2	0.0	1.1	0.0
Withers height / croup height	1.0	0.0	1.1	0.0
Legginess index	1.2	0.1	1.2	0.1

Forelimb	♀		♂	
Number of dogs	6		4	
	Proportion of total length			
	%	SD	%	SD
Shoulder blade	26	0.5	27	0.7
Upper arm	26	0.8	26	0.6
Lower arm	32	0.8	31	0.6
Pastern	15	0.6	16	1.6

Hindlimb	♀		♂	
Number of dogs	6		4	
	Proportion of total length			
	%	SD	%	SD
Thigh	38	2.1	38	0.8
Lower leg	36	2.0	36	3.0
Hind pastern	26	0.3	27	2.4

Gait parameters

Forelimb

	Stride length				Stride duration				Stance length				Stance duration			
	m				s				m				s			
MV	0.76	1.17	1.89	1.81	0.72	0.49	0.40	0.40	0.48	0.54	0.58	0.64	0.47	0.24	0.13	0.15
SD	0.1	0.2	0.1	0.2	0.1	0.0	0.0	0.0	0.1	0.1	0.1	0.1	0.1	0.0	0.0	0.0
Number of strides					93	145	85	46	All numbers equal							

	Swing length				Swing duration				Dutyfactor				Speed			
	m				s								m/s			
MV	0.28	0.63	1.32	1.17	0.25	0.25	0.27	0.25	65	49	33	37	1.04	2.36	4.75	4.52
SD	0.1	0.1	0.1	0.2	0.0	0.0	0.0	0.0	4	4.0	3	3.7	0.2	0.3	0.3	0.4
Number of strides	All numbers equal															

Hindlimb

	Stride length				Stride duration				Stance length				Stance duration			
	m				s				m				s			
MV	0.79	1.24	1.72	1.90	0.72	0.49	0.39	0.39	0.50	0.50	0.55	0.57	0.47	0.21	0.13	0.13
SD	0.1	0.2	0.2	0.1	0.1	0.0	0.0	0.0	0.1	0.1	0.1	0.0	0.0	0.0	0.0	0.0
Number of strides					104	171	84	83	All numbers equal							

	Swing length				Swing duration				Dutyfactor				Speed			
	m				s								m/s			
MV	0.29	0.74	1.17	1.33	0.25	0.28	0.26	0.27	65	43	34	32	1.10	2.51	4.35	4.78
SD	0.1	0.1	0.2	0.1	0.0	0.0	0.0	0.0	2.9	3.9	3.0	2.9	0.1	0.3	0.7	0.2
Number of strides	All numbers equal															

Kinematics

Forelimb

Forelimb	Shoulder blade				Shoulder joint				Upper arm				Elbow joint			
Touch down	*62*	*59*	*60*	*65*	145	144	142	147	76	81	81	77	140	143	145	133
Lift off	*95*	*100*	*113*	*103*	122	125	137	130	29	25	24	26	137	143	160	152
Mid-stance	-	-	-	-	131	131	136	134	-	-	-	-	145	135	138	134
Effective motion	*33*	*45*	*53*	*38*	23	19	5	17	47	56	57	50	3	0	15	19
Stance amplitude	*33*	*45*	*60*	*46*	24	24	27	25	48	56	64	57	32	37	44	36
Swing amplitude	*35*	*47*	*60*	*44*	32	40	38	51	51	62	78	61	48	65	70	72
Number of strides	93	145	85	47	Values in italics are marker-based values corrected to reflect the X-Ray video and position data.											

Forelimb	Lower arm				Carpal joint				Pastern			
Touch down	64	62	65	57	198	201	212	201	133	140	148	145
Lift off	108	118	136	126	130	136	167	159	22	18	31	33
Mid-stance	-	-	-	-	198	201	212	201	-	-	-	-
Effective motion	44	56	71	69	67	65	45	42	111	122	117	112
Stance amplitude	67	67	72	71	70	74	56	60	109	117	111	110
Swing amplitude	66	87	103	106	94	117	103	126	120	153	151	176

Hindlimb

Hindlimb	Pelvis				Hip joint				Thigh				Stifle joint			
Touch down	63	58	55	63	114	116	118	108	52	59	63	45	129	142	128	140
Lift off	53	48	50	39	152	153	147	152	99	105	97	113	121	132	139	141
Mid-stance	-	-	-	-	114	116	111	108	-	-	-	-	129	142	141	140
Effective motion	9	9	5	23	38	37	29	44	47	46	58	67	8	10	11	1
Stance amplitude	17	12	20	27	44	38	43	47	53	46	58	67	13	18	34	26
Swing amplitude	10	15	22	28	46	44	54	58	51	54	67	77	33	56	52	54
Number of strides	104	171	84	83	Values in italics are marker-based values corrected to reflect the X-Ray video and position data.											

Hindlimb	Lower leg				Ankle joint				Hind pastern			
Touch down	77	83	65	95	114	128	127	130	37	45	62	35
Lift off	22	27	42	28	140	152	152	155	118	125	109	127
Mid-stance	-	-	-	-	115	128	136	131	-	-	-	-
Effective motion	56	56	22	67	26	24	25	25	82	80	79	92
Stance amplitude	55	55	74	68	50	53	62	73	91	82	84	100
Swing amplitude	67	80	98	98	43	62	68	64	82	86	101	107

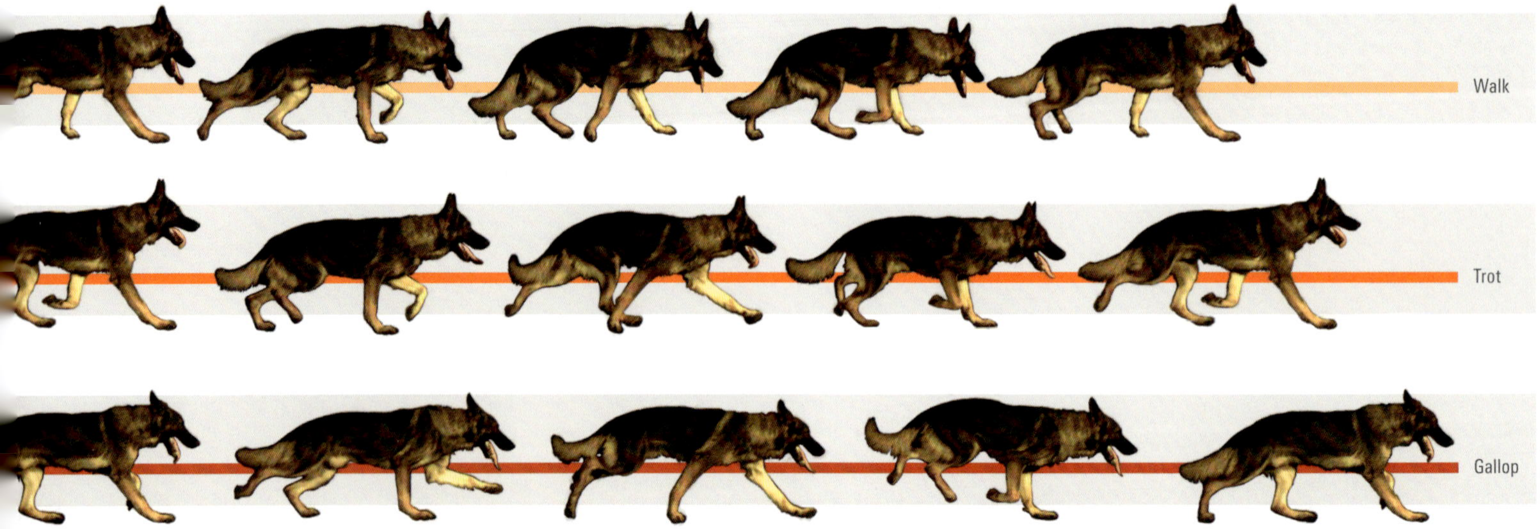

GERMAN SHEPHERD (Working)

DESCRIPTION OF BREED CONTINUED

The division of Germany after 1945 led to different breeding programs being developed in different directions. Both of the lines which emerged, however, clearly preserved the characteristics of the working dog. The differences mainly lay in the appearance of the line of the back. In the years after reunification the populations were allowed to mix, although some breeders did attempt to preserve the differences between the lines. Today, a distinction is made between two lines, one of which is bred for showing and the other for working purposes. However, the same standard applies to both lines and demands a height at the withers of 60 – 65 centimeters in dogs and 55 – 60 centimeters in bitches. Trunk length should exceed height at the withers by 10-17% and the depth of the chest should amount to between 45 and 48 percent of the height at the withers. The shoulder blade and upper arm should be approximately the same length, as should the upper and lower leg. The length of the front pastern should be approximately one third of the forearm. Ideally, the shoulder blade and upper arm should form a 90° angle and the knee joint an angle of approximately 120°.

PROPORTIONS

The only sex-specific difference observed in this line of German Shepherds is the ratio between trunk length and height at the withers. In bitches, trunk length very slightly exceeds height at the withers, while in dogs, height at the withers very slightly exceeds croup height. The shoulder blade is slightly longer and the forearm slightly shorter in this breed than the cross-breed average. The low level of variability of all the forelimb segments is striking, though the proportional length of the front pastern does vary somewhat in males.

GAIT PARAMETERS

Comparatively speaking, the relative stride length achieved by the forelimbs is middle-of-the-range in the walk but long in all other gaits. Hindlimb stride length is long in all gaits.

The forward excursion of the forelimbs is lower in the walk than in the trot and the gallop. No difference is observed in forward excursion in these two latter gaits. All the forelimb segments play a role here. Backwards excursion is also lower in the walk than in the trot or the leading limb in the gallop, and greatest in the trailing limb in the gallop. The main factors at play here are the motion of the shoulder blade and upper arm, though the position of the forearm also plays a role in the gallop. Forward excursion in the hindlimbs is lowest in the trot. It is slightly higher in the walk and much higher in the gallop, especially in the leading limb. Backwards excursion is almost the same in the walk and the trot but greater in both limbs in the gallop. These differences are caused by differences in the movement of all the joints.

In a cross-breed comparison, the excursion of the forelimbs is situated in the middle of the range in all gaits except the gallop, where the excursion of the leading limb is very short. Backwards excursion is only middle-of-the range in the walk, and otherwise falls near the top of the range. With the exception of the trailing limb, the forward excursion of the hindlimbs is always middle-of-the-range. Backwards excursion is always situated towards the top of the range.

The 90° angle between the shoulder blade and the upper arm required by the breed standard is not observed in any of the motion parameters, and nor is the 120° angle of the knee joint. Both joints exceed the specification during the stance phase.

Proportions

Body	♀		♂	
Number of dogs	8		5	
	MV	SD	MV	SD
Withers height [cm]	37	1.4	42	1.3
Body length / Withers height	1.2	0.0	2.0	0.0
Withers height / croup height	1.1	0.0	1.1	0.0
Legginess index	1.0	0.1	1.3	0.1

Forelimb	♀		♂	
Number of dogs	8		5	
	Proportion of total length			
	%	SD	%	SD
Shoulder blade	29	0.7	29	1.5
Upper arm	27	0.8	26	2.4
Lower arm	28	2.3	28	1.8
Pastern	16	2.2	17	1.9

Hindlimb	♀		♂	
Number of dogs	8		5	
	Proportion of total length			
	%	SD	%	SD
Thigh	37	1.2	37	1.1
Lower leg	36	1.1	35	1.5
Hind pastern	27	1.4	28	2.3

Gait parameters

Forelimb

	Stride length				Stride duration				Stance length				Stance duration			
	m				s				m				s			
MV	0.50	0.77	1.22	1.07	0.59	0.41	0.32	0.32	0.30	0.35	0.39	0.37	0.36	0.19	0.11	0.12
SD	0.1	0.1	0.1	0.1	0.0	0.0	0.0	0.0	0.0	0.0	0.0	0.0	0.0	0.0	0.0	0.0
Number of strides					114	164	29	47	All numbers equal							

	Swing length				Swing duration				Dutyfactor				Speed			
	m				s								m/s			
MV	0.21	0.42	0.83	0.70	0.23	0.22	0.21	0.20	62	47	35	37	0.86	1.88	3.75	3.28
SD	0.0	0.1	0.1	0.1	0.0	0.0	0.0	0.0	5	4.0	5	5.1	0.1	0.4	0.5	0.4
Number of strides	All numbers equal															

Hindlimb

	Stride length				Stride duration				Stance length				Stance duration			
	m				s				m				s			
MV	0.51	0.77	1.21	1.06	0.59	0.41	0.32	0.32	0.31	0.31	0.30	0.34	0.38	0.17	0.09	0.11
SD	0.0	0.1	0.2	0.1	0.0	0.0	0.0	0.0	0.0	0.0	0.0	0.0	0.0	0.0	0.0	0.0
Number of strides					111	160	52	71	All numbers equal							

	Swing length				Swing duration				Dutyfactor				Speed			
	m				s								m/s			
MV	0.20	0.46	0.90	0.72	0.21	0.23	0.23	0.21	64	43	28	34	0.86	1.91	3.71	3.37
SD	0.0	0.1	0.1	0.1	0.0	0.0	0.0	0.0	3.9	5.7	3.3	2.9	0.1	0.4	0.5	0.3
Number of strides	All numbers equal															

Kinematics

Forelimb

Forelimb	Shoulder blade				Shoulder joint				Upper arm				Elbow joint			
Touch down	*59*	*60*	*58*	*65*	130	134	143	141	71	74	85	76	131	127	141	128
Lift off	*90*	*105*	*113*	*105*	126	133	135	133	36	28	22	28	132	129	140	131
Mid-stance	-	-	-	-	131	135	145	151	-	-	-	-	134	127	139	127
Effective motion	*31*	*45*	*55*	*40*	16	11	8	8	42	46	63	49	1	3	1	2
Stance amplitude	*31*	*47*	*62*	*46*	24	16	21	15	43	48	66	55	30	29	26	20
Swing amplitude	*32*	*47*	*69*	*50*	28	24	30	16	45	49	70	54	48	63	66	67
Number of strides	113	164	29	47	Values in italics are marker-based values corrected to reflect the X-Ray video and position data.											

Forelimb	Lower arm				Carpal joint				Pastern			
Touch down	60	52	56	52	182	188	191	199	122	135	135	148
Lift off	96	101	118	103	138	127	157	141	42	26	39	39
Mid-stance	-	-	-	-	181	188	191	199	-	-	-	-
Effective motion	48	49	62	51	45	61	34	58	93	110	96	109
Stance amplitude	62	59	62	57	49	70	52	81	90	108	93	112
Swing amplitude	67	80	92	93	82	111	107	123	114	145	156	160

Hindlimb

Hindlimb	Pelvis				Hip joint				Thigh				Stifle joint			
Touch down	47	39	33	50	104	97	90	97	58	58	57	47	148	146	154	155
Lift off	48	34	42	26	*141*	*139*	*148*	*138*	*93*	*105*	*106*	*112*	122	127	127	134
Mid-stance	-	-	-	-	105	99	91	97	-	-	-	-	147	145	154	155
Effective motion	4	5	9	24	*37*	*42*	*58*	*41*	*44*	*47*	*49*	*65*	27	19	26	21
Stance amplitude	16	11	22	26	*43*	*42*	*58*	*51*	*47*	*47*	*48*	*66*	28	24	44	40
Swing amplitude	10	13	24	33	*44*	*46*	*63*	*53*	*46*	*53*	*57*	*74*	49	61	61	66
Number of strides	112	161	52	71	Values in italics are marker-based values corrected to reflect the X-Ray video and position data.											

Hindlimb	Lower leg				Ankle joint				Hind pastern			
Touch down	90	88	96	108	140	130	140	141	49	42	43	32
Lift off	38	32	31	32	150	153	145	142	111	121	114	110
Mid-stance	-	-	-	-	138	130	140	141	-	-	-	-
Effective motion	61	56	65	76	16	23	16	14	77	79	71	78
Stance amplitude	62	57	67	77	43	59	56	68	85	82	71	88
Swing amplitude	69	78	96	106	49	76	67	73	84	92	103	105

PROPORTIONS

In the male dogs which took part in this study, trunk length was slightly longer than in bitches in relation to height at the withers, and chests were slightly less deep. The proportional length of the upper arm seemed to be much more variable in these dogs than in the bitches we investigated. One of the males in our study was slightly larger than specified by the standard and had a proportionally longer upper arm than the other animals. One of the bitches we investigated displayed a much longer forearm than the others, despite complying with the standard in terms of height.

GAIT PARAMETERS

Comparatively speaking, relative stride length in this breed is high in almost all gaits, though the relative stride length of the forelimbs in the walk is more middle-of-the range. In the forelimbs, forward excursion prior to touchdown is lowest in the walk and does not differ between the trot and the gallop. The difference between the walk and the trot is mainly due to differences in the motion of the forearm and upper arm. Backwards excursion does not differ between the walk and the trot, but is greater in the gallop, with higher values displayed by the trailing forelimb than the leading forelimb. Differences can be put down to the size of the movements which occur in the shoulder blade and upper arm. Forwards excursion in the hindlimbs is lowest in the trot and highest in the leading limb in the gallop. There is no difference in the forward excursion displayed by the trailing hindlimb in the gallop and the hindlimbs in the walk. The low degree of excursion observed in the trot is the result of the positioning of the pelvis, and the high level of excursion observed in the leading limb in the gallop is caused by the positioning of the femur and lower leg. Backwards excursion is lowest in the trailing hindlimb in the gallop and highest in the leading hindlimb in the gallop and in the trot. In a cross-breed comparison, the forward excursion of the forelimbs prior to touchdown is only high in the leading limb in the gallop. In all other gaits the values are middle-of-the range. The backwards excursion of the forelimbs prior to liftoff is, like the forwards excursion of the hindlimbs, high. The values observed for the hindlimbs when it comes to forwards excursion in the gallop are the highest obtained for any breed. The backwards excursion of the hindlimbs is moderate in symmetrical gaits and very short in the gallop.

ENGLISH COCKER SPANIEL

DESCRIPTION OF BREED

English Cocker Spaniels were originally bred as gun dogs where their role was to flush game out of dense brush. Nowadays they are popular family pets. The first references to the breed are found in a Welsh statute book dating from the 10th century. The first English book about dogs, which was written in the 14th century, claims that the Spaniel originated in Spain. Another theory is that the name derives not from "Spain" but from the old French word "espanir", meaning to crouch or flatten. The Cocker part comes from "woodcock", one of the birds that the dog was trained to hunt. Until well into the 18th century no distinction was made between Spaniels and Setters, but when the Kennel Club was founded in 1873, Field, Springer and Cocker Spaniels were all registered as separate breeds. The first breed standard was formulated in 1892. The first English Cocker Spaniels were imported to Germany in the early years of the 20th century (Beyersdorf 2007). In 1907, the two Spaniel Clubs which then existed in Germany merged to form the Hunting Spaniel Club, which coordinated breeding activities in Germany.
According to the current breed standard, height at the withers should be 39 – 41 centimeters in dogs and 38 or 39 centimeters in bitches. Movement should be expansive.

Proportions

Body	♀		♂	
Number of dogs	5		4	
	MV	SD	MV	SD
Withers height [cm]	45	2.3	50	2.1
Body length / Withers height	1.7	0.1	1.2	0.1
Withers height / croup height	1.0	0.0	1.0	0.0
Legginess index	1.3	0.1	1.2	0.2

Forelimb	♀		♂	
Number of dogs	5		4	
	Proportion of total length			
	%	SD	%	SD
Shoulder blade	31	2.1	29	3.0
Upper arm	27	1.2	26	2.3
Lower arm	27	2.0	29	2.6
Pastern	15	1.3	16	2.3

Hindlimb	♀		♂	
Number of dogs	5		4	
	Proportion of total length			
	%	SD	%	SD
Thigh	41	1.6	39	2.9
Lower leg	35	2.3	35	1.4
Hind pastern	24	2.7	25	2.6

Gait parameters

Forelimb

	Stride length				Stride duration				Stance length				Stance duration				Swing length				Swing duration				Dutyfactor				Speed			
	m				s				m				s				m				s								m/s			
MV	0.66	0.90	1.36	1.29	0.59	0.43	0.35	0.35	0.41	0.44	0.43	0.45	0.39	0.22	0.12	0.13	0.24	0.47	0.93	0.84	0.21	0.21	0.23	0.22	65	51	34	37	1.12	2.10	3.86	3.71
SD	0.1	0.1	0.2	0.2	0.0	0.0	0.0	0.0	0.0	0.0	0.0	0.1	0.0	0.0	0.0	0.0	0.1	0.1	0.2	0.2	0.0	0.0	0.0	0.0	3.6	3.2	4.9	4.5	0.2	0.2	0.5	0.5
Number of strides					109	143	79	87	All numbers equal																							

Hindlimb

	Stride length				Stride duration				Stance length				Stance duration				Swing length				Swing duration				Dutyfactor				Speed			
	m				s				m				s				m				s								m/s			
MV	0.69	0.94	1.38	1.37	0.59	0.43	0.35	0.35	0.41	0.38	0.44	0.44	0.37	0.19	0.12	0.12	0.28	0.56	0.93	0.93	0.23	0.24	0.23	0.23	62	44	35	34	1.16	2.17	3.93	3.90
SD	0.1	0.1	0.2	0.2	0.0	0.0	0.0	0.0	0.0	0.0	0.1	0.0	0.0	0.0	0.0	0.0	0.1	0.1	0.2	0.2	0.0	0.0	0.0	0.0	4.7	6.4	3.4	3.3	0.2	0.3	0.5	0.5
Number of strides					119	153	88	96	All numbers equal																							

Kinematics

Forelimb

Forelimb	Shoulder blade				Shoulder joint				Upper arm				Elbow joint				Lower arm				Carpal joint				Pastern			
Touch down	64	58	58	67	138	136	128	131	74	78	70	64	131	135	134	120	57	57	64	56	195	197	202	200	138	140	139	145
Lift off	*96*	*96*	*106*	*100*	*122*	*125*	*129*	*128*	27	29	23	28	128	138	149	147	101	109	126	119	120	128	158	153	19	19	32	35
Mid-stance	-	-	-	-	128	124	119	120	-	-	-	-	128	128	131	122	-	-	-	-	195	197	202	200	-	-	-	-
Effective motion	*32*	*38*	*48*	*33*	*15*	*11*	*1*	*3*	47	49	47	36	3	3	15	27	44	52	62	63	76	69	44	47	119	121	108	110
Stance amplitude	*35*	*41*	*49*	*38*	*17*	*16*	*15*	*13*	49	51	51	45	30	35	28	39	64	63	65	67	76	77	57	61	111	112	100	103
Swing amplitude	*31*	*38*	*55*	*43*	*21*	*25*	*24*	*24*	48	52	65	57	48	64	55	48	67	84	92	85	106	116	106	110	137	153	153	151
Number of strides	91	123	79	87	Values in italics are marker-based values corrected to reflect the X-Ray video and position data.																							

Hindlimb

Hindlimb	Pelvis				Hip joint				Thigh				Stifle joint				Lower leg				Ankle joint				Hind pastern			
Touch down	37	36	38	40	93	94	85	87	56	58	53	47	142	143	142	141	87	85	89	94	134	132	136	134	46	47	46	39
Lift off	32	32	23	18	130	130	125	128	98	99	106	110	126	129	133	144	28	31	28	34	146	151	151	158	118	121	123	123
Mid-stance	-	-	-	-	92	94	91	87	-	-	-	-	143	144	142	141	-	-	-	-	134	133	137	135	-	-	-	-
Effective motion	5	4	15	22	37	36	40	41	42	40	53	63	17	14	9	3	59	55	61	59	14	19	16	24	72	73	78	84
Stance amplitude	10	11	20	22	39	36	39	43	44	41	51	62	20	18	28	20	57	54	62	63	43	53	66	74	82	78	86	96
Swing amplitude	7	10	19	25	41	39	50	50	44	44	59	71	46	57	67	58	73	79	107	98	45	62	71	66	77	83	93	100
Number of strides	99	133	89	96	Values in italics are marker-based values corrected to reflect the X-Ray video and position data.																							

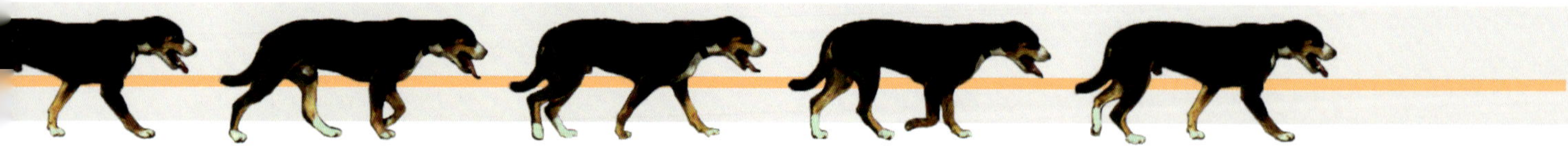
Walk

Trot

Gallop

ENTLEBUCHER MOUNTAIN DOG

DESCRIPTION OF BREED

The Entlebucher Mountain Dog is the smallest of the Swiss Mountain Dog breeds. Its name reflects its geographic origins in the Entlebuch valley. The breed is first mentioned in an article from 1889, though at that time the animals were not yet being pure-bred and were only rarely distinguished from Appenzell Mountain Dogs, which fulfilled the same role as herding, sheep and guard dogs but in a different region of the Alps. The first Entlebucher Mountain Dogs were shown in Switzerland in 1913, when they were also entered into the Swiss Stud Book. This breed developed much more slowly than the other Mountain Dog breeds, with the Swiss Entlebucher Mountain Dog Club not being set up until 1926, and the first standard laid out in 1927. As is the case with the Appenzell Mountain Dog, today's population stems from a very small number of individuals.

According to the standard, height at the withers should be 44 – 52 centimeters in dogs and 42 – 50 centimeters in bitches. The ratio of height at withers to the length of the body should be 8:10, the chest should be deep. The upper arm should be of equal length or slightly shorter than the shoulder blade and together they should form an angle of 110 – 120 degrees. The upper and lower thigh should be of approximately equal length and form an open angle. Movement is described as expansive.

PROPORTIONS

On average, trunk length in relation to height at the withers is greater in bitches than in dogs. Other than that, no sex-specific differences were observed. The shoulder blade is slightly longer and the forearm slightly shorter than the cross-breed average, while in the hindlimbs, the femur is slightly longer and the lower leg slightly shorter.

GAIT PARAMETERS

Comparatively speaking, stride length is long. This is caused in all gaits by a long stance phase. The swing phase falls into the middle of the range in a cross-breed comparison. Forwards excursion prior to touchdown is smallest in the leading forelimb in the gallop and greatest in the trot. This difference is caused by differences in the movements of the proximal limb segments. The forwards excursion of the forelimb in the walk does not differ from that of the trailing forelimb in the gallop. Backwards excursion is lowest in the leading forelimb in the gallop and greatest in the trailing forelimb in the same gait. Backwards excursion is greater in the trot than the walk. The high degree of excursion observed in the trailing forelimb in the gallop is contributed to by all the limb segments. The differences between the excursion observed in the walk and that seen in the trot are caused by the motion of the forearm. When it comes to the hindlimbs, forward excursion is lowest in the trot and highest in the leading limb in the gallop. There is no difference between the degree of excursion observed in the walk and that exhibited by the trailing limb in the gallop. The differences in the forward excursion of the hindlimbs can be put down to differences in the movements of the proximal limb segments. When it comes to backwards excursion, there is a clear difference between the symmetrical gaits and the gallop, with both hindlimbs displaying greater excursion in the gallop. Here too, the reason for this lies in the pelvis and femur. Excursion in both directions is long in most gaits for both the forelimbs and the hindlimbs. The only exception to this is the gallop, in which forelimb excursion in both directions is low.

The angle of 110 – 120° between the shoulder blade and upper arm required by the breed standard is, according to our data, always exceeded during the movement which occurs during the stance phase. The long stride lengths and very high degree of excursion observed in some gaits/limbs indicate that this breed is, as required, good at covering ground.

Proportions

Body	♀		♂	
Number of dogs	7		5	
	MV	SD	MV	SD
Withers height [cm]	51	2.3	55	2.0
Body length / Withers height	1.1	0.1	1.1	0.1
Withers height / croup height	1.0	0.0	1.0	0.0
Legginess index	1.3	0.1	1.2	0.1

Forelimb	♀		♂	
Number of dogs	7		5	
	Proportion of total length			
	%	SD	%	SD
Shoulder blade	27	1.7	28	1.1
Upper arm	26	1.6	28	1.2
Lower arm	30	0.8	28	1.1
Pastern	17	0.7	16	1.5

Hindlimb	♀		♂	
Number of dogs	7		5	
	Proportion of total length			
	%	SD	%	SD
Thigh	37	0.7	35	0.4
Lower leg	37	1.2	37	1.0
Hind pastern	27	1.5	27	1.1

Gait parameters

Forelimb

	Stride length				Stride duration				Stance length				Stance duration			
	m				s				m				s			
MV	0.70	0.99	-	-	0.67	0.45	-	-	0.41	0.44	-	-	0.41	0.21	-	-
SD	0.1	0.1	-	-	0.0	0.0	-	-	0.0	0.0	-	-	0.0	0.0	-	-
Number of strides					122	208	-	-								

	Swing length				Swing duration				Dutyfactor				Speed			
	m				s								m/s			
MV	0.28	0.55	-	-	0.26	0.24	-	-	61	46	-	-	1.04	2.19	-	-
SD	0.0	0.1	-	-	0.0	0.0	-	-	3.4	3.5	-	-	0.1	0.2	-	-
Number of strides	All numbers equal															

Hindlimb

	Stride length				Stride duration				Stance length				Stance duration			
	m				s				m				s			
MV	0.71	1.01	-	-	0.67	0.45	-	-	0.40	0.35	-	-	0.39	0.17	-	-
SD	0.1	0.1	-	-	0.0	0.0	-	-	0.0	0.0	-	-	0.0	0.0	-	-
Number of strides					140	209	-	-								

	Swing length				Swing duration				Dutyfactor				Speed			
	m				s								m/s			
MV	0.31	0.66	-	-	0.28	0.28	-	-	58	37	-	-	1.05	2.23	-	-
SD	0.0	0.1	-	-	0.0	0.0	-	-	3.4	3.9	-	-	0.1	0.2	-	-
Number of strides	All numbers equal															

Kinematics

Forelimb

Forelimb	Shoulder blade				Shoulder joint				Upper arm				Elbow joint			
Touch down	60	63	-	-	146	153	-	-	86	90	-	-	153	155	-	-
Lift off	*94*	100	-	-	*128*	132	-	-	34	32	-	-	151	150	-	-
Mid-stance	-	-	-	-	*142*	151	-	-	-	-	-	-	148	154	-	-
Effective motion	*34*	37	-	-	*18*	21	-	-	52	58	-	-	1	5	-	-
Stance amplitude	*36*	39	-	-	*17*	24	-	-	51	58	-	-	25	33	-	-
Swing amplitude	*35*	38	-	-	*23*	36	-	-	54	61	-	-	58	73	-	-
Number of strides	122	208	0	0	Values in italics are marker-based values corrected to reflect the X-Ray video and position data.											

Forelimb	Lower arm				Carpal joint				Pastern			
Touch down	67	65	-	-	188	187	-	-	121	122	-	-
Lift off	117	118	-	-	159	164	-	-	42	46	-	-
Mid-stance	-	-	-	-	188	187	-	-	-	-	-	-
Effective motion	50	53	-	-	28	22	-	-	78	76	-	-
Stance amplitude	56	54	-	-	28	37	-	-	75	76	-	-
Swing amplitude	77	90	-	-	84	107	-	-	108	141	-	-

Hindlimb

Hindlimb	Pelvis				Hip joint				Thigh				Stifle joint			
Touch down	-	-	-	-	-	-	-	-	60	66	-	-	140	149	-	-
Lift off	-	-	-	-	-	-	-	-	*111*	*111*	-	-	138	143	-	-
Mid-stance	-	-	-	-	-	-	-	-	-	-	-	-	141	149	-	-
Effective motion	-	-	-	-	-	-	-	-	*51*	*45*	-	-	3	5	-	-
Stance amplitude	-	-	-	-	-	-	-	-	*51*	*44*	-	-	12	15	-	-
Swing amplitude	-	-	-	-	-	-	-	-	*53*	*52*	-	-	35	51	-	-
Number of strides	140	209	0	0	Values in italics are marker-based values corrected to reflect the X-Ray video and position data.											

Hindlimb	Lower leg				Ankle joint				Hind pastern			
Touch down	81	83	-	-	135	144	-	-	54	61	-	-
Lift off	37	42	-	-	162	162	-	-	125	120	-	-
Mid-stance	-	-	-	-	136	144	-	-	-	-	-	-
Effective motion	44	40	-	-	27	19	-	-	72	59	-	-
Stance amplitude	43	41	-	-	44	46	-	-	75	64	-	-
Swing amplitude	60	69	-	-	48	56	-	-	75	71	-	-

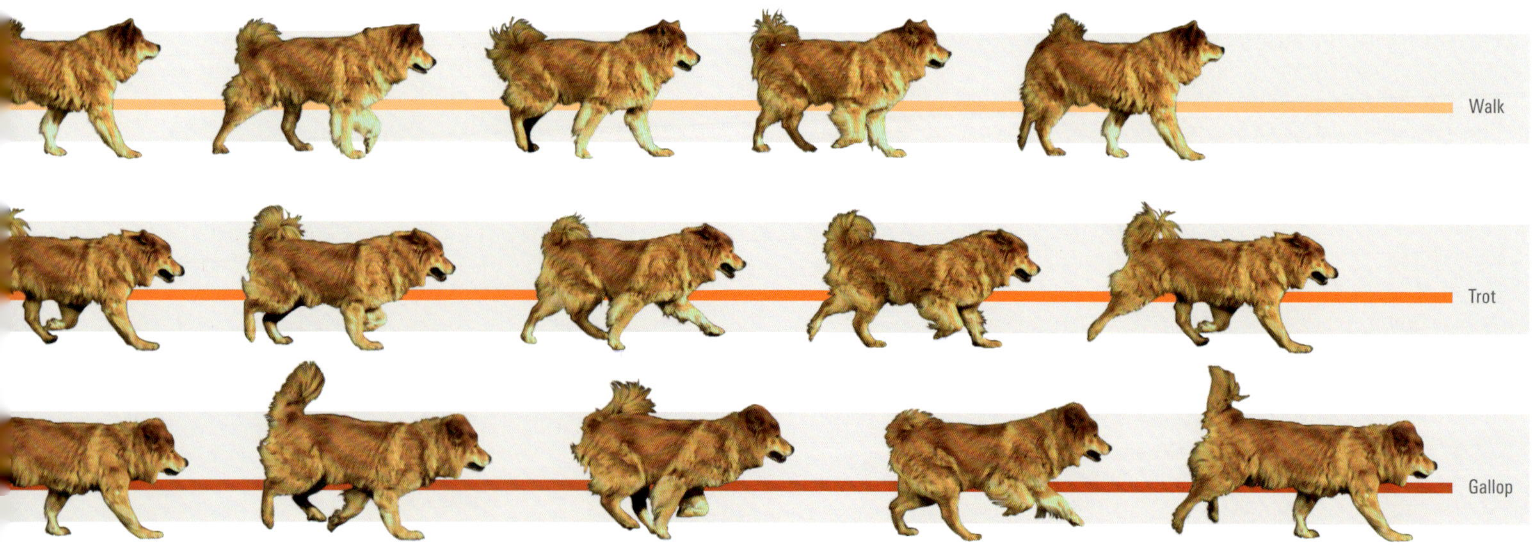

PROPORTIONS

In all the animals we investigated, trunk length exceeded height at the withers by a small amount. Sex-specific differences occur in the proportional lengths of the upper and forearm. In dogs, the upper arm and the forearm are the same length, as required by the breed standard, while in bitches, the forearm is slightly longer than the upper arm. In the hindlimbs, the proportional lengths of the femur and the lower leg are equal in bitches, while in dogs, the lower leg is slightly longer than the femur. Overall, the variability of the various segment lengths is very low. In bitches, the proportional lengths of the front pastern and the femur hardly vary at all, and in dogs there is extremely little variability in the proportional length of the lower leg.

GAIT PARAMETERS

Comparatively speaking, relative stride length in the forelimbs is middle-of-the-range in the walk and slightly longer in the trot, and in the hindlimbs the situation is reversed.
There is no difference between the walk and the trot in the degree of excursion in either direction observed in the forelimbs. In the hindlimbs, excursion in both directions is lower in the trot than in the walk. The difference in forward excursion between these two gaits is caused by differences in the movements of the proximal limb segments, while the difference in backward excursion is caused by a difference in the position in which the forearm is held. The pelvis, too, probably plays a role.
In a cross-breed comparison, forwards excursion in the forelimbs is middle-of-the range in the walk and the trot, while backwards excursion is higher up the range. Forwards excursion in the hindlimbs is very low in both the walk and the trot, while backwards excursion is comparatively high.
The angulation of the knee does not deviate at touchdown from that observed in other breeds of a similar size, while at takeoff the joint angle is larger than in other breeds.

EURASIER

DESCRIPTION OF BREED

The Eurasier is one of the youngest breeds of dog. Breeding began in 1960 by crossing Chow Chow dogs with Keeshond bitches, and the initial name of the breed, chosen by founder Julius Wipfel, was actually Wolf Chow (the German for Keeshond being "Wolfspitz"). In the early days, the development of the breed was significantly influenced by the breeder Charlotte Baldamus. The aim of the program was to create a Husky-like dog that would be suitable as a family pet. In 1972, the animals were crossed with Samoyeds. This was followed by more of the original Chow Chow/Keeshond crosses and, later, renewed Samoyed crosses. As a result of the Samoyed influence, the original name of the breed was no longer justified (it was also deemed to be too close to those of the founding breeds), and in 1973 it was changed at the suggestion of its founder to Eurasier. That same year, the breed was officially recognized by both the FCI and the VDH. Coloration can vary greatly in Eurasiers, from gray to red to dun, black or black with patches of color. In terms of temperament, Eurasiers are calm, confident dogs which behave neutrally towards strangers and have no hunting instinct.
According to the standard, height at the withers should be 52 – 60 centimeters in dogs and 48 – 56 centimeters in bitches. Trunk length should be slightly greater than height at the withers. The upper arm should be of approximately equal length to the forearm and the upper and lower thigh should also be approximately the same length. The angle of the stifle joint should not be too open, and motion should be expansive with good drive and good reach in front.

Proportions

Body	♀		♂	
Number of dogs	4		6	
	MV	SD	MV	SD
Withers height [cm]	31	3.3	33	3.0
Body length / Withers height	1.3	0.1	1.4	0.2
Withers height / croup height	1.0	0.0	1.0	0.0
Leggginess index	0.7	0.1	0.7	0.1

Forelimb	♀		♂	
Number of dogs	4		6	
	Proportion of total length			
	%	SD	%	SD
Shoulder blade	32	1.4	31	2.0
Upper arm	28	1.6	29	2.1
Lower arm	26	2.1	26	1.3
Pastern	15	1.0	14	1.0

Hindlimb	♀		♂	
Number of dogs	4		6	
	Proportion of total length			
	%	SD	%	SD
Thigh	36	2.2	38	3.5
Lower leg	37	2.8	36	2.5
Hind pastern	27	2.4	26	1.5

Gait parameters

Forelimb

	Stride length				Stride duration				Stance length				Stance duration				Swing length				Swing duration				Dutyfactor				Speed			
	m				s				m				s				m				s								m/s			
MV	0.34	0.49	-	-	0.65	0.39	-	-	0.24	0.26	-	-	0.46	0.22	-	-	0.11	0.23	-	-	0.18	0.16	-	-	72	57	-	-	0.54	1.27	-	-
SD	0.1	0.1	-	-	0.1	0.1	-	-	0.0	0.0	-	-	0.1	0.1	-	-	0.0	0.1	-	-	0.0	0.0	-	-	3.5	6.1	-	-	0.1	0.3	-	-
Number of strides					134	243	-	-	All numbers equal																							

Hindlimb

	Stride length				Stride duration				Stance length				Stance duration				Swing length				Swing duration				Dutyfactor				Speed			
	m				s				m				s				m				s								m/s			
MV	0.35	0.50	-	-	0.65	0.39	-	-	0.23	0.24	-	-	0.44	0.20			0.12	0.26	-	-	0.21	0.19	-	-	68	51	-	-	0.55	1.30	-	-
SD	0.1	0.1	-	-	0.1	0.1	-	-	0.0	0.0	-	-	0.1	0.1	-	-	0.1	0.1	-	-	0.0	0.0	-	-	4.4	5.7	-	-	0.1	0.3	-	-
Number of strides					141	243	-	-	All numbers equal																							

Kinematics

Forelimb	Shoulder blade				Shoulder joint				Upper arm				Elbow joint				Lower arm				Carpal joint				Pastern			
Touch down	*63*	*62*	-	-	139	141	-	-	71	74	-	-	133	136	-	-	62	62	-	-	199	200	-	-	136	139	-	-
Lift off	*95*	*102*	-	-	123	129	-	-	33	32	-	-	134	136	-	-	102	104	-	-	133	117	-	-	32	13	-	-
Mid-stance	-	-	-	-	138	139	-	-	-	-	-	-	133	135	-	-	-	-	-	-	199	200	-	-	-	-	-	-
Effective motion	*32*	*40*	-	-	16	12	-	-	38	42	-	-	1	0	-	-	39	42	-	-	65	83	-	-	-104	-126	-	-
Stance amplitude	*34*	*42*	-	-	17	17	-	-	40	45	-	-	25	25	-	-	60	57	-	-	68	90	-	-	99	120	-	-
Swing amplitude	*38*	*46*	-	-	20	21	-	-	40	46	-	-	35	43	-	-	55	66	-	-	87	107	-	-	116	145	-	-
Number of strides	134	243	0	0	Values in italics are marker-based values corrected to reflect the X-Ray video and position data.																							

Hindlimb	Pelvis				Hip joint				Thigh				Stifle joint				Lower leg				Ankle joint				Hind pastern			
Touch down	45	40	-	-	109	107	-	-	64	67	-	-	161	161	-	-	97	95	-	-	153	150	-	-	56	54	-	-
Lift off	39	32	-	-	*136*	*134*	-	-	*97*	*102*	-	-	136	139	-	-	44	42	-	-	161	159	-	-	117	118	-	-
Mid-stance	-	-	-	-	109	107	-	-	-	-	-	-	161	161	-	-	-	-	-	-	153	149	-	-	-	-	-	-
Effective motion	6	8	-	-	*27*	*28*	-	-	*32*	*35*	-	-	25	22	-	-	53	53	-	-	-7	-10	-	-	-61	-63	-	-
Stance amplitude	14	12	-	-	*30*	*30*	-	-	*35*	*37*	-	-	24	22	-	-	51	51	-	-	28	46	-	-	66	68	-	-
Swing amplitude	7	11	-	-	*30*	*31*	-	-	*34*	*38*	-	-	42	55	-	-	60	69	-	-	32	52	-	-	64	69	-	-
Number of strides	141	243	0	0	Values in italics are marker-based values corrected to reflect the X-Ray video and position data.																							

PROPORTIONS

All the animals we investigated complied with the breed standard in terms of height at the withers. However, withers height and croup height are identical, so although the breed standard calls for slightly longer hindlimbs, our measurements do not indicate that this is the case. The low legginess index value shows that the chest in this breed is very deep. Comparatively speaking, the proportional length of the upper arm is average, the forearm slightly shorter than average and the shoulder blade slightly longer than average to make up for it. The proportional lengths of the hindlimb segments all correspond to those observed in most other breeds. Overall, no sex-specific differences were noted.

GAIT PARAMETERS

In a cross-breed comparison, relative stride lengths are short and are composed of short stance phases and short swing phases in every gait except the trot, where the relative length of the stance phase of the forelimbs is middle-of-the-range.

Horizontal excursion in both directions is more or less the same in the walk and the trot. The backward excursion of the hindlimbs, however, is greater in the trot than the walk due to differences in the position of the pelvis and femur at liftoff.

Although the stance phases are all very short in comparative terms, all the values for limb excursion are, relatively speaking, middle-of-the-range.

FRENCH BULLDOG

DESCRIPTION OF BREED

The history of the French Bulldog before it began to be pure-bred is identical to that of the English Bulldog. The term bulldog is first found in documents dating from 1631. Bulldogs were kept as bull-biters, until bull-baiting was banned by the English parliament in 1802, after which they were kept as fighting dogs. For this purpose it was desirable that they were smaller and more agile, which was achieved by crossing the old bulldog stock with terriers. When dog-fighting, too, was forbidden in England in 1858, bulldogs became the poor man's companion. However, the miniature versions of the old fighting dogs soon found fans abroad, especially in France and Belgium. Small bulldogs arrived in France around the turn of the century with the migration of English workers to Normandy. In Normandy, the dogs were once again crossed with terriers and, for the first time, pugs, and their characteristic sticking up ("bat") ears were secured. The first stud book was published in 1885 and the first breed charactistics in 1888. The breed was recognized in 1898 and the "bat ears", which had been the cause of much argument, were finally incorporated into the breed standard. After the 1870/1871 war, French Bulldogs were introduced to Germany by a soldier.

The current standard requires the hindlimbs to be slightly longer than the forelimbs, the chest to be very deep, and the upper arm and forearm, like the hind pastern to be short. Height at the withers can be up to 35 centimeters.

Proportions

Body	♀		♂	
Number of dogs	6		4	
	MV	SD	MV	SD
Withers height [cm]	55	3.2	58	1.4
Body length / Withers height	1.2	0.1	1.2	0.0
Withers height / croup height	1.0	0.0	1.0	0.0
Legginess index	1.2	0.1	1.2	0.1

Forelimb	♀		♂	
Number of dogs	6		4	
	Proportion of total length			
	%	SD	%	SD
Shoulder blade	28	1.3	28	3.3
Upper arm	28	1.8	26	1.2
Lower arm	28	1.5	30	2.6
Pastern	17	2.2	15	2.4

Hindlimb	♀		♂	
Number of dogs	6		4	
	Proportion of total length			
	%	SD	%	SD
Thigh	36	2.5	38	2.3
Lower leg	37	3.2	37	2.3
Hind pastern	27	1.3	26	1.8

Gait parameters

Forelimb

	Stride length (m)				Stride duration (s)				Stance length (m)				Stance duration (s)				Swing length (m)				Swing duration (s)				Dutyfactor				Speed (m/s)			
MV	0.69	0.97	1.36	1.33	0.68	0.45	0.38	0.38	0.43	0.46	0.48	0.48	0.44	0.23	0.14	0.15	0.26	0.51	0.88	0.85	0.24	0.23	0.24	0.23	64	50	38	39	1.06	2.12	3.50	3.44
SD	0.1	0.1	0.2	0.2	0.1	0.0	0.0	0.0	0.1	0.0	0.1	0.1	0.1	0.0	0.0	0.0	0.1	0.1	0.2	0.2	0.0	0.0	0.0	0.0	5.5	4.5	4.3	4.7	0.4	0.1	0.3	0.3
Number of strides					135	178	116	120	All numbers equal																							

Hindlimb

	Stride length (m)				Stride duration (s)				Stance length (m)				Stance duration (s)				Swing length (m)				Swing duration (s)				Dutyfactor				Speed (m/s)			
MV	0.70	0.96	1.33	1.35	0.67	0.46	0.38	0.38	0.44	0.46	0.51	0.51	0.44	0.23	0.16	0.15	0.26	0.50	0.82	0.84	0.23	0.23	0.23	0.23	65	50	41	40	1.08	2.07	3.46	3.50
SD	0.2	0.1	0.1	0.1	0.1	0.0	0.0	0.0	0.1	0.0	0.1	0.1	0.1	0.0	0.0	0.0	0.1	0.1	0.2	0.1	0.0	0.0	0.0	0.0	5.8	4.1	5.3	5.6	0.4	0.2	0.3	0.2
Number of strides					136	183	116	121	All numbers equal																							

Kinematics

Forelimb

	Shoulder blade				Shoulder joint				Upper arm				Elbow joint				Lower arm				Carpal joint				Pastern			
Touch down	*61*	*61*	*60*	*62*	147	148	143	136	81	82	78	69	144	144	144	128	62	63	65	59	192	194	197	192	130	133	134	135
Lift off	*91*	*98*	*110*	*97*	120	130	140	125	29	32	30	33	124	135	152	143	95	104	122	110	126	125	150	138	31	21	28	28
Mid-stance	-	-	-	-	142	144	139	131	-	-	-	-	141	140	141	123	-	-	-	-	192	194	197	192	-	-	-	-
Effective motion	*30*	*37*	*50*	*35*	27	18	3	11	52	50	48	36	20	9	8	15	32	41	56	51	66	69	47	55	99	112	106	107
Stance amplitude	*30*	*38*	*51*	*36*	27	26	23	24	52	54	54	44	32	34	35	27	55	54	60	56	68	80	65	69	98	112	105	102
Swing amplitude	*32*	*40*	*58*	*43*	33	33	28	34	55	58	66	52	52	62	62	53	58	77	88	88	88	107	94	102	114	141	141	152
Number of strides	135	178	116	122	Values in italics are marker-based values corrected to reflect the X-Ray video and position data.																							

Hindlimb

	Pelvis				Hip joint				Thigh				Stifle joint				Lower leg				Ankle joint				Hind pastern			
Touch down	51	48	47	53	103	106	101	96	52	57	54	44	143	143	141	143	90	86	88	99	142	140	143	147	50	53	54	47
Lift off	46	39	36	30	*137*	*140*	*144*	*136*	*91*	*101*	*109*	*106*	117	127	133	139	32	31	30	38	147	157	164	168	116	125	135	131
Mid-stance	-	-	-	-	103	106	101	96	-	-	-	-	143	144	141	143	-	-	-	-	142	140	143	147	-	-	-	-
Effective motion	5	9	11	23	*33*	*34*	*44*	*40*	*39*	*44*	*55*	*62*	25	16	8	4	59	55	58	62	13	17	21	21	65	72	80	85
Stance amplitude	17	14	22	29	*39*	*37*	*45*	*46*	*44*	*44*	*54*	*62*	27	22	25	22	59	55	59	65	40	57	70	80	81	83	91	102
Swing amplitude	8	17	20	25	*38*	*42*	*52*	*53*	*40*	*49*	*57*	*68*	48	57	53	50	70	77	84	88	48	65	67	69	70	83	91	104
Number of strides	136	183	116	121	Values in italics are marker-based values corrected to reflect the X-Ray video and position data.																							

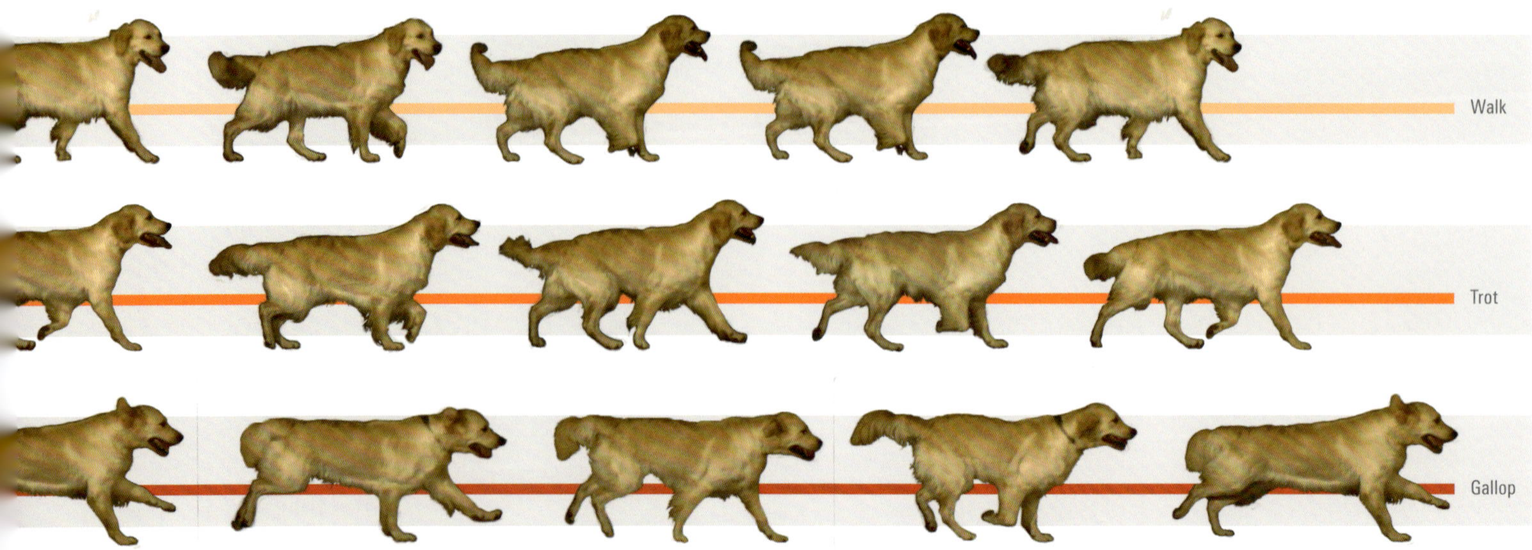

PROPORTIONS
The limb segment proportions of the Golden Retriever are essentially average for the breeds in this study. Of the ten Golden Retrievers we measured, two bitches were slightly taller than the standard allows. In one of them, this did not cause the limb segment proportions to deviate in any way, while the other had a longer femur and shorter lower leg than the other individuals. As a rule, the shoulder blade and the upper arm are the same length in bitches, while in dogs the shoulder blade is slightly longer than the upper arm.

GAIT PARAMETERS
Comparatively speaking, relative stride length in most gaits is middle-of-the-range, though the relative swing phase length is usually shorter than that of the stance phase. The relative stance phase lengths are particularly long in the hindlimbs.
In the forelimbs, forward excursion prior to touchdown is always the same, regardless of the gait. The backward excursion of the forelimbs is equal in the trot and the gallop (leading limb), lower in the walk and highest of all in the trailing limb in the gallop. The same gradation is observed in the movement of the shoulder blade. The forward excursion of the hindlimbs is equally low in the trot and the gallop (trailing limb), slightly higher in the walk and highest of all in the leading limb in the gallop. These differences are brought about by differences in the position of the pelvis and forearm. The backwards excursion of the hindlimbs is lowest in the walk and highest in the gallop, where the trailing limb displays greater excursion values than the leading. These differences can be put down to differences in the movements of the proximal limb segments.
Comparatively speaking, the forward excursion of the forelimbs is moderate in the walk and the trot and low in the gallop. Backwards excursion is always middle-of-the-range. The forward excursion of the hindlimbs is high in the walk but middle-of-the-range in all the other gaits. Backwards excursion is situated in the bottom third of the range in the walk, and otherwise in the middle of the range. As this comparison of excursion values shows, the
forward reach in this breed is not significantly larger than that observed in other breeds of a similar size.

GOLDEN RETRIEVER

DESCRIPTION OF BREED
Retrievers are hunting dogs whose original purpose was to retrieve game birds after they had been shot. The name retriever pre-dates today's Retriever breeds by a long time. As well as enjoying the act of retrieving, Retrievers need to be eager to follow the commands of their handlers, have a very good nose, a love of water, be good at swimming, have a thick, insulating coat, and be steady and trainable. English breeders identified these qualities in the St. John's Water Dog of New Foundland and the Tweed Water Spaniel, two breeds which are extinct today. Later, Black Labradors, Irish Setters and sandy-colored Bloodhounds were also included in the breeding program. Golden Retrievers were recognized as a breed by the English Kennel Club in 1911, and in 1912 they are mentioned for the first time as successful participants in English hunting dog trials. Two World Wars interrupted the breeding program, but it flourished again in England after 1947. The first German litter was registered in the VDH in 1962.
According to the breed standard, height at the withers should be 56 – 61 centimeters in dogs and 51 – 56 centimeters in bitches. The ribs should be deep. The shoulder blade should be long and the upper arm the same length. The requirements made of the hindlimbs are more general: the stifles should be well bent and the shank a good length. Powerful movement with a good forward reach and good drive is desired.

Proportions

Body	♀		♂	
Number of dogs	6		4	
	MV	SD	MV	SD
Withers height [cm]	61	2.1	65	2.2
Body length / Withers height	1.0	0.0	1.0	0.0
Withers height / croup height	1.0	0.0	1.0	0.0
Legginess index	1.6	0.3	1.5	0.1

Forelimb	♀		♂	
Number of dogs	6		4	
	Proportion of total length			
	%	SD	%	SD
Shoulder blade	26	0.7	26	0.8
Upper arm	28	1.3	28	0.9
Lower arm	30	2.5	31	1.2
Pastern	16	1.7	15	1.3

Hindlimb	♀		♂	
Number of dogs	6		4	
	Proportion of total length			
	%	SD	%	SD
Thigh	36	0.8	35	1.8
Lower leg	38	1.0	39	1.9
Hind pastern	27	0.8	26	0.4

Gait parameters

Forelimb

	Stride length				Stride duration				Stance length				Stance duration				Swing length				Swing duration				Dutyfactor				Speed			
	m				s				m				s				m				s								m/s			
MV	0.83	1.26	1.99	2.00	0.76	0.48	0.39	0.40	0.51	0.53	0.56	0.61	0.48	0.21	0.12	0.13	0.33	0.73	1.43	1.39	0.28	0.27	0.27	0.27	63	45	30	33	1.12	2.63	5.11	4.94
SD	0.1	0.2	0.1	0.1	0.1	0.0	0.0	0.0	0.1	0.0	0.0	0.0	0.1	0.0	0.0	0.0	0.1	0.2	0.1	0.1	0.0	0.0	0.0	0.0	5	5	1	1	0.2	0.4	0.3	0.2
Number of strides					66	89	25	24	All numbers equal																							

Hindlimb

	Stride length				Stride duration				Stance length				Stance duration				Swing length				Swing duration				Dutyfactor				Speed			
	m				s				m				s				m				s								m/s			
MV	0.83	1.21	1.96	2.00	0.76	0.48	0.39	0.40	0.50	0.48	0.48	0.52	0.48	0.21	0.10	0.11	0.33	0.72	1.48	1.48	0.28	0.27	0.29	0.29	63	43	26	28	1.11	2.46	4.98	4.99
SD	0.1	0.2	0.1	0.1	0.1	0.0	0.0	0.0	0.0	0.0	0.1	0.0	0.1	0.0	0.0	0.0	0.1	0.2	0.1	0.1	0.0	0.0	0.0	0.0	6	7	3	2	0.2	0.4	0.3	0.4
Number of strides					68	94	28	34	All numbers equal																							

Kinematics

Forelimb	Shoulder blade				Shoulder joint				Upper arm				Elbow joint				Lower arm				Carpal joint				Pastern			
Touch down	65	64	63	72	153	152	153	164	88	88	90	92	147	148	155	148	59	60	65	56	187	191	204	192	128	133	140	138
Lift off	95	99	115	115	128	134	145	153	33	35	29	38	140	150	159	160	106	115	129	121	140	156	174	170	33	40	45	48
Mid-stance	-	-	-	-	140	139	139	149	-	-	-	-	140	136	141	134	-	-	-	-	187	191	204	192	-	-	-	-
Effective motion	30	35	52	43	25	18	8	11	55	53	61	54	7	2	4	12	47	55	64	65	47	35	30	23	94	92	96	90
Stance amplitude	34	39	53	47	25	23	22	20	56	55	64	60	27	32	30	30	61	58	62	64	48	45	51	55	91	89	91	93
Swing amplitude	31	37	61	45	31	37	45	58	57	62	84	70	51	68	76	78	68	88	104	111	88	112	107	111	118	148	144	161
Number of strides	66	89	25	24	Values in italics are marker-based values corrected to reflect the X-Ray video and position data.																							

Hindlimb	Pelvis				Hip joint				Thigh				Stifle joint				Lower leg				Ankle joint				Hind pastern			
Touch down	57	53	33	49	110	112	96	104	53	60	68	55	145	147	160	153	92	87	93	98	142	140	148	143	50	52	55	43
Lift off	52	47	34	29	148	141	134	139	96	102	112	110	130	136	149	148	34	34	37	38	159	162	162	162	124	128	125	125
Mid-stance	-	-	-	-	110	112	96	104	-	-	-	-	145	147	161	154	-	-	-	-	143	141	149	145	-	-	-	-
Effective motion	5	6	-1	20	38	36	38	35	43	42	45	55	14	11	11	5	57	53	56	60	16	22	17	20	74	76	70	82
Stance amplitude	21	10	11	23	43	37	36	36	48	42	43	54	16	20	27	25	57	53	56	61	40	53	56	68	81	79	71	89
Swing amplitude	11	14	12	27	44	42	57	52	45	51	57	72	36	61	64	76	64	77	94	99	37	60	58	65	77	85	98	108
Number of strides	68	94	28	34	Values in italics are marker-based values corrected to reflect the X-Ray video and position data.																							

PROPORTIONS

All the individuals we investigated were the size required by the breed standard, except for one male animal, which was slightly too tall. The upper arm was very slightly longer in this individual, but otherwise there were no differences in limb segment proportions. No sex-specific differences were observed except in the variability of the limb segment proportions. All the limb segments except the shoulder blade, femur and lower leg are more variable in bitches than in dogs.

GAIT PARAMETERS

For their functional lengths, both the forelimbs and the hindlimbs produce very long strides in almost every gait. Stride length in the trot, however, is only middle-of-the-range, as is that of the leading limb in the gallop. In the trot, the stance phase is short and is not compensated by the medium-length swing phase.

The forwards excursion of the forelimbs prior to touchdown does not differ between the walk, the trot and the gallop (leading limb), but is lower in the trailing limb in the gallop. The most striking difference between the gallop and the other gaits lies in the position of the forearm. The backwards excursion of the forelimbs does not differ between the walk and the trot. It increases in the gallop, thanks to a significant increase in the rotation of the shoulder blade, and the excursion of the trailing limb is greater than that of the leading limb. The degree of backwards excursion observed in the hindlimbs is low in the trot and the gallop (trailing limb), slightly greater in the walk and highest of all in the leading limb in the gallop. Backwards excursion is only greater in the gallop (leading limb) than in the other gaits. This can be put down to the position of the pelvis and a summation of all the other segment movements.

Compared to other breeds, forelimb excursion is moderate in Large Münsterländers, except in the walk and the trot, where forward excursion is high. The backwards excursion of the hindlimbs is high in the walk and in the leading limb in the gallop, but the forwards excursion of this limb is, comparatively speaking, low. The long bound demanded by the standard is therefore present in the symmetrical gaits but not observable in the gallop.

LARGE MÜNSTERLÄNDER

DESCRIPTION OF BREED

The Large Münsterländer has its roots in the Westphalia/Lower Saxony area of Germany and descends, like all long-haired German setters, from a group of animals known as "bird dogs" (or alternatively hawk or quail dogs). Although the breed is originally German, it is possible that some crossing with English and French hunting dogs took place too (most probably with the black and white English Setter and the Epagneul bleu de Picard, according to Vornholt (2004)). When the breed standard for the German Longhaired Pointer was formulated in 1878, dogs with the black and white coloring were excluded, which soon led to the idea of turning dogs of this coloring which also displayed excellent hunting abilities into a separate breed. The first breed club was founded in 1919. Emphasis was placed on dogs with a calm disposition but which were spirited and powerful during the hunt and which were quick learners.

Height at the withers should be 60 – 65 centimeters in dogs and 58 – 63 centimeters in bitches. Body length should ideally be equal to height at the withers but may exceed it by up to two centimeters. The fore- and hindlimbs should be correctly angulated, and the dog's gait should be long-striding with good forward reach. Movement should be springy in the walk and the trot.

Proportions

Body	♀		♂	
Number of dogs	9		2	
	MV	SD	MV	SD
Withers height [cm]	62	2.8	69	0.7
Body length / Withers height	1.1	0.0	1.1	0.0
Withers height / croup height	1.0	0.0	1.0	0.0
Legginess index	1.5	0.4	1.5	0.1

Forelimb	♀		♂	
Number of dogs	9		2	
	Proportion of total length			
	%	SD	%	SD
Shoulder blade	27	1.6	28	0.8
Upper arm	27	1.6	27	1.0
Lower arm	31	2.3	32	0.7
Pastern	16	1.6	13	0.5

Hindlimb	♀		♂	
Number of dogs	9		2	
	Proportion of total length			
	%	SD	%	SD
Thigh	38	3.2	37	1.6
Lower leg	36	2.7	38	3.0
Hind pastern	25	1.0	25	1.4

Gait parameters

Forelimb

	Stride length				Stride duration				Stance length				Stance duration			
	m				s				m				s			
MV	0.86	1.22	1.34	-	0.73	0.46	0.40	-	0.52	0.55	0.47	-	0.46	0.22	0.15	-
SD	0.1	0.1	0.1	-	0.1	0.1	0.0	-	0.1	0.0	0.0	-	0.0	0.0	0.0	-
Number of strides					75	114	20	-	All numbers equal							

	Swing length				Swing duration				Dutyfactor				Speed			
	m				s								m/s			
MV	0.33	0.67	0.65	-	0.27	0.24	0.25	-	63	47	37	-	1.17	2.71	3.35	-
SD	0.1	0.1	0.4	-	0.0	0.0	0.0	-	3	4	2	-	0.2	0.7	0.1	-
Number of strides	All numbers equal															

Hindlimb

	Stride length				Stride duration				Stance length				Stance duration			
	m				s				m				s			
MV	0.88	1.23	1.44	1.97	0.73	0.47	0.40	0.39	0.50	0.48	0.53	0.61	0.44	0.19	0.16	0.13
SD	0.1	0.1	0.1	0.1	0.1	0.1	0.0	0.0	0.0	0.1	0.0	0.0	0.1	0.0	0.0	0.0
Number of strides					89	83	24	10	All numbers equal							

	Swing length				Swing duration				Dutyfactor				Speed			
	m				s								m/s			
MV	0.38	0.75	0.90	0.39	0.29	0.28	0.25	0.26	60	41	39	33	1.21	2.63	3.56	5.01
SD	0.1	0.1	0.1	0.6	0.0	0.0	0.0	0.0	5	5	2	1	0.2	0.5	0.1	0.1
Number of strides	All numbers equal															

Kinematics

Forelimb

	Shoulder blade				Shoulder joint				Upper arm				Elbow joint			
Touch down	*63*	*65*	*56*	*71*	147	155	146	171	84	90	90	100	150	153	159	153
Lift off	*95*	*105*	*104*	*106*	*128*	*136*	*145*	*139*	33	31	41	33	133	145	167	154
Mid-stance	-	-	-	-	137	142	126	111	-	-	-	-	143	147	154	133
Effective motion	*32*	*40*	*48*	*35*	*19*	*19*	*1*	*32*	51	59	49	68	17	8	8	1
Stance amplitude	*32*	*44*	*54*	*39*	*21*	*28*	*19*	*57*	55	61	61	81	30	36	27	28
Swing amplitude	*34*	*43*	*58*	*59*	*25*	*37*	*27*	*42*	57	64	63	68	57	74	59	85
Number of strides	86	114	20	2	Values in italics are marker-based values corrected to reflect the X-Ray video and position data.											

	Lower arm				Carpal joint				Pastern			
Touch down	66	63	69	54	197	195	189	185	130	133	120	130
Lift off	101	114	126	122	146	150	151	142	46	36	25	20
Mid-stance	-	-	-	-	197	195	189	185	-	-	-	-
Effective motion	34	51	57	68	51	46	38	42	85	97	95	109
Stance amplitude	63	56	63	69	50	53	49	72	92	93	89	110
Swing amplitude	67	88	85	107	88	111	68	107	104	147	113	163

Hindlimb

	Pelvis				Hip joint				Thigh				Stifle joint			
Touch down	61	57	62	-	115	114	120	-	53	57	58	39	142	143	137	132
Lift off	55	45	57	-	154	148	173	-	98	101	117	131	132	137	144	169
Mid-stance	-	-	-	-	116	115	120	-	-	-	-	-	142	141	138	132
Effective motion	7	11	5	-	39	34	53	-	45	44	59	93	10	6	7	37
Stance amplitude	20	14	15	-	40	33	55	-	45	43	57	89	17	17	25	58
Swing amplitude	9	18	12	-	44	41	57	-	47	52	65	99	36	54	47	73
Number of strides	89	83	24	10	Values in italics are marker-based values corrected to reflect the X-Ray video and position data.											

	Lower leg				Ankle joint				Hind pastern			
Touch down	89	86	80	93	143	143	141	139	54	56	61	46
Lift off	34	36	28	37	158	157	162	170	124	121	135	133
Mid-stance	-	-	-	-	143	140	142	140	-	-	-	-
Effective motion	55	50	52	55	15	15	21	31	70	65	74	87
Stance amplitude	54	49	56	71	39	48	62	100	76	69	76	105
Swing amplitude	64	74	81	91	44	60	55	75	73	78	84	102

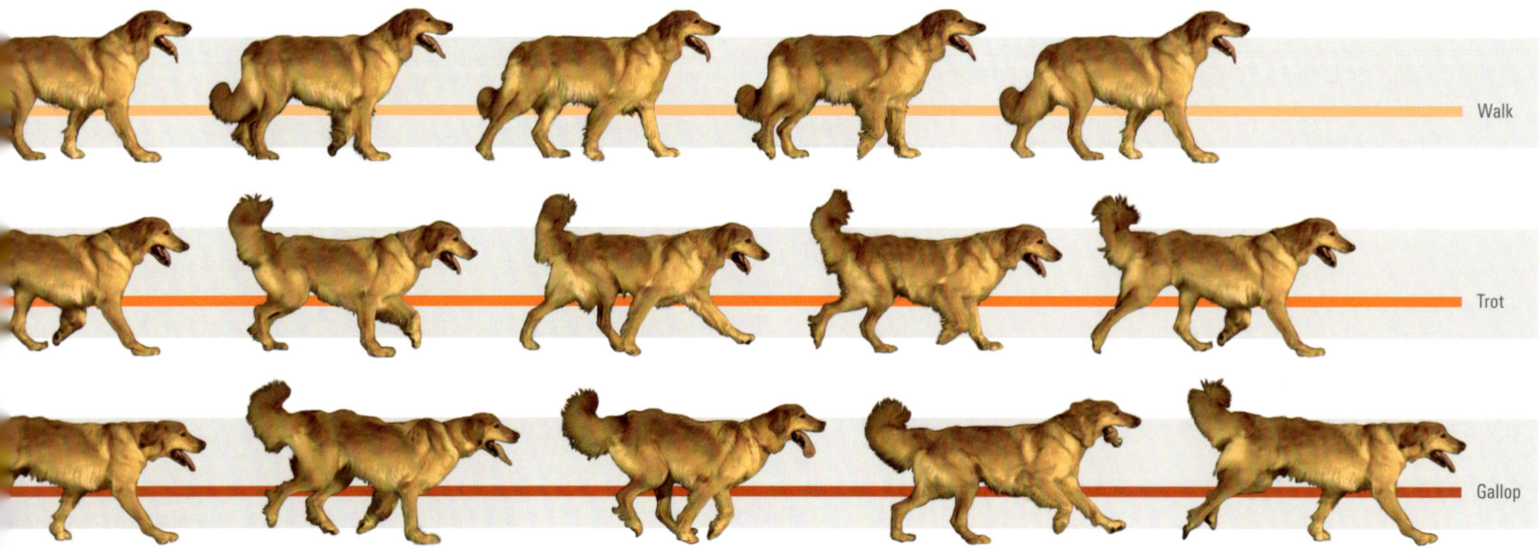

PROPORTIONS

As required by the standard, trunk length exceeded height at the withers in all the Hovawarts we investigated. A legginess index value of 1.5 points ties in with the relatively high chest observed in this breed. No sex-specific differences were detected. Overall, the limb segment proportions displayed by this breed were average in comparison with the other breeds in the study. The proportional length of the upper arm varies significantly within Hovawarts.

GAIT PARAMETERS

Relative stride length is long in almost every gait. The stride length of the trailing limb in the gallop, however, is shorter than that observed in many other breeds. With the exception of the gallop, where it is greater in the leading limb, the forwards excursion of the forelimbs does not differ between the gaits. Backward excursion differs slightly between the symmetrical gaits and the gallop, a phenomenon based on the movement of the forearm. The excursion of the hindlimbs is less uniform. Forward excursion is lowest in the trailing limb in the gallop and highest in the leading limb in the same gait. It is lower in the trot than the walk. The differences are caused by the accumulated movements of all the limb segments. Backwards excursion is greater in the gallop than in the walk and the trot, where the values do not differ. The difference is caused by a difference in the size of the movement of the femur.

Comparatively speaking, forelimb excursion is high. The value for the trailing forelimb in the gallop, however, is only middle-of-the-range. The forward excursion of the hindlimbs is low compared to that of other breeds, and very low when it comes to the trailing limb in the gallop. Backwards excursion, on the other hand, is fairly high, with values in all gaits falling into the top third of the range. The ground-covering gait demanded by the standard is observable in the forelimbs.

HOVAWART

DESCRIPTION OF BREED

The Hovawart was first included in a stud book in 1922, though the first attempts at breeding by Kurt Friedrich König began slightly earlier than that, just after the World War I. The Hovawart was recognized as an independent breed in 1937, and as a working dog in 1964. The Hovawart is thus a relatively young breed, though the name stems from the Middle High German. A reference to similar dogs is found in the Schwabenspiegel, a legal code dating from 1275, at which point Hovawarts were used as farm guard dogs. The breeding program which began after World War I focused on large, long to rough-haired farm dogs from the Harz region of Germany, crossed on several occasions with Alsatians, Kuvaszok, Leonbergers, Leonberger/Swiss Mountain Dog crosses and a Newfoundland, with the aim of obtaining a large, robust dog with long hair and lop ears. The correct colors are black, black and gold, or blond.

Wither's height is 63-70 cm in males and 58-65 cm in bitches. According to the standard, body length should exceed height at the withers by approximately 10 – 15 percent. The chest should be deep and the shoulder blade and upper arm long. The gait should be effortless and ground-covering.

Proportions

Body	♀		♂	
Number of dogs	6		2	
	MV	SD	MV	SD
Withers height [cm]	52	2.3	55	0.7
Body length / Withers height	1.1	0.0	1.1	0.0
Withers height / croup height	1.0	0.0	1.0	0.0
Legginess index	1.4	0.1	1.2	-

Forelimb	♀		♂	
Number of dogs	6		2	
	Proportion of total length			
	%	SD	%	SD
Shoulder blade	27	1.1	27	1.3
Upper arm	27	1.6	29	0.2
Lower arm	30	1.8	29	1.4
Pastern	16	1.8	14	2.9

Hindlimb	♀		♂	
Number of dogs	6		2	
	Proportion of total length			
	%	SD	%	SD
Thigh	37	2.3	37	0.8
Lower leg	39	2.3	39	2.7
Hind pastern	25	1.3	24	1.9

Gait parameters

Forelimb

	Stride length				Stride duration				Stance length				Stance duration				Swing length				Swing duration				Dutyfactor				Speed			
	m				s				m				s				m				s								m/s			
MV	0.79	1.02	1.58	1.58	0.63	0.47	0.38	0.37	0.46	0.47	0.48	0.52	0.38	0.22	0.13	0.13	0.33	0.55	0.84	1.03	0.25	0.25	0.25	0.24	60	47	33	35	1.26	2.16	4.15	4.32
SD	0.1	0.1	0.2	0.2	0.0	0.0	0.0	0.0	0.1	0.1	0.1	0.1	0.0	0.0	0.0	0.0	0.0	0.1	0.5	0.2	0.0	0.0	0.0	0.0	4	3	6	4	0.1	0.2	0.7	0.5
Number of strides					50	47	48	40	All numbers equal																							

Hindlimb

	Stride length				Stride duration				Stance length				Stance duration				Swing length				Swing duration				Dutyfactor				Speed			
	m				s				m				s				m				s								m/s			
MV	0.81	1.06	1.58	1.56	0.62	0.47	0.38	0.38	0.45	0.41	0.47	0.47	0.36	0.19	0.12	0.12	0.36	0.65	0.82	1.09	0.26	0.28	0.26	0.26	58	40	32	32	1.29	2.25	4.19	4.12
SD	0.1	0.1	0.2	0.2	0.0	0.0	0.0	0.0	0.1	0.0	0.0	0.1	0.0	0.0	0.0	0.0	0.1	0.1	0.5	0.2	0.0	0.0	0.0	0.0	6	3	4	4	0.1	0.2	0.7	0.6
Number of strides					46	46	43	45	All numbers equal																							

Kinematics

Forelimb	Shoulder blade				Shoulder joint				Upper arm				Elbow joint				Lower arm				Carpal joint				Pastern			
Touch down	*60*	*59*	*60*	*66*	145	147	142	149	75	78	72	74	128	133	131	125	54	54	59	51	189	191	199	189	135	135	142	139
Lift off	*103*	*103*	*110*	*114*	128	131	144	143	25	28	23	29	138	145	153	155	113	117	130	126	150	161	165	164	36	43	35	39
Mid-stance	-	-	-	-	136	136	133	140	-	-	-	-	127	122	127	124	-	-	-	-	189	191	199	189	-	-	-	-
Effective motion	*43*	*44*	*50*	*48*	17	16	-1	6	50	51	49	44	10	12	22	30	60	63	71	75	39	30	35	25	99	92	107	100
Stance amplitude	*46*	*47*	*54*	*54*	18	21	19	21	51	52	54	54	31	36	38	39	68	65	70	75	37	43	51	54	92	89	101	99
Swing amplitude	*44*	*47*	*60*	*53*	28	27	26	35	55	55	70	58	52	62	66	69	81	92	99	104	102	115	112	111	134	155	157	168
Number of strides	50	47	48	40	Values in italics are marker-based values corrected to reflect the X-Ray video and position data.																							

Hindlimb	Pelvis				Hip joint				Thigh				Stifle joint				Lower leg				Ankle joint				Hind pastern			
Touch down	46	42	45	42	102	107	107	96	57	65	62	53	146	148	151	149	89	82	89	96	135	134	141	138	45	51	51	40
Lift off	40	38	31	22	142	143	141	132	102	105	111	111	137	140	144	149	35	35	33	38	164	162	172	166	129	126	138	128
Mid-stance	-	-	-	-	104	107	107	95	-	-	-	-	145	148	152	150	-	-	-	-	133	135	142	139	-	-	-	-
Effective motion	6	4	14	21	41	36	34	37	45	40	48	57	9	7	8	1	54	47	56	58	29	27	31	28	84	75	87	88
Stance amplitude	16	10	20	22	43	35	37	38	47	39	47	57	18	19	29	22	54	46	56	60	50	53	67	70	86	76	87	94
Swing amplitude	10	11	19	26	47	42	49	47	48	47	52	66	44	58	55	68	70	75	91	100	53	63	62	64	88	85	100	105
Number of strides	46	46	43	45	Values in italics are marker-based values corrected to reflect the X-Ray video and position data.																							

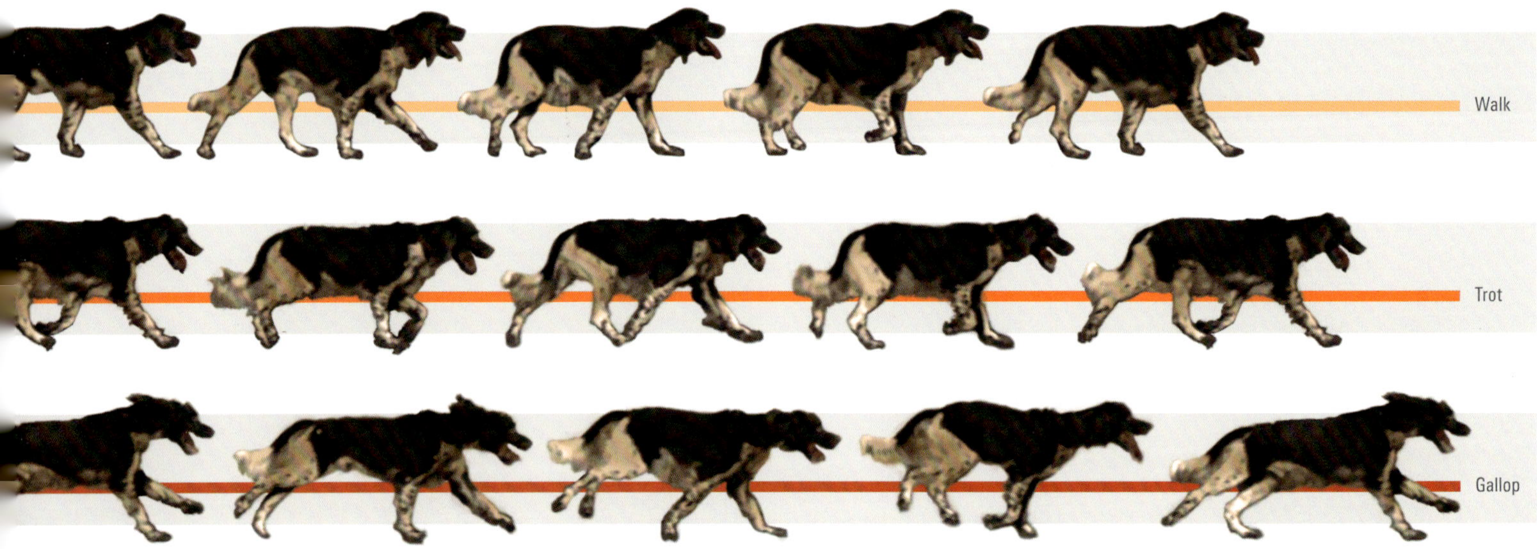

PROPORTIONS

Two of the animals we investigated were (according to the standard) too long in their trunk for their height at the withers, but this did not affect their limb segment proportions. No sex-specific differences were detected, and although the legginess index values indicate that the chest is less deep in bitches, this must be interpreted with caution as we were only able to take the relevant measurements for one male. The same applies to the length of the upper arm, which appears to be slightly greater in dogs than in bitches: the values here are based on measurements from just two dogs. All in all, the limb segment proportions of Small Münsterländers are average for the breeds involved in this study, though the upper arm in male dogs appears to be slightly longer. The length of the upper arm is much more variable in bitches than in dogs.

GAIT PARAMETERS

In the walk, this breed displays the greatest relative stride length of all the breeds we investigated. Its relative stride length in the trot is also high. The stride length of the trailing limbs in the gallop is middle-of-the-range, while that of the leading limbs is fairly high up the range. The degree of forward excursion of the forelimbs prior to touchdown does not differ between the gaits. Backwards excursion is equal in the walk and the trot and increases in the gallop thanks to the movement of the shoulder blade and upper arm. The backwards excursion of the trailing limb is greater than that of the leading limb. The same pattern is observed in the backwards excursion of the hindlimbs, this time as the result of the movement of the pelvis and femur. The forwards excursion of the hindlimbs is not quite so uniform. It is lowest in the trot, and in the gallop it is lower in the trailing limb than the leading limb. The forwards excursion of the leading limb in the gallop is greater than that observed in the walk. These differences are caused by differences in the movements of all the limb segments.
Compared to other breeds, the excursion of the forelimbs in the walk is high. In the gallop, the forwards excursion of the forelimbs is middle-of-the-range. The forwards excursion of the hindlimbs is moderate to low (trot), while backwards excursion is always high, except in the trot where it is middle-of-the-range. As is the case in all other breeds, the angle of the shoulder is never 90° during locomotion, but always exceeds this value. The ground-covering gait is clearly observed in the forelimbs during the walk and the trot.

SMALL MÜNSTERLÄNDER

DESCRIPTION OF BREED

Small Münsterländers are enthusiastic hunting dogs whose ancestors are still the subject of some debate. The most commonly held view is that the breed stems from the centuries old French breed Epagneul Breton, which was brought to Westphalia by officers in Napoleon's army. Small Münsterländers began to be purebred in the Münsterland region of Germany in the second half of the nineteenth century. Today's dogs can be traced back to two main lines, the Heitmann line and the Dorstener line, which had already existed for decades without being crossed with any other dogs. The Small Münsterländer Setter Club was set up in Germany in 1912, but the first breed standard not formulated until 1921.
According to the standard, height at the withers must be between 52 and 56 centimeters for dogs, while bitches can be two centimeters shorter. Body length should not exceed height at the withers by more than 5 centimeters, the upper arm should be as long as possible, meeting the shoulder blade at an angle of 90°, and movement should be expansive.

Proportions

Body	♀		♂	
Number of dogs	7		2	
	MV	SD	MV	SD
Withers height [cm]	70	1.7	73	1.1
Body length / Withers height	1.2	0.1	1.3	0.1
Withers height / croup height	1.0	0.0	1.0	0.0
Legginess index	1.2	0.1	1.2	0.2

Forelimb	♀		♂	
Number of dogs	7		2	
	Proportion of total length			
	%	SD	%	SD
Shoulder blade	28	1.4	31	1.7
Upper arm	28	1.2	26	2.0
Lower arm	29	2.0	27	1.0
Pastern	15	0.9	16	1.3

Hindlimb	♀		♂	
Number of dogs	7		2	
	Proportion of total length			
	%	SD	%	SD
Thigh	37	1.5	32	2.4
Lower leg	37	1.3	41	1.8
Hind pastern	25	1.8	27	0.5

Gait parameters

Forelimb

	Stride length (m)				Stride duration (s)				Stance length (m)				Stance duration (s)			
	1	2	3	4	1	2	3	4	1	2	3	4	1	2	3	4
MV	0.81	1.14	-	1.30	0.85	0.54	-	0.46	0.55	0.58	-	0.58	0.59	0.29	-	0.22
SD	0.1	0.1	-	0.2	0.1	0.0	-	0.0	0.1	0.1	-	0.0	0.1	0.0	-	0.0
Number of strides					104	132	0	16	All numbers equal							

	Swing length (m)				Swing duration (s)				Dutyfactor				Speed (m/s)			
	1	2	3	4	1	2	3	4	1	2	3	4	1	2	3	4
MV	0.27	0.56	-	0.72	0.27	0.25	-	0.24	69	53	-	48	0.96	2.11	-	2.82
SD	0.1	0.1	-	0.2	0.0	0.0	-	0.0	2.7	5.3	-	6.1	0.2	0.4	-	0.3
Number of strides	All numbers equal															

Hindlimb

	Stride length (m)				Stride duration (s)				Stance length (m)				Stance duration (s)			
	1	2	3	4	1	2	3	4	1	2	3	4	1	2	3	4
MV	0.85	1.16	1.31	1.24	0.86	0.54	0.47	0.46	0.56	0.55	0.51	0.57	0.58	0.27	0.20	0.22
SD	0.1	0.1	0.1	0.1	0.1	0.0	0.0	0.0	0.1	0.1	0.0	0.0	0.1	0.0	0.0	0.0
Number of strides					96	162	12	15	All numbers equal							

	Swing length (m)				Swing duration (s)				Dutyfactor				Speed (m/s)			
	1	2	3	4	1	2	3	4	1	2	3	4	1	2	3	4
MV	0.30	0.60	0.81	0.66	0.28	0.27	0.27	0.24	67	50	42	48	1.00	2.12	2.78	2.68
SD	0.1	0.1	0.2	0.1	0.0	0.0	0.0	0.0	2.9	5.2	4.3	3.6	0.1	0.4	0.2	0.2
Number of strides	All numbers equal															

Kinematics

Forelimb

	Shoulder blade				Shoulder joint				Upper arm				Elbow joint			
	1	2	3	4	1	2	3	4	1	2	3	4	1	2	3	4
Touch down	*65*	*68*	-	*72*	*142*	*147*	-	*153*	77	79	-	76	134	136	-	121
Lift off	*92*	*103*	-	*88*	*120*	*133*	-	*111*	28	30	-	28	125	133	-	135
Mid-stance	-	-	-	-	129	135	-	125	-	-	-	-	144	137	-	132
Effective motion	*27*	*30*	-	*16*	*22*	*14*	-	*41*	49	49	-	48	9	3	-	14
Stance amplitude	*30*	*37*	-	*34*	*22*	*20*	-	*42*	50	51	-	57	29	27	-	28
Swing amplitude	*28*	*36*	-	*33*	*26*	*25*	-	*47*	51	53	-	51	46	54	-	38
Number of strides	104	132	0	16	Values in italics are marker-based values corrected to reflect the X-Ray video and position data.											

	Lower arm				Carpal joint				Pastern			
	1	2	3	4	1	2	3	4	1	2	3	4
Touch down	57	58	-	45	184	187	-	187	128	131	-	143
Lift off	97	103	-	106	116	119	-	143	19	17	-	37
Mid-stance	-	-	-	-	184	187	-	187	-	-	-	-
Effective motion	40	45	-	62	68	68	-	44	109	115	-	106
Stance amplitude	66	59	-	67	76	84	-	71	106	113	-	109
Swing amplitude	64	75	-	74	88	103	-	104	121	141	-	152

Hindlimb

	Pelvis				Hip joint				Thigh				Stifle joint			
	1	2	3	4	1	2	3	4	1	2	3	4	1	2	3	4
Touch down	48	46	39	51	101	103	93	95	53	57	54	44	141	145	138	136
Lift off	48	41	44	37	*141*	*139*	*143*	*137*	*93*	*98*	*99*	*100*	120	128	125	134
Mid-stance	-	-	-	-	103	104	93	101	-	-	-	-	141	143	138	132
Effective motion	0	5	-5	14	*40*	*42*	*50*	*42*	*40*	*42*	*46*	*56*	21	17	14	2
Stance amplitude	16	11	10	18	*50*	*42*	*51*	*51*	*46*	*43*	*46*	*58*	20	19	30	27
Swing amplitude	8	11	13	18	*49*	*40*	*55*	*47*	*42*	*46*	*46*	*58*	40	55	54	39
Number of strides	96	162	13	14	Values in italics are marker-based values corrected to reflect the X-Ray video and position data.											

	Lower leg				Ankle joint				Hind pastern			
	1	2	3	4	1	2	3	4	1	2	3	4
Touch down	89	88	85	92	149	151	156	139	59	62	71	47
Lift off	32	35	30	39	153	161	170	167	121	127	140	128
Mid-stance	-	-	-	-	148	147	155	125	-	-	-	-
Effective motion	57	53	55	53	8	14	14	28	62	64	68	81
Stance amplitude	55	52	55	58	38	50	60	87	76	74	77	103
Swing amplitude	66	74	75	75	40	56	45	62	65	75	77	88

PROPORTIONS

There are more sex-specific differences in the limb segment proportions of Mastiffs than is the case for any other breed. The only segment for which proportional length does not differ between dogs and bitches is the front pastern. Bitches display a slightly shorter shoulder blade, a shorter lower leg and a shorter hind pastern, but a proportionally longer upper arm, forearm and femur than dogs. The differences are more striking in the femur and lower legs than in the forelimbs. Overall, the limb proportions seen in the bitches of this breed are closer to the cross-breed average than those of Mastiff males.

GAIT PARAMETERS

Relative stride lengths in this breed are, comparatively speaking, short, regardless of the gait. The forelimb stance phase is always middle-of-the-range but is canceled out by a relatively short swing phase. Both the stance and the swing phase are short in the hindlimbs. The forwards excursion of the forelimbs prior to touchdown only deviates, by being higher, in the leading limb in the gallop, in all the other gaits it is the same. This difference is down to differences in the movement of the forearm. Backwards excursion prior to lift off is equally low in the walk and in the leading limb in the gallop, slightly greater in the trot, thanks to increased movement in the shoulder blade, and greater still in the trailing limb in the gallop. When it comes to the hindlimbs, forwards excursion is lowest in the trailing limb in the gallop, slightly greater in the trot, greater still in the walk and greatest of all in the leading limb in the gallop. The same gradation is observed in the position of the pelvis and forearm. Backwards excursion is equal in the walk, the trot and the leading limb in the gallop, but is greater in the trailing limb in the gallop, mainly because the movement of the pelvis is also greater.

In comparative terms, the excursion of the forelimbs is generally middle-of-the-range, though backwards excursion is very short in the walk and, in particular, in the leading limb in the gallop. The excursion of the hindlimbs is always situated towards the bottom of the range, especially in the gallop where it is low in both directions.

MASTIFF

DESCRIPTION OF BREED

The ancestors of the Mastiff were probably brought to the British Isles with various waves of immigrants. Both the Celts and the Teutons kept large, mastiff-like dogs, and there is a very early written reference to mastiffs in the Welsh Laws, a statute book dating from the 9th century. At the beginning of the 11th century, mastiffs were being used for bear-baiting, and documents from the 16th century also refer to mastiffs in connection with the "pleasant sports" of bear-hunting and bear-baiting. Later, mastiffs were put to bait lions to entertain the gentry, and used in bull-fighting for the amusement of the common people. When dog-fighting became popular, however, mastiffs were deemed unsuitable because of their size. The dogs being bred today are almost all descendants of animals bred by one particular family which is said to have kept purebred Mastiffs for over 300 years. However, opinions differ as to the pedigree of these original dogs. It is documented to have started in either 1820 or 1825, and began with the crossing of a Mastiff and a Saint Bernard/Mastiff cross. The Mastiff Club was founded in 1872. Later on, to avoid inbreeding, Mastiffs were crossed with smooth-haired Saint Bernards, which, to dampen concerns about breed purity, were quickly termed Alpine Mastiffs.

The shoulder height is at minmum 76 cm for males and 69 cm for females. The Mastiff should move effortlessly, with long strides.

Proportions

Body	♀		♂	
Number of dogs	7		1	
	MV	SD	MV	SD
Withers height [cm]	45	1.7	50	-
Body length / Withers height	1.1	0.1	1.1	-
Withers height / croup height	1.0	0.0	1.1	-
Legginess index	1.1	0.1	1.3	-

Forelimb	♀		♂	
Number of dogs	7		1	
	Proportion of total length			
	%	SD	%	SD
Shoulder blade	28	1.4	28	-
Upper arm	27	2.2	26	-
Lower arm	30	0.9	32	-
Pastern	15	1.6	15	-

Hindlimb	♀		♂	
Number of dogs	7		1	
	Proportion of total length			
	%	SD	%	SD
Thigh	36	1.7	35	-
Lower leg	38	2.0	38	-
Hind pastern	25	2.9	27	-

Gait parameters

Forelimb

	Stride length (m)				Stride duration (s)				Stance length (m)				Stance duration (s)				Swing length (m)				Swing duration (s)				Dutyfactor				Speed (m/s)			
MV	0.59	0.82	1.08	-	0.61	0.43	0.35	-	0.34	0.35	0.39	-	0.36	0.20	0.14	-	0.25	0.47	0.69	-	0.25	0.24	0.22	-	59	46	38	-	0.97	1.89	3.00	-
SD	0.1	0.1	0.0	-	0.1	0.0	0.0	-	0.0	0.0	0.0	-	0.1	0.0	0.0	-	0.1	0.1	0.1	-	0.0	0.0	0.0	-	3.8	5.6	2.5	-	0.1	0.2	0.1	-
Number of strides					97	104	26	0	All numbers equal																							

Hindlimb

	Stride length (m)				Stride duration (s)				Stance length (m)				Stance duration (s)				Swing length (m)				Swing duration (s)				Dutyfactor				Speed (m/s)			
MV	0.58	0.80	0.99	0.98	0.60	0.44	0.36	0.36	0.34	0.31	0.37	0.37	0.36	0.18	0.14	0.14	0.24	0.48	0.62	0.62	0.24	0.25	0.22	0.22	60	42	40	40	0.97	1.82	2.73	2.73
SD	0.1	0.1	0.1	0.1	0.1	0.0	0.0	0.0	0.0	0.1	0.0	0.0	0.0	0.0	0.0	0.0	0.1	0.1	0.1	0.1	0.0	0.1	0.0	0.0	5.1	9.3	4.7	5.3	0.1	0.2	0.3	0.4
Number of strides					122	122	53	51	All numbers equal																							

Kinematics

Forelimb

Forelimb	Shoulder blade				Shoulder joint				Upper arm				Elbow joint				Lower arm				Carpal joint				Pastern			
Touch down	*58*	*62*	*60*	-	148	153	162	-	80	81	92	-	145	142	160	-	65	60	68	-	191	189	201	-	127	130	132	-
Lift off	92	99	104	-	131	136	148	-	39	37	44	-	158	154	180	-	118	117	136	-	165	161	173	-	47	44	37	-
Mid-stance	-	-	-	-	144	149	159	-	-	-	-	-	148	144	159	-	-	-	-	-	191	189	201	-	-	-	-	-
Effective motion	*34*	*37*	*44*	-	17	17	14	-	22	26	-47	-	13	12	20	-	53	56	68	-	26	29	28	-	38	46	-95	-
Stance amplitude	*36*	*39*	*46*	-	19	19	27	-	41	46	50	-	24	24	28	-	59	60	69	-	34	47	45	-	77	84	90	-
Swing amplitude	*35*	*39*	*45*	-	24	24	32	-	44	49	57	-	52	61	65	-	78	86	95	-	80	104	65	-	114	144	128	-
Number of strides	97	104	26	0	Values in italics are marker-based values corrected to reflect the X-Ray video and position data.																							

Hindlimb

Hindlimb	Pelvis				Hip joint				Thigh				Stifle joint				Lower leg				Ankle joint				Hind pastern			
Touch down	49	46	28	39	101	102	68	71	53	56	56	45	145	144	143	146	92	87	87	101	143	140	138	142	50	52	50	39
Lift off	43	38	22	24	*139*	*134*	*115*	*128*	*98*	*96*	*93*	*104*	133	132	116	140	40	41	28	41	160	160	157	159	121	119	128	118
Mid-stance	-	-	-	-	103	102	68	71	-	-	-	-	145	144	143	146	-	-	-	-	140	140	138	143	-	-	-	-
Effective motion	6	8	6	15	*38*	*32*	*31*	*44*	*45*	*40*	*37*	*59*	12	12	27	6	53	47	59	61	5	6	-19	-17	12	9	-78	-79
Stance amplitude	16	13	11	18	*39*	*34*	*33*	*43*	*46*	*40*	*37*	*59*	17	17	29	31	53	47	59	62	39	48	58	76	75	69	83	96
Swing amplitude	11	15	10	17	*41*	*40*	*32*	*49*	*47*	*46*	*40*	*63*	39	53	56	45	62	71	78	85	44	54	41	44	75	80	81	83
Number of strides	122	122	53	51	Values in italics are marker-based values corrected to reflect the X-Ray video and position data.																							

PON - POLSKI OWCZAREK NIZINNY

DESCRIPTION OF BREED

The Polish Lowland Sheepdog originates from the Bydogoszez area of Poland, where it was used as a guard dog as well as a sheep dog. Originally, it occurred in three sizes, but it was decided at the beginning of the pure-breeding program to concentrate on the medium-sized variety. PONs are distinguished by their stamina, weather-hardiness and aptitude for learning. Various authors claim that the breed is descended from Asian breeds. They base this claim on the similarity between the PON and the Tibetan Terrier and other Asian sheepdogs. Other argue that this similarity is the result of the dogs performing similar work under similar climatic conditions. It is known that the first time a PON was shown at a Polish pedigree dog show was in 1924. Pedigree dogs were something of a rarity in Poland after World War II, which caused the country's cynologists to dedicate more attention to native farm dogs. From then on, PONs appeared more and more frequently at dog shows, and in 1953 a standard was formulated on the basis of a male dog which the judges found to be optimum in both appearance and temperament. The first PON kennel was set up in Poland in 1946, and the breed was recognized by the FCI in 1963. Shortly after the World Dog Show in 1965, the first PONS were bred in East Germany, and around ten years later in West Germany too.
According to the breed standard, height at the withers is between 42 and 47 centimeters in bitches and between 45 and 52 centimeters in dogs. Body length should exceed height by a small amount (the ration of withers height to body length is ideally 9:10), the chest must be deep and the gait long and expansive.

PROPORTIONS

All the PONs we measured complied with the standard in terms of height at the withers. As only one male animal took part in the study, we are not able to draw any conclusions regarding sex-specific differences. Limb segment proportions in bitches are average for the breeds involved in this study. With regard to segment length variability within the breed, it can be said that the forearm, not the upper arm, is the least variable segment. Indeed, the proportional length of the upper arm varies the most of all the forelimb segments.

GAIT PARAMETERS

Compared to other breeds, the PON's relative stride lengths are middle-of-the-range in the walk and the trot and rather short in the gallop. Interestingly, the relative stance phase lengths seen in the symmetrical gaits are short while swing phase lengths are longer, and in the gallop the situation is reversed.
There is no difference between the gaits in the degree of forward excursion displayed by the forelimbs, but backwards excursion is greater in the trailing limb in the gallop than in either the trot or the walk, between which no difference is observed. The greatest backwards excursion of the forelimbs is brought about by the rotation of the shoulder blade and the movement of the forearm. In the hindlimbs, forwards excursion is lowest in the trot, slightly greater in the walk and the gallop (trailing limb) and highest of all in the leading limb in the gallop. These differences can be put down to differences in the movements of all the limb segments. Backwards excursion is greatest in the trailing limb in the gallop – there is no difference in the degree of backwards excursion observed in the other gaits.
Comparatively speaking, excursion in PONs is generally middle-of-the-range, though the backwards excursion of the hindlimbs is lower than average in the trot and the gallop. Our data do not indicate that the PON gait is any more expansive ground-covering than that of other breeds.

Proportions

Body	♀		♂	
Number of dogs	4		9	
	MV	SD	MV	SD
Withers height [cm]	48	2.1	49	2.3
Body length / Withers height	1.1	0.0	1.1	0.0
Withers height / croup height	1.0	0.0	1.0	0.0
Legginess index	1.2	0.1	1.2	0.1

Forelimb	♀		♂	
Number of dogs	4		9	
	Proportion of total length			
	%	SD	%	SD
Shoulder blade	27	2.1	27	1.2
Upper arm	27	0.9	27	0.8
Lower arm	29	0.8	29	1.7
Pastern	16	1.2	17	1.6

Hindlimb	♀		♂	
Number of dogs	4		9	
	Proportion of total length			
	%	SD	%	SD
Thigh	38	2.2	37	2.0
Lower leg	38	2.3	36	1.4
Hind pastern	24	1.1	27	1.0

Gait parameters

Forelimb

	Stride length				Stride duration				Stance length				Stance duration			
	m				s				m				s			
MV	0.66	0.89	1.35	1.31	0.54	0.41	0.36	0.35	0.37	0.40	0.44	0.46	0.33	0.20	0.12	0.13
SD	0.2	0.2	0.1	0.1	0.1	0.1	0.0	0.0	0.1	0.0	0.0	0.0	0.1	0.1	0.0	0.0
Number of strides					174	204	108	157	All numbers equal							

	Swing length				Swing duration				Dutyfactor				Speed			
	m				s								m/s			
MV	0.29	0.49	0.91	0.85	0.21	0.21	0.23	0.22	60	48	35	38	1.28	2.18	3.75	3.69
SD	0.2	0.1	0.1	0.1	0.0	0.0	0.0	0.0	8.2	7.5	2.3	3.0	0.6	0.5	0.3	0.3
Number of strides	All numbers equal															

Hindlimb

	Stride length				Stride duration				Stance length				Stance duration			
	m				s				m				s			
MV	0.67	0.85	1.24	1.31	0.53	0.41	0.35	0.34	0.37	0.35	0.42	0.44	0.32	0.18	0.13	0.12
SD	0.2	0.1	0.1	0.1	0.1	0.1	0.0	0.0	0.1	0.1	0.0	0.0	0.1	0.1	0.0	0.0
Number of strides					162	207	159	145	All numbers equal							

	Swing length				Swing duration				Dutyfactor				Speed			
	m				s								m/s			
MV	0.30	0.50	0.82	0.87	0.22	0.23	0.22	0.22	59	44	37	36	1.33	2.12	3.55	3.76
SD	0.1	0.1	0.1	0.1	0.0	0.0	0.0	0.0	8.1	8.7	4.4	3.0	0.6	0.4	0.5	0.3
Number of strides	All numbers equal															

Kinematics

Forelimb

Forelimb	Shoulder blade				Shoulder joint				Upper arm				Elbow joint			
Touch down	*67*	*62*	*61*	*65*	*148*	*147*	*143*	*151*	80	80	82	86	142	140	144	141
Lift off	98	104	116	105	134	139	147	145	36	35	31	40	140	146	158	155
Mid-stance	-	-	-	-	144	144	143	148	-	-	-	-	137	134	143	135
Effective motion	*31*	*42*	*55*	*40*	*14*	*8*	*4*	*6*	35	34	23	28	2	6	14	14
Stance amplitude	*34*	*40*	*56*	*47*	*15*	*12*	*13*	*17*	46	47	56	60	26	26	25	35
Swing amplitude	*31*	*39*	*60*	*48*	*20*	*20*	*20*	*34*	49	54	69	60	46	56	57	54
Number of strides	174	204	108	157	Values in italics are marker-based values corrected to reflect the X-Ray video and position data.											

Forelimb	Lower arm				Carpal joint				Pastern			
Touch down	62	60	62	55	190	192	198	197	128	133	137	142
Lift off	105	111	127	116	131	137	164	151	28	26	37	36
Mid-stance	-	-	-	-	191	194	198	197	-	-	-	-
Effective motion	43	51	65	60	59	56	34	46	81	80	49	75
Stance amplitude	58	58	67	63	60	63	60	64	98	102	98	104
Swing amplitude	65	79	95	93	87	105	87	106	119	138	128	147

Hindlimb

Hindlimb	Pelvis				Hip joint				Thigh				Stifle joint			
Touch down	41	39	37	48	103	105	104	99	62	66	67	55	153	155	156	159
Lift off	38	36	25	22	*140*	*141*	*136*	*132*	*101*	*106*	*110*	*114*	133	139	134	147
Mid-stance	-	-	-	-	110	110	110	123	-	-	-	-	145	148	144	143
Effective motion	4	3	12	25	*36*	*36*	*32*	*33*	*39*	*40*	*43*	*59*	20	15	22	11
Stance amplitude	11	7	21	28	*40*	*37*	*36*	*43*	*44*	*41*	*44*	*59*	19	19	26	25
Swing amplitude	7	9	18	28	*39*	*40*	*44*	*44*	*41*	*44*	*48*	*68*	43	55	61	70
Number of strides	162	207	158	145	Values in italics are marker-based values corrected to reflect the X-Ray video and position data.											

Hindlimb	Lower leg				Ankle joint				Hind pastern			
Touch down	91	89	89	104	144	144	139	147	52	55	50	43
Lift off	37	38	29	38	156	159	165	167	118	121	136	129
Mid-stance	-	-	-	-	131	129	118	107	-	-	-	-
Effective motion	54	51	60	65	0	5	-14	-19	66	65	86	86
Stance amplitude	53	49	59	63	38	48	69	80	72	69	90	99
Swing amplitude	64	73	93	100	40	55	60	72	70	77	93	102

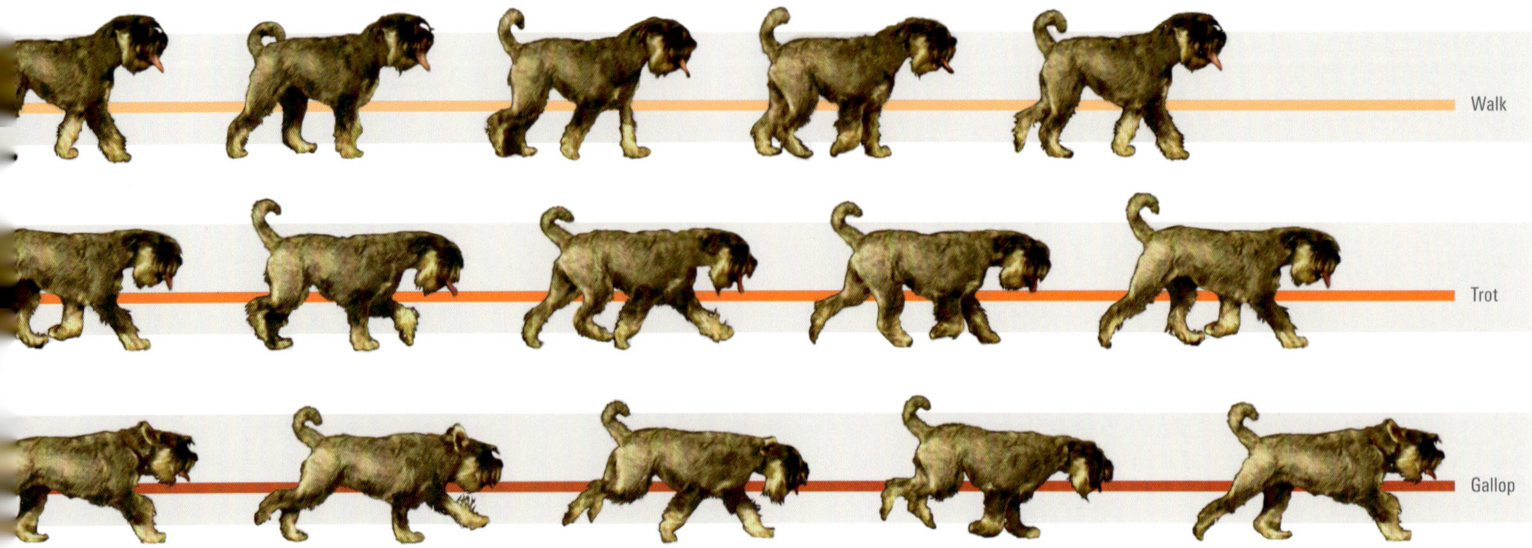

STANDARD SCHNAUZER

DESCRIPTION OF BREED

Standard Schnauzer originated in southern Germany and were originally only distinguishable from Pinschers by the length of their topcoat. Indeed, for a long time they were known alternatively as Rough-haired Pinschers. Now, however, the two breeds are quite distinct. Schnauzer were originally kept as stable dogs, to chase rats and mice, and by wagon drivers as guard dogs. Wagoners favoured rough-haired animals as they were better suited to keeping watch over the wagon even in the winter and in bad weather. It is not clear when exactly the name Schnauzer began to be used, but it first appears in written documents dating from 1842. Schnauzer were first shown at dog shows in 1879, the first standard was formulated in 1880, and five years later organized breeding began with the establishment of the Pinscher Club. The stable dog thus became a pedigree dog, though its name was not officially changed from Pinscher to Schnauzer until 1917.

Schnauzer should be square-built, and height at the withers should be between 45 and 50 centimeters. The shoulder blade should slope at an angle of 50° to the horizontal, forming an angle of 95 – 105° with the upper arm. The upper thigh should be moderately long, the shank long. The forelegs should swing as far forward as possible when the animal is moving.

PROPORTIONS

All the Standard Schnauzer we investigated complied with the breed standard in terms of height at the withers. The same cannot be said, however, for the ratio of trunk length to withers height. Trunk length exceeded height at the withers in all the individuals we measured. The femur and lower leg were equal in length, as required by the standard. The only sex-specific difference we observed in limb segment proportions relates to the hind pastern, which is slightly longer in dogs than in bitches. The proportional length of the upper arm hardly varies at all, and in bitches, that of the forearm is also very constant. The least variable limb segment in the hindlimbs is the hind pastern.

GAIT PARAMETERS

Comparatively speaking, the relative stride length of this breed is middle-of-the-range in almost all gaits. However, the relative stride length of the forelimbs in the trot is in the top third of the range, and that of the hindlimbs in the walk is in the bottom part of the range. The relative stride length of the leading hindlimb in the gallop is also low.

The forwards excursion in the forelimbs is greatest in the gallop (more so in the leading limb than the trailing limb) and lowest in the walk. All the joints – in various constellations – play a role in this. Backwards excursion in the forelimbs is lowest in the walk and highest in the trailing limb in the gallop. There is no difference between the degree of backwards excursion observed in the trot and that observed in the leading forelimb in the gallop. The gradation is caused by differences in the movements of the shoulder blade and forearm. When it comes to the hindlimbs, forwards excursion is lowest in the trot and in the trailing limb in the gallop. It is slightly greater in the walk and highest of all in the leading limb in the gallop. These differences can be put down to differences in the position of the pelvis and in the movements of the femur and lower leg. Backwards excursion is lowest in the walk, slightly higher in the trot and highest in the gallop, where the trailing limb is extended even further backwards than the leading limb.

In comparative terms, the relative excursion displayed by this breed falls into the middle of the range in all gaits. As in other breeds, the angle of the shoulder joint during locomotion always exceeds the 95 – 105° required by the breed standard.

Proportions

Body	♀		♂	
Number of dogs	8		3	
	MV	SD	MV	SD
Withers height [cm]	25	1.4	27	1.0
Body length / Withers height	1.7	0.1	1.6	0.1
Withers height / croup height	1.0	0.0	1.0	0.0
Legginess index	0.5	0.1	0.5	0.1

Forelimb	♀		♂	
Number of dogs	8		3	
	Proportion of total length			
	%	SD	%	SD
Shoulder blade	32	3.4	35	1.8
Upper arm	26	2.2	26	4.1
Lower arm	25	2.8	23	1.3
Pastern	16	2.2	16	1.7

Hindlimb	♀		♂	
Number of dogs	8		3	
	Proportion of total length			
	%	SD	%	SD
Thigh	38	4.5	35	4.2
Lower leg	33	2.8	36	2.8
Hind pastern	29	3.5	29	1.9

Gait parameters

Forelimb

	Stride length				Stride duration				Stance length				Stance duration				Swing length				Swing duration				Dutyfactor				Speed			
	m				s				m				s				m				s								m/s			
MV	0.69	0.90	-	-	0.41	0.32	-	-	0.41	0.42	-	-	0.25	0.16	-	-	0.28	0.48	-	-	0.16	0.16	-	-	61	49	-	-	1.65	2.79	-	-
SD	0.1	0.1	-	-	0.1	0.0	-	-	0.1	0.1	-	-	0.0	0.0	-	-	0.1	0.1	-	-	0.0	0.0	-	-	4.8	5.6	-	-	0.2	0.4	-	-
Number of strides					197	242	0	0	All numbers equal																							

Hindlimb

	Stride length				Stride duration				Stance length				Stance duration				Swing length				Swing duration				Dutyfactor				Speed			
	m				s				m				s				m				s								m/s			
MV	0.76	1.00	-	-	0.41	0.32	-	-	0.45	0.43	-	-	0.25	0.14	-	-	0.31	0.57	-	-	0.16	0.17	-	-	61	46	-	-	1.81	3.11	-	-
SD	0.1	0.1	-	-	0.1	0.0	-	-	0.1	0.1	-	-	0.0	0.0	-	-	0.1	0.1	-	-	0.0	0.0	-	-	6.0	7.9	-	-	0.2	0.4	-	-
Number of strides					218	275	0	0	All numbers equal																							

Kinematics

Forelimb	Shoulder blade				Shoulder joint				Upper arm				Elbow joint				Lower arm				Carpal joint				Pastern			
Touch down	68	68	-	-	138	136	-	-	71	69	-	-	143	143	-	-	73	74	-	-	207	209	-	-	134	135	-	-
Lift off	*94*	*96*	-	-	*143*	*143*	-	-	49	47	-	-	153	154	-	-	104	107	-	-	134	133	-	-	31	26	-	-
Mid-stance	-	-	-	-	*131*	*131*	-	-	-	-	-	-	161	157	-	-	-	-	-	-	207	209	-	-	-	-	-	-
Effective motion	*26*	*29*	-	-	*15*	*13*	-	-	21	21	-	-	10	11	-	-	31	33	-	-	73	76	-	-	104	108	-	-
Stance amplitude	*33*	*34*	-	-	*24*	*21*	-	-	29	28	-	-	26	25	-	-	49	45	-	-	82	84	-	-	106	107	-	-
Swing amplitude	*30*	*35*	-	-	*30*	*29*	-	-	34	37	-	-	32	36	-	-	55	61	-	-	91	100	-	-	130	141	-	-
Number of strides	197	242	0	0	Values in italics are marker-based values corrected to reflect the X-Ray video and position data.																							

Hindlimb	Pelvis				Hip joint				Thigh				Stifle joint				Lower leg				Ankle joint				Hind pastern			
Touch down	33	30	-	-	*114*	*112*	-	-	*61*	*63*	-	-	175	172	-	-	94	89	-	-	134	137	-	-	*59*	*67*	-	-
Lift off	26	20	-	-	*143*	*142*	-	-	*97*	*101*	-	-	140	146	-	-	43	45	-	-	154	161	-	-	111	116	-	-
Mid-stance	-	-	-	-	*126*	*124*	-	-	-	-	-	-	169	169	-	-	-	-	-	-	132	133	-	-	-	-	-	-
Effective motion	7	9	-	-	*29*	*30*	-	-	*36*	*39*	-	-	35	26	-	-	51	44	-	-	22	25	-	-	*52*	*49*	-	-
Stance amplitude	12	15	-	-	*30*	*29*	-	-	*41*	*41*	-	-	35	25	-	-	51	44	-	-	41	51	-	-	*59*	*55*	-	-
Swing amplitude	10	19	-	-	*35*	*39*	-	-	*40*	*45*	-	-	50	57	-	-	66	74	-	-	41	49	-	-	*63*	*72*	-	-
Number of strides	218	275	0	0	Values in italics are marker-based values corrected to reflect the X-Ray video and position data.																							

SCOTTISH TERRIER

DESCRIPTION OF BREED

Scottish Terriers as we know them today have only been around since about 1879. Not much is known about the origins of the breed, except that it comes from the Scottish Highlands and Islands. Scotland's four terrier breeds (Scottish, Cairn, Skye and West Highland White) are closely related, and the original Scottish Terriers were large and medium-sized dogs. The first mention in the literature of a short-legged Scottish terrier dates from 1847. From the description, however, the animal in question actually appears to be a Dandie Dinmont Terrier. A Scottish Terrier is mentioned again in 1888, and in this case the description fits today's breed very well. The breed standard was formulated the same year. The dogs were originally bred for fox and badger hunting. The first Scottish Terrier Club was founded in 1883 and the first Scottish Terriers shown in Germany in 1903. The breed was recognized in Germany in 1906.
According the standard, Scottish Terriers should have a deep chest and a good forward reach when moving. Height at the withers should not exceed 28 centimeters.

PROPORTIONS

None of the Scottish Terriers to take part in the study were larger than specified by the standard. The low legginess index values we obtained indicate a very deep chest, as is required by the standard. Certain sex-specific differences do exist: proportionally, the shoulder blade and lower leg are slightly longer in dogs than in bitches, while the femur is slightly longer in bitches than in dogs. Overall, segment proportions are highly variable. This applies in particular to the proportional length of the femur in both sexes, that of the shoulder blade in bitches and that of the upper arm in dogs.

GAIT PARAMETERS

Compared to the other breeds involved in this study, relative stride length in Scottish Terriers is short in all gaits. In the walk, the relative length of the forelimb stance phase is short while that of the swing phase is middle-of-the-range, and in the hindlimbs the situation is reversed. The relative stance phase, swing phase and stride lengths observed in the trot are all short.
There is no difference between the walk and the trot in terms of the degree of forwards excursion displayed by the forelimbs prior to touchdown. Backwards excursion is greater in the trot than the walk, and this is the result of the accumulated movements of all the limb segments. In the hindlimbs, forwards excursion is greater in the walk than the trot, a phenomenon in which the movements of all the limb segments play a role. Backwards excursion is greater in the trot than the walk, and again, all the limb segments play a role.
Like the relative stride length, the relative excursion displayed by this breed is also fairly low in comparative terms. This applies to both the forelimbs and the hindlimbs when it comes to forward excursion. The forwards excursion of the forelimb is particularly low in the trot.

Proportions: Body	♀		♂	
Number of dogs	5		4	
	MV	SD	MV	SD
Withers height [cm]	49	2.3	48	4.3
Body length / Withers height	1.1	0.0	1.2	0.1
Withers height / croup height	1.0	0.0	1.0	0.0
Legginess index	1.1	0.1	1.1	0.0

Forelimb	♀		♂	
Number of dogs	5		4	
	Proportion of total length			
	%	SD	%	SD
Shoulder blade	28	1.2	28	0.7
Upper arm	27	1.0	27	2.5
Lower arm	31	1.4	30	1.2
Pastern	14	1.5	15	1.4

Hindlimb	♀		♂	
Number of dogs	5		4	
	Proportion of total length			
	%	SD	%	SD
Thigh	38	0.8	39	2.0
Lower leg	36	1.6	35	3.3
Hind pastern	26	0.8	26	1.3

Gait parameters

Forelimb	Stride length				Stride duration				Stance length				Stance duration				Swing length				Swing duration				Dutyfactor				Speed			
	m				s				m				s				m				s								m/s			
MV	0.66	0.89	1.36	1.40	0.63	0.45	0.36	0.37	0.39	0.42	0.43	0.46	0.38	0.22	0.12	0.13	0.27	0.47	0.93	0.94	0.24	0.23	0.24	0.24	61	49	34	35	1.05	1.96	3.74	3.75
SD	0.0	0.1	0.1	0.1	0.1	0.0	0.0	0.0	0.0	0.0	0.0	0.0	0.0	0.0	0.0	0.0	0.0	0.0	0.1	0.1	0.0	0.0	0.0	0.0	3	3	3	3	0.1	0.2	0.2	0.3
Number of strides					109	160	84	38	All numbers equal																							

Hindlimb	Stride length				Stride duration				Stance length				Stance duration				Swing length				Swing duration				Dutyfactor				Speed			
	m				s				m				s				m				s								m/s			
MV	0.66	0.87	1.35	1.39	0.63	0.45	0.36	0.36	0.37	0.34	0.40	0.41	0.36	0.19	0.12	0.11	0.29	0.53	0.95	0.98	0.26	0.27	0.25	0.25	58	41	32	32	1.06	1.90	3.69	3.80
SD	0.0	0.0	0.1	0.1	0.1	0.0	0.0	0.0	0.0	0.0	0.1	0.0	0.0	0.0	0.0	0.0	0.0	0.1	0.2	0.1	0.0	0.0	0.0	0.0	3.0	4.3	5.8	3.7	0.1	0.1	0.3	0.3
Number of strides					104	142	88	80	All numbers equal																							

Kinematics

Forelimb	Shoulder blade				Shoulder joint				Upper arm				Elbow joint				Lower arm				Carpal joint				Pastern			
Touch down	*67*	*62*	*60*	*64*	150	150	150	155	78	83	90	75	138	141	152	127	60	58	62	51	189	185	191	178	129	128	129	127
Lift off	*102*	*105*	*111*	*109*	131	137	139	133	28	33	28	24	145	151	153	141	116	118	126	117	162	164	174	171	46	46	48	55
Mid-stance	-	-	-	-	139	140	139	133	-	-	-	-	138	134	145	127	-	-	-	-	189	185	191	178	-	-	-	-
Effective motion	*35*	*43*	*51*	*45*	20	13	11	21	50	50	62	51	7	10	2	14	56	60	64	66	27	21	17	7	83	82	81	73
Stance amplitude	*37*	*45*	*51*	*39*	21	22	22	25	49	52	63	55	24	29	23	22	61	62	63	65	29	43	41	45	78	82	80	76
Swing amplitude	*35*	*44*	*55*	*40*	28	22	27	40	53	53	71	61	55	60	52	55	80	89	87	94	92	103	90	118	121	134	136	156
Number of strides	152	160	84	38	Values in italics are marker-based values corrected to reflect the X-Ray video and position data.																							

Hindlimb	Pelvis				Hip joint				Thigh				Stifle joint				Lower leg				Ankle joint				Hind pastern			
Touch down	42	36	33	38	106	105	102	96	65	69	67	58	152	156	157	156	88	88	89	99	145	147	152	150	56	59	62	50
Lift off	37	32	36	25	141	135	154	141	105	104	114	117	144	146	146	154	39	43	32	37	163	163	156	157	124	120	125	120
Mid-stance	-	-	-	-	109	105	102	96	-	-	-	-	152	156	157	156	-	-	-	-	142	147	152	150	-	-	-	-
Effective motion	5	4	-3	13	35	31	53	45	41	35	47	59	8	10	11	2	49	45	58	62	19	16	11	12	68	61	63	70
Stance amplitude	14	9	15	16	35	31	54	53	41	34	47	59	14	18	26	21	48	45	58	63	53	57	64	76	81	69	68	88
Swing amplitude	9	15	22	21	37	35	58	52	41	41	50	65	45	62	69	68	68	78	102	105	53	65	60	64	75	77	96	96
Number of strides	117	162	88	80	Values in italics are marker-based values corrected to reflect the X-Ray video and position data.																							

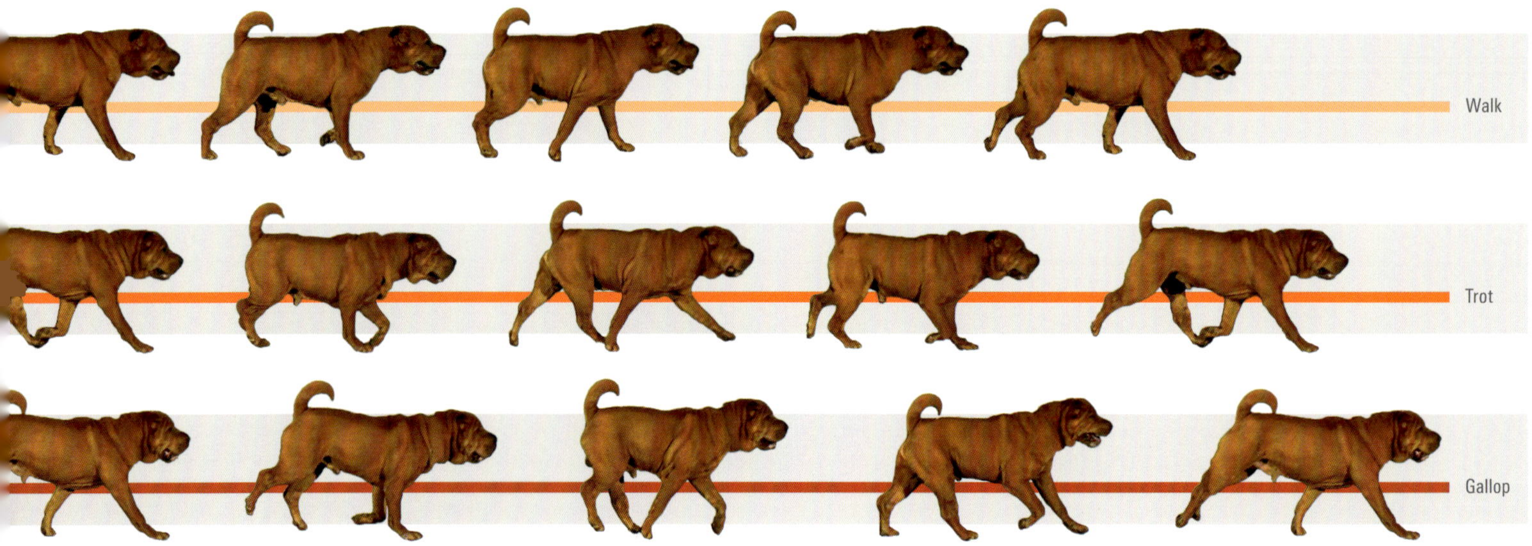

PROPORTIONS

One of the individuals we investigated was slightly smaller than specified by the standard, but this did not affect its limb segment proportions in any way. Limb segment and body proportions do not differ between the sexes. Overall, Shar Pei proportions were average for the breeds involved in this study. There is significant variability in the length of the upper arm in male dogs, while the length of the shoulder blade remains almost constant.

GAIT PARAMETERS

Comparatively speaking, relative stride length is middle-of-the-range in most gaits. That of the forelimbs in the walk and of the leading forelimb in the gallop is comparatively long, while the hindlimbs only display top-of-the-range values in the walk.

The forwards excursion of the forelimbs is lowest in the walk and slightly greater in the trot and the gallop. The differences are caused by a combination of movements in the various segments – no clear gradation is observed from segment to segment. Backwards excursion in the forelimbs is greater in the gallop than in the walk or the trot. The increase in excursion in the gallop is caused by a greater backwards rotation of the shoulder blade. Forwards excursion in the hindlimbs is greatest in the leading limb in the gallop (thanks to movement of the femur and lower leg), but apart from that does not differ at all between the gaits. In comparative terms, relative excursion values differ widely from gait to gait and from limb to limb in this breed. Forwards excursion in the forelimbs is middle-of-the-range, except in the leading limb in the gallop where it falls lower down the range. Backwards excursion is fairly high in all gaits. The forwards excursion of the hindlimbs is low, while backwards excursion is fairly high in the walk, fairly low in the trot and middle-of-the-range in the gallop.

SHAR PEI

DESCRIPTION OF BREED

Opinion is divided as to the origin of this breed. Some authors claim that the Shar Pei is a Chinese breed which has been around for thousands of years, but because no descriptions corresponding to these dogs appear in early texts or travel reports, most dog experts believe that the breed is much younger. Genetic investigations by Parker et al. (2004), on the other hand, support the assumption that it is indeed a very old breed. Shar Peis had many uses in China, and were put to work in the hunt for wild boar, as sheep dogs and as guard dogs. The ancestors of the Shar Pei were also used for dog fighting, as their bristly, very loose coats made it easy for them to escape the jaws of their opponents when bitten. However, they were too small to have much success against the much larger Bulldog crosses that were also used in the "sport", so their popularity as fighting dogs did not last long (Metz, 2010). The breed was almost wiped out in the 1950s when such a high dog tax was introduced that Shar Peis became a status symbol of the rich and were thus sought out and killed by the disgruntled poor. Only a few animals survived, in isolated villages in southern China, Macao, Hong Kong and Taiwan. In the 1970s, two breeders from Hong Kong decided to take their dogs to the USA to ensure that the breed survived. Numerous breeding programs were started in the USA, and in 1979 the first Shar Peis were introduced to Germany. The FCI recognized an interim standard in 1981, and a definitive standard in 1994.

Height at the withers should be between 44 and 51 centimeters, the chest deep and movement long and expansive.

Proportions

Body	♀		♂	
Number of dogs	7		1	
	MV	SD	MV	SD
Withers height [cm]	24	1.1	27	-
Body length / Withers height	1.9	0.1	1.8	-
Withers height / croup height	1.0	0.0	1.0	-
Legginess index	0.4	0.1	0.5	-

Forelimb	♀		♂	
Number of dogs	7		1	
	Proportion of total length			
	%	SD	%	SD
Shoulder blade	32	1.3	29	-
Upper arm	27	2.2	29	-
Lower arm	25	1.8	26	-
Pastern	16	2.0	15	-

Hindlimb	♀		♂	
Number of dogs	7		1	
	Proportion of total length			
	%	SD	%	SD
Thigh	39	3.0	37	-
Lower leg	31	1.8	33	-
Hind pastern	30	2.0	30	-

Gait parameters

Forelimb	Stride length				Stride duration				Stance length				Stance duration				Swing length				Swing duration				Dutyfactor				Speed			
	m				s				m				s				m				s								m/s			
MV	0.30	0.42	-	-	0.44	0.33	-	-	0.19	0.21	-	-	0.29	0.17	-	-	0.08	0.12	-	-	0.15	0.16	-	-	66	53	-	-	0.66	1.26	-	-
SD	0.1	0.1	-	-	0.0	0.0	-	-	0.0	0.0	-	-	0.0	0.0	-	-	0.1	0.1	-	-	0.0	0.0	-	-	3.7	4.2	-	-	0.1	0.2	-	-
Number of strides					71	48	0	0	All numbers equal																							

Hindlimb	Stride length				Stride duration				Stance length				Stance duration				Swing length				Swing duration				Dutyfactor				Speed			
	m				s				m				s				m				s								m/s			
MV	0.32	0.41	0.59	0.56	0.44	0.33	0.26	0.26	0.20	0.18	0.22	0.21	0.28	0.16	0.10	0.10	0.12	0.23	0.35	0.33	0.16	0.17	0.16	0.15	63	48	39	40	0.71	1.23	2.24	2.13
SD	0.1	0.1	0.0	0.0	0.0	0.0	0.0	0.0	0.0	0.0	0.0	0.0	0.0	0.0	0.0	0.0	0.0	0.1	0.1	0.1	0.0	0.0	0.0	0.0	3.9	6.9	1.8	1.9	0.1	0.3	0.1	0.1
Number of strides					106	120	30	34	All numbers equal																							

Kinematics

Forelimb	Shoulder blade				Shoulder joint				Upper arm				Elbow joint				Lower arm				Carpal joint				Pastern			
Touch down	62	59	-	-	140	136	-	-	77	77	-	-	145	139	-	-	*62*	*58*	-	-	207	211	-	-	141	149	-	-
Lift off	*96*	*98*	-	-	*140*	*137*	-	-	44	39	-	-	150	153	-	-	106	113	-	-	122	122	-	-	16	8	-	-
Mid-stance	-	-	-	-	135	133	-	-	-	-	-	-	147	138	-	-	-	-	-	-	207	211	-	-	-	-	-	-
Effective motion	*34*	*39*	-	-	*0*	*1*	-	-	33	37	-	-	6	14	-	-	*44*	*55*	-	-	85	90	-	-	125	141	-	-
Stance amplitude	*39*	*41*	-	-	*4*	*9*	-	-	42	47	-	-	30	33	-	-	62	66	-	-	91	103	-	-	124	137	-	-
Swing amplitude	*38*	*46*	-	-	*11*	*21*	-	-	40	47	-	-	44	50	-	-	63	74	-	-	110	137	-	-	140	175	-	-
Number of strides	71	48	0	0	Values in italics are marker-based values corrected to reflect the X-Ray video and position data.																							

Hindlimb	Pelvis				Hip joint				Thigh				Stifle joint				Lower leg				Ankle joint				Hind pastern			
Touch down	36	36	32	48	97	98	96	94	*61*	*62*	*64*	*56*	138	136	135	130	77	73	72	84	127	126	137	125	49	51	65	40
Lift off	30	32	24	31	*129*	*132*	*131*	*131*	*99*	*100*	*107*	*99*	106	111	108	112	22	25	11	22	145	148	141	144	123	123	130	122
Mid-stance	-	-	-	-	100	99	106	106	-	-	-	-	135	134	124	123	-	-	-	-	124	122	120	115	-	-	-	-
Effective motion	6	3	8	17	*32*	*34*	*35*	*37*	*38*	*38*	*44*	*54*	32	25	27	18	55	48	61	62	18	24	14	18	74	72	65	82
Stance amplitude	10	9	15	23	*37*	*37*	*37*	*41*	*38*	*41*	*43*	*55*	31	26	34	23	54	48	61	62	37	46	58	58	79	74	70	89
Swing amplitude	8	9	14	22	*38*	*39*	*39*	*47*	*42*	*44*	*45*	*58*	49	54	54	44	67	69	78	83	44	62	48	42	82	89	86	104
Number of strides	106	121	30	34	Values in italics are marker-based values corrected to reflect the X-Ray video and position data.																							

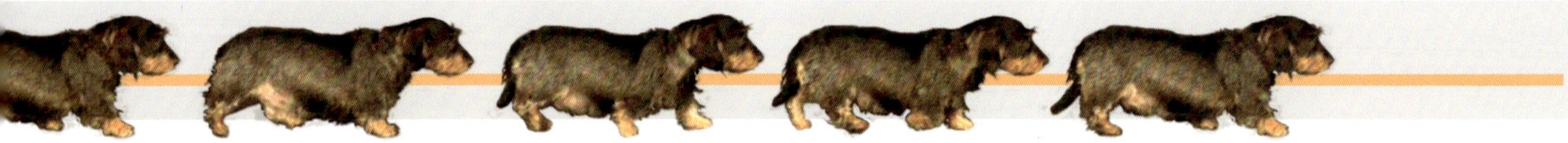
Walk

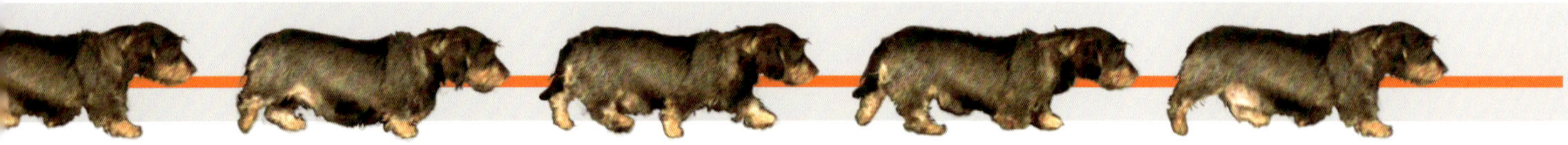
Trot

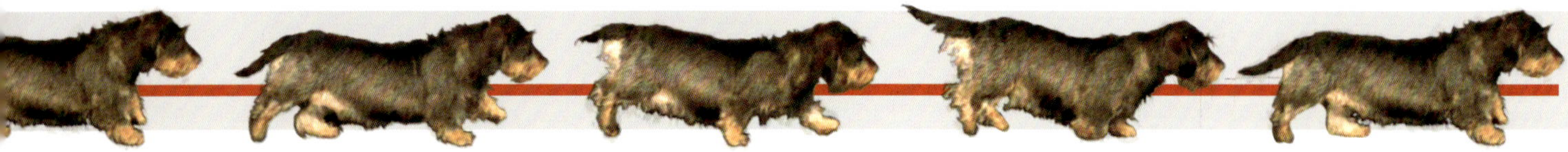
Gallop

DACHSHUND

DESCRIPTION OF BREED

The Dachshund is a hunting dog which was bred specifically to flush animals out of underground sets or dens but which also performs extremely well above ground. Descriptions of short-legged dogs being used in this way are found in French literature dating back to 1562, and the German text "The Perfect German Hunter" from 1727 contains a description of a short-legged dog very similar to a Dachshund which, like today's Dachshunds, descended from short-legged hounds. Although mentioned in very early texts from both England and France, it is thought that the Dachshund originated in Germany. The first German stud book of 1840 contains 54 Dachshunds, though Dachshund breeding was actually on the decline at this point in Germany. The systematic breeding of purebreds was eventually resumed by hunting enthusiasts among the nobility, and the first breed standard was laid down in 1879. The German Dachshund Club was founded in 1888. There are nine breeds of Dachshund, and the dogs are separated according to size (Standard, Dwarf and Kaninchen (rabbit)) and coat type (Short-haired, Wire-haired and Long-haired).

According to the standard, ground clearance should be around one third of the height at withers and body length should be in harmonious relation to height at withers, about 1 to 1.7 – 1.8. The shoulder blade should be equal in length to the upper arm, and the forearm short. In the hindquarters, the upper thigh and the pastern should be long, but the lower thigh short.

PROPORTIONS

As only one male Dachshund participated in the study, no conclusions can be drawn about sex-specific differences. All the individuals we investigated complied with the breed standard in terms of ground clearance, and the ratio of trunk length to height at the withers was, at an average of 1:9, also very close to the requirement of the standard. The proportional length of the upper arm is, however, smaller than that of the shoulder blade, and the forearm is even shorter, something observed in very few breeds. Compared to other breeds the Dachshund does not have a particularly long femur, but it does have a much shorter lower leg and a longer hind pastern.

GAIT PARAMETERS

Compared to that observed in other breeds, relative stride length in the Dachshund is fairly long in the walk and the trot but fairly short in the gallop.

The forwards excursion of the forelimbs prior to touchdown does not differ at all between the walk and the trot, while backwards excursion prior to liftoff is greater in the trot. This is caused by differences in the movements of the upper arm and forearm. In the hindlimbs, forwards excursion is equally low in the trot and the gallop (trailing limb), slightly greater in the walk and greatest of all in the leading limb in the gallop. The increase in excursion observed in the walk is down to the movement of the lower leg, while that observed in the gallop is caused by the movements of all the limb segments. Backwards excursion in the hindlimbs does not differ between the walk and the trot but is greater in the gallop, where the trailing limb extends even further backwards than the leading limb. The differences are down to differences in the movement of all the limb segments. In a cross-breed comparison, relative excursion in Dachshunds falls into the middle of the range in all gaits.

Proportions

Body	♀		♂	
Number of dogs	7		4	
	MV	SD	MV	SD
Withers height [cm]	40	2.4	43	2.9
Body length / Withers height	1.1	0.1	1.1	0.1
Withers height / croup height	1.0	0.0	1.0	0.0
Legginess index	1.3	0.2	1.4	0.1

Forelimb	♀		♂	
Number of dogs	7		4	
	Proportion of total length			
	%	SD	%	SD
Shoulder blade	27	1.0	26	1.0
Upper arm	26	1.1	27	0.2
Lower arm	30	2.8	31	2.1
Pastern	16	2.0	16	1.3

Hindlimb	♀		♂	
Number of dogs	7		4	
	Proportion of total length			
	%	SD	%	SD
Thigh	37	1.4	37	1.4
Lower leg	37	1.2	38	2.0
Hind pastern	26	1.6	25	1.8

Gait parameters

Forelimb

	Stride length				Stride duration				Stance length				Stance duration			
	m				s				m				s			
MV	0.51	0.68	1.08	-	0.63	0.41	0.32	-	0.31	0.33	0.40	-	0.39	0.20	0.13	-
SD	0.1	0.1	0.1	-	0.1	0.0	0.0	-	0.0	0.0	0.0	-	0.1	0.0	0.0	-
Number of strides					149	222	25	0	All numbers equal							

	Swing length				Swing duration				Dutyfactor				Speed			
	m				s								m/s			
MV	0.66	0.70	0.68	-	0.24	0.21	0.19	-	62	50	40	-	0.82	1.65	3.33	-
SD	0.2	0.2	0.1	-	0.0	0.0	0.0	-	3.5	3.8	3.4	-	0.1	0.2	0.4	-

Hindlimb

	Stride length				Stride duration				Stance length				Stance duration			
	m				s				m				s			
MV	0.52	0.68	1.13	-	0.63	0.41	0.32	-	0.31	0.29	0.37	-	0.38	0.18	0.11	-
SD	0.1	0.1	0.1	-	0.1	0.0	0.0	-	0.0	0.0	0.0	-	0.0	0.0	0.0	-
Number of strides					147	228	26	0	All numbers equal							

	Swing length				Swing duration				Dutyfactor				Speed			
	m				s								m/s			
MV	0.64	0.75	0.76	-	0.24	0.23	0.21	-	61	44	35	-	0.83	1.65	3.44	-
SD	0.3	0.3	0.1	-	0.0	0.0	0.0	-	2.7	4.5	1.9	-	0.1	0.2	0.3	-

Kinematics

Forelimb	Shoulder blade				Shoulder joint				Upper arm				Elbow joint			
Touch down	*62*	*62*	56	-	151	151	173	-	79	79	117	-	142	141	183	-
Lift off	*97*	*103*	100	-	132	139	161	-	35	36	61	-	142	144	187	-
Mid-stance	-	-	-	-	140	144	172	-	-	-	-	-	147	140	183	-
Effective motion	*35*	*41*	44	-	20	12	12	-	44	43	56	-	0	3	3	-
Stance amplitude	*38*	*44*	44	-	20	17	46	-	44	44	68	-	23	30	42	-
Swing amplitude	*37*	*44*	45	-	24	22	33	-	47	48	65	-	43	56	67	-
Number of strides	149	222	25	0	Values in italics are marker-based values corrected to reflect the X-Ray video and position data.											

Forelimb	Lower arm				Carpal joint				Pastern			
Touch down	63	62	67	-	189	189	200	-	127	128	133	-
Lift off	108	108	126	-	146	138	133	-	38	30	7	-
Mid-stance	-	-	-	-	189	189	200	-	-	-	-	-
Effective motion	45	46	59	-	44	51	67	-	89	97	126	-
Stance amplitude	58	56	63	-	42	60	102	-	84	94	140	-
Swing amplitude	69	79	100	-	87	106	114	-	115	138	157	-

Hindlimb	Pelvis				Hip joint				Thigh				Stifle joint			
Touch down	59	52	54	-	117	116	111	-	58	64	57	-	145	145	142	-
Lift off	47	41	34	-	147	145	145	-	100	103	111	-	134	138	138	-
Mid-stance	-	-	-	-	118	117	111	-	-	-	-	-	144	145	140	-
Effective motion	12	10	20	-	30	29	34	-	41	39	54	-	11	7	4	-
Stance amplitude	15	11	26	-	31	29	39	-	43	39	53	-	15	14	28	-
Swing amplitude	12	15	20	-	35	32	46	-	44	45	58	-	40	53	63	-
Number of strides	147	228	26	0	Values in italics are marker-based values corrected to reflect the X-Ray video and position data.											

Hindlimb	Lower leg				Ankle joint				Hind pastern			
Touch down	87	81	85	-	142	139	153	-	54	58	67	-
Lift off	35	35	27	-	158	160	171	-	123	125	144	-
Mid-stance	-	-	-	-	141	137	151	-	-	-	-	-
Effective motion	52	46	57	-	17	21	19	-	70	67	77	-
Stance amplitude	52	46	58	-	35	47	69	-	73	70	83	-
Swing amplitude	67	75	96	-	43	57	47	-	75	79	95	-

TIBETAN TERRIER

DESCRIPTION OF BREED

Whatever its name may suggest, the Tibetan Terrier is not a terrier but a true herding dog. In its country of origin, Tibet, the breed was often kept on farms and in monasteries. Originally, the distinction between Tibetan Terriers and Lhasa Apsos was very blurred. The first breed standard was formulated around 1900, though the breed was still relatively unknown in Europe at that time. In 1929, the main difference between the Lhasa Apso and the Tibetan Terrier was one of size alone. Nowadays, however, the two breeds are clearly separate. The Tibetan Terrier population in Europe can be traced back to two dogs which an English doctor received as a gift from a Tibetan in India and brought back to the continent. The breed was recognized by the English Kennel Club in 1931.

The standard is very general when it comes to limb proportions and angulation. Height at the withers should be between 36 and 41 centimeters for dogs, while bitches should be slightly smaller.

PROPORTIONS

One of the Tibetan Terriers in the study was significantly larger than permitted by the standard, and unlike in most dogs which violate the size requirements of their breed standard, we observed some clear differences in the limb segment proportions of this individual compared to those of other Tibetan Terriers. The animal in question had a proportionally shorter forearm and a proportionally shorter lower leg than its fellows. The limb segment proportions of the other dogs were average in a cross-breed comparison, no idiosyncrasies were detected and no sex-specific differences were observed. The proportional length of the upper arm is very constant in male Tibetan Terriers.

GAIT PARAMETERS

Comparatively speaking, the relative stride length values for this breed are middle-of-the-range in the walk (both for the forelimbs and the hindlimbs) and lie in the bottom third of the range for all the other gaits.

The forwards excursion of the forelimbs does not differ between the walk and the trot but increases in the gallop due to the position of the shoulder blade and upper arm. Backwards excursion is greater in the walk and the trot than it is in the gallop, and this is due to differences in the movements of the upper and forearm. Forwards excursion in the hindlimbs is greatest in the walk and lowest in the trot. All the segments play a role in bringing about this difference. Backwards excursion in the hindlimbs is equal in the walk and the trot but increases in the gallop thanks to an increase in the movement of all the limb segments.

Comparatively speaking, relative excursion values differ significantly from limb to limb and from gait to gait. The backwards excursion of the forelimbs is low in the walk and the trot but very high in the gallop. Backwards excursion is only very low in the gallop, otherwise values fall into the middle of the range. Forwards excursion in the hindlimbs is always very low, while backwards excursion is always middle-of-the-range.

Proportions

Body	♀		♂	
Number of dogs	11		3	
	MV	SD	MV	SD
Withers height [cm]	60	2.2	67	2.5
Body length / Withers height	1.2	0.0	1.2	0.0
Withers height / croup height	1.0	0.0	1.0	0.0
Legginess index	1.2	0.1	1.2	0.1

Forelimb	♀		♂	
Number of dogs	11		3	
	Proportion of total length			
	%	SD	%	SD
Shoulder blade	27	1.4	27	0.9
Upper arm	27	1.1	27	1.4
Lower arm	31	1.2	32	0.5
Pastern	16	1.5	13	0.2

Hindlimb	♀		♂	
Number of dogs	11		3	
	Proportion of total length			
	%	SD	%	SD
Thigh	37	2.2	37	1.5
Lower leg	37	1.8	39	1.3
Hind pastern	26	1.8	25	0.8

Gait parameters

Forelimb

	Stride length				Stride duration				Stance length				Stance duration			
	m				s				m				s			
MV	0.79	1.17	1.77	1.64	0.75	0.48	0.41	0.40	0.51	0.58	0.68	0.67	0.50	0.25	0.16	0.17
SD	0.1	0.1	0.2	0.4	0.1	0.0	0.0	0.0	0.1	0.1	0.0	0.1	0.1	0.0	0.0	0.0
Number of strides					105	173	22	21	All numbers equal							

	Swing length				Swing duration				Dutyfactor				Speed			
	m				s								m/s			
MV	0.27	0.59	1.09	0.96	0.25	0.23	0.25	0.23	67	52	40	43	1.05	2.45	4.34	4.05
SD	0.0	0.1	0.2	0.3	0.0	0.0	0.0	0.0	2.1	3.2	2.3	4.8	0.1	0.4	0.6	0.8
Number of strides	All numbers equal															

Hindlimb

	Stride length				Stride duration				Stance length				Stance duration			
	m				s				m				s			
MV	0.79	1.17	1.76	1.87	0.76	0.47	0.41	0.40	0.53	0.54	0.62	0.64	0.52	0.23	0.15	0.15
SD	0.1	0.2	0.1	0.4	0.1	0.0	0.0	0.0	0.1	0.1	0.0	0.1	0.1	0.0	0.0	0.0
Number of strides					94	121	20	50	All numbers equal							

	Swing length				Swing duration				Dutyfactor				Speed			
	m				s								m/s			
MV	0.26	0.62	1.14	1.23	0.24	0.24	0.26	0.26	69	49	36	37	1.04	2.46	4.28	4.58
SD	0.1	0.1	0.1	0.3	0.0	0.0	0.0	0.0	2.8	3.2	1.4	3.4	0.1	0.5	0.3	0.9
Number of strides	All numbers equal															

Kinematics

Forelimb

Forelimb	Shoulder blade				Shoulder joint				Upper arm				Elbow joint			
Touch down	*61*	*62*	63	58	138	144	157	144	76	82	94	86	136	137	155	129
Lift off	*93*	*96*	125	84	116	122	152	119	29	26	28	35	138	144	158	155
Mid-stance	-	-	-	-	126	123	148	116	-	-	-	-	138	122	151	126
Effective motion	*32*	*34*	61	26	22	22	5	25	48	56	67	50	2	7	3	26
Stance amplitude	*34*	*38*	62	35	21	29	33	31	47	57	72	60	31	37	33	40
Swing amplitude	*31*	*34*	65	36	26	35	36	54	48	58	71	58	46	66	67	65
Number of strides	106	173	22	21	Values in italics are marker-based values corrected to reflect the X-Ray video and position data.											

Forelimb	Lower arm				Carpal joint				Pastern			
Touch down	60	55	61	44	183	189	190	183	123	133	130	139
Lift off	110	118	131	120	133	140	153	141	25	23	23	21
Mid-stance	-	-	-	-	183	189	190	183	-	-	-	-
Effective motion	50	62	69	76	49	48	38	42	98	110	108	118
Stance amplitude	69	69	70	78	56	58	64	51	97	105	112	108
Swing amplitude	68	90	98	101	81	108	103	120	109	143	149	178

Hindlimb

Hindlimb	Pelvis				Hip joint				Thigh				Stifle joint			
Touch down	55	53	43	62	101	103	86	101	45	49	43	35	128	135	132	132
Lift off	49	44	33	39	141	144	126	143	91	99	93	103	114	127	118	133
Mid-stance	-	-	-	-	101	103	86	102	-	-	-	-	128	135	132	132
Effective motion	6	9	10	23	40	41	41	42	45	50	50	69	14	8	14	0
Stance amplitude	18	14	28	26	43	42	51	54	49	50	49	69	16	20	30	30
Swing amplitude	7	14	22	28	42	45	49	52	46	54	55	78	34	53	48	52
Number of strides	94	121	20	50	Values in italics are marker-based values corrected to reflect the X-Ray video and position data.											

Hindlimb	Lower leg				Ankle joint				Hind pastern			
Touch down	83	86	89	97	130	136	150	138	47	50	60	39
Lift off	23	28	25	29	143	154	165	160	120	126	139	131
Mid-stance	-	-	-	-	129	134	148	139	-	-	-	-
Effective motion	59	57	64	68	14	22	15	22	73	77	79	91
Stance amplitude	59	57	64	72	43	56	59	84	82	81	85	109
Swing amplitude	63	75	84	91	38	62	49	58	74	83	92	102

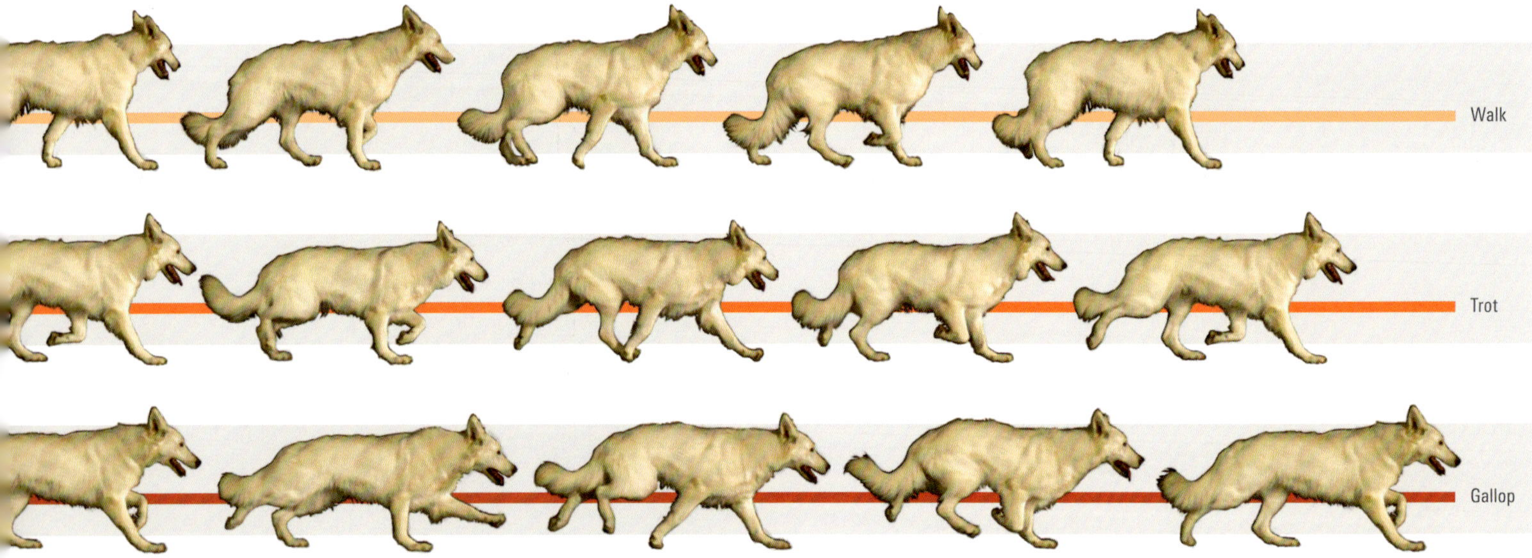

PROPORTIONS

One bitch and one dog were larger than permitted by the standard, but their limb segment proportions were not affected. Trunk length exceeded height at the withers in all individuals and, as the legginess index shows, on average, chest depth was slightly less than half of withers height. The shoulder blade and upper arm are not longer in this breed than in other breeds, but absolutely average. No obvious sex-specific differences were observed except for the fact that the front pastern is proportionally longer in bitches than in dogs, while the lower leg is slightly shorter. However, it should be pointed out that more bitches than dogs were represented in the study. Interestingly, the upper arm is the least variable forelimb segment in bitches and the most variable in dogs.

GAIT PARAMETERS

Comparatively speaking, relative stride length is moderate in the walk and long in the other gaits. The relative stride length of the hindlimbs in the gallop is very long.

Forelimb excursion in both directions is greater in the trot than the walk and greater still in the gallop. Forwards excursion is equal in the trot and the trailing limb in the gallop, while backwards excursion is greater in the leading limb in the gallop than it is in the trot, and is greatest of all in the trailing limb in the gallop. These differences are mainly due to differences in the movements of the proximal limb segments. In the hindlimbs, forwards excursion does not differ between the walk and the trot but is greater in the gallop, with the highest values being obtained by the trailing limb. Backwards excursion is also approximately equal in the walk and the trot, and when it comes to the gallop the trailing limb is extended further backwards than the leading limb.

Compared to that observed in other breeds, relative excursion in White Swiss Shepherds is high in both the forelimbs and the hindlimbs in all gaits. The breed standard requires the front legs to reach out far, and this is certainly the case.

WHITE SWISS SHEPHERD DOG

DESCRIPTION OF BREED

White Swiss Shepherd Dogs are descended from German Shepherds. White Shepherds are not albinos, and neither do they have a weaker constitution than German Shepherds. In the early days of the German Shepherd breeding program white individuals occurred on a regular basis. After years of debate, the color white became grounds for disqualification under the breed standard for German Shepherds in Germany, with the American Kennel Club eventually following the example of its German counterpart. The only Kennel Club which continued to recognize white as a standard color was that of Canada. In 1970, the first White Shepherds returned to Europe via Switzerland. In 1984 a renewed attempt was made to establish the breed in Switzerland, and in 1991 the "White Shepherd Dogs Switzerland" society was accepted as a breed club by the Swiss Cynological Society. In order to create a solid breeding base for the program, out of 40 dogs being evaluated on criteria of appearance and temperament, 36 were entered into the stud Book. As Switzerland was the first country to boast 8 independent breeding lines, it was Switzerland which petitioned the FCI to recognize the breed, which in turn is how the dogs got their name. The breed has been provisionally recognized by the FCI since 2002.

Body length should be slightly greater than height (12:10) in White Swiss Shepherds, with the chest reaching to approximately half of withers height. Both the shoulder blade and upper arm should be long, and the dogs should display a good forward reach. Height at the withers should be between 60 and 66 centimeters in dogs and between 55 and 61 centimeters in bitches.

Proportions

Body	♀		♂	
Number of dogs	6		5	
	MV	SD	MV	SD
Withers height [cm]	28	1.2	30	1.9
Body length / Withers height	1.8	0.0	1.8	0.1
Withers height / croup height	1.0	0.1	1.1	0.0
Legginess index	0.6	0.1	0.6	0.0

Forelimb	♀		♂	
Number of dogs	6		5	
	Proportion of total length			
	%	SD	%	SD
Shoulder blade	34	1.8	33	4.7
Upper arm	28	3.3	29	1.3
Lower arm	24	2.8	23	2.8
Pastern	15	2.9	14	3.6

Hindlimb	♀		♂	
Number of dogs	6		5	
	Proportion of total length			
	%	SD	%	SD
Thigh	38	2.0	38	3.1
Lower leg	33	2.7	33	2.7
Hind pastern	29	2.2	29	3.3

Gait parameters

Forelimb

	Stride length				Stride duration				Stance length				Stance duration			
	m				s				m				s			
MV	0.39	0.50	0.69	0.72	0.53	0.38	0.28	0.28	0.24	0.24	0.25	0.27	0.33	0.19	0.11	0.11
SD	0.0	0.1	0.1	0.1	0.1	0.0	0.0	0.0	0.0	0.0	0.0	0.0	0.0	0.0	0.0	0.0
Number of strides					96	161	65	68	All numbers equal							

	Swing length				Swing duration				Dutyfactor				Speed			
	m				s								m/s			
MV	0.10	0.21	0.26	0.34	0.20	0.20	0.17	0.17	63	49	40	40	0.74	1.30	2.42	2.55
SD	0.1	0.1	0.2	0.2	0.1	0.0	0.0	0.0	9	7.0	5	5.6	0.1	0.1	0.4	0.4
Number of strides	All numbers equal															

Hindlimb

	Stride length				Stride duration				Stance length				Stance duration			
	m				s				m				s			
MV	0.39	0.48	0.71	0.76	0.54	0.38	0.29	0.28	0.24	0.22	0.25	0.27	0.34	0.19	0.11	0.11
SD	0.1	0.1	0.1	0.1	0.0	0.0	0.0	0.0	0.0	0.0	0.0	0.0	0.0	0.0	0.0	0.0
Number of strides					124	198	70	89	All numbers equal							

	Swing length				Swing duration				Dutyfactor				Speed			
	m				s								m/s			
MV	0.15	0.26	0.37	0.40	0.20	0.20	0.18	0.17	63	49	37	38	0.72	1.23	2.46	2.69
SD	0.1	0.1	0.2	0.2	0.1	0.0	0.0	0.0	7.6	5.7	5.9	5.2	0.1	0.2	0.6	0.4
Number of strides	All numbers equal															

Kinematics

Forelimb

Forelimb	Shoulder blade				Shoulder joint				Upper arm				Elbow joint			
Touch down	*64*	*63*	60	71	137	136	134	137	*78*	*78*	73	66	128	130	149	121
Lift off	*95*	*97*	106	102	121	128	129	133	*31*	*31*	24	31	145	137	144	138
Mid-stance	-	-	-	-	131	133	129	134	-	-	-	-	141	133	150	131
Effective motion	*31*	*34*	46	31	16	9	4	4	*37*	*37*	30	1	16	7	5	16
Stance amplitude	*32*	*35*	47	47	19	20	23	21	*50*	*51*	57	43	35	28	29	30
Swing amplitude	*34*	*37*	52	47	24	29	28	28	*54*	*56*	63	57	48	49	47	28
Number of strides	96	161	65	68												

Forelimb	Lower arm				Carpal joint				Pastern			
Touch down	60	62	76	55	195	197	212	196	136	136	136	141
Lift off	113	106	121	107	146	134	140	133	33	28	20	27
Mid-stance	-	-	-	-	196	197	212	196	-	-	-	-
Effective motion	53	44	45	51	50	64	72	63	-86	-60	-64	6
Stance amplitude	68	56	55	64	56	76	82	83	99	107	116	111
Swing amplitude	75	73	77	76	86	95	92	93	121	131	139	140

Values in italics are marker-based values corrected to reflect the X-Ray video and position data.

Hindlimb

Hindlimb	Pelvis				Hip joint				Thigh				Stifle joint			
Touch down	42	43	48	46	*101*	*105*	*94*	*91*	*58*	*59*	*45*	*45*	*143*	*144*	*126*	*130*
Lift off	34	32	22	22	*137*	*139*	*125*	*136*	*98*	*102*	*98*	*109*	*119*	*124*	*116*	*128*
Mid-stance	-	-	-	-	*112*	*104*	*97*	*114*	-	-	-	-	*125*	*130*	*103*	*112*
Effective motion	8	12	26	24	*36*	*34*	*31*	*45*	*40*	*43*	*53*	*64*	*24*	*20*	*10*	*2*
Stance amplitude	17	16	28	25	*35*	*29*	*29*	*48*	*44*	*45*	*53*	*65*	33	30	24	21
Swing amplitude	12	17	27	31	*37*	*35*	*35*	*49*	*42*	*46*	*55*	*71*	56	60	55	53
Number of strides	124	198	70	86												

Hindlimb	Lower leg				Ankle joint				Hind pastern			
Touch down	85	84	80	85	122	129	131	122	37	44	50	36
Lift off	26	31	23	24	140	146	147	146	114	115	124	121
Mid-stance	-	-	-	-	126	125	115	114	-	-	-	-
Effective motion	59	53	58	61	-6	-7	6	-4	-26	-36	32	-8
Stance amplitude	58	53	57	60	39	51	51	71	84	76	77	94
Swing amplitude	73	73	81	89	47	59	38	48	83	83	90	103

Values in italics are marker-based values corrected to reflect the X-Ray video and position data.

PROPORTIONS

All the Pembroke Welsh Corgis we investigated complied with the standard in terms of size, and they all also had a deep chest, as the low legginess index value indicates. No sex-specific differences in limb segment proportions were found. The proportional length of the upper arm is average for the breeds in this study, while the shoulder blade is significantly longer and the forearm significantly shorter than in most breeds. In the hindlimbs, the lower leg is shorter and the hind pastern longer. The lengths of almost all the limb segments vary between individuals, with the most uniform segment being the upper arm in male dogs.

GAIT PARAMETERS

Compared to other breeds, the relative stride length displayed by Pembroke Welsh Corgis is high in the walk. In the trot and the gallop the relative stride length of the forelimbs is middle-of-the-range while that of the hindlimbs is fairly low.

The forwards excursion of the forelimbs does not differ between the walk and the trot but is greater in the gallop, where no difference is seen in the values obtained by the trailing and leading limbs. Backwards excursion is lowest in the walk, greater (and equal) in the trot and in the leading limb in the gallop, and greatest in the trailing limb in the gallop. All the joints are involved in increasing excursion. The forwards excursion of the hindlimbs is lowest in the trailing limb in the gallop, greater (and equal) in the trot and the walk, and greatest in the leading limb in the gallop. Once again, the movement of all the limb segments contributes to these differences. The backwards excursion of the hindlimbs is equal in the walk and the trot, and greater in the gallop, where the values are the same for the trailing and the leading limb. The increase is due mainly to a difference in the position of the pelvis.

Comparatively speaking, relative excursion values vary widely from limb to limb and gait to gait. The forwards excursion of the forelimbs is low in all gaits, while backwards excursion values are situated in the middle or top third of the range. In the hindlimbs, forwards excursion is very high in the walk and the trot and middle-of-the-range in the gallop. The backwards excursion of the hindlimbs is low in all gaits.

During locomotion, the angle of the shoulder always exceeds 90°.

PEMBROKE WELSH CORGI

DESCRIPTION OF BREED

Pembroke Welsh Corgis originated, as the name suggests, in Pembrokeshire, Wales, where they were originally used as herding dogs for cattle and ponies, and to keep small predators away from the farmyard. Opinions differ as to the earlier origins of the breed: one school of thought is that Pembroke Corgis were brought to the British Isles by the Vikings in the 9th or 10th century, while Cardigan Corgis are said to have been brought to Wales by the Celts. Räber (1993) holds both of these theories to be fairly improbable, proposing that Corgis are more likely to be descended from small Welsh farm dogs. Welsh Corgis were first shown in England in 1925, when they immediately caught the attention of the press. Three years later, the breed was added to the Kennel Club Register, though at this point no distinction was made between Pembrokes and Cardigans. In 1943, following a dispute at a dog show during which judges were accused of being biased towards one or the other variety, the two lines were separated and the differences set down officially. In the early days of the breeding program, Pembroke Welsh Corgis were not particularly popular, though this soon changed when Princess Elizabeth was given one as a gift in 1933 and then procured herself another, events which were widely reported in the press. Following this, Corgis more or less stopped being used as working dogs altogether. Today's population can be traced back to four different bitches and five dogs.

According to the standard, height at the withers should be between 25.4 and 30.5 centimeters for both sexes. The chest should be deep and the shoulder blade meet the upper arm at an angle of 90 degrees.

Proportions

Body	♀		♂	
Number of dogs	7		2	
	MV	SD	MV	SD
Withers height [cm]	47	1.5	50	5.3
Body length / Withers height	1.1	0.0	1.0	0.0
Withers height / croup height	1.0	0.0	1.0	0.0
Legginess index	1.3	0.1	1.3	0.1

Forelimb	♀		♂	
Number of dogs	7		2	
	Proportion of total length			
	%	SD	%	SD
Shoulder blade	26	0.9	25	0.0
Upper arm	27	1.0	28	0.4
Lower arm	33	0.6	33	0.9
Pastern	14	1.1	14	0.5

Hindlimb	♀		♂	
Number of dogs	7		2	
	Proportion of total length			
	%	SD	%	SD
Thigh	36	0.9	38	1.7
Lower leg	39	1.2	38	1.5
Hind pastern	25	1.4	25	0.2

Gait parameters

Forelimb

	Stride length				Stride duration				Stance length				Stance duration			
	m				s				m				s			
MV	0.58	0.84	1.58	1.52	0.57	0.41	0.33	0.33	0.32	0.33	0.42	0.44	0.33	0.17	0.10	0.11
SD	0.1	0.1	0.2	0.2	0.1	0.0	0.0	0.0	0.0	0.0	0.0	0.0	0.0	0.0	0.0	0.0
Number of strides					132	160	128	112	All numbers equal							

	Swing length				Swing duration				Dutyfactor				Speed			
	m				s								m/s			
MV	0.26	0.50	1.15	1.08	0.24	0.24	0.23	0.22	58	42	29	32	1.01	2.02	4.71	4.57
SD	0.0	0.1	0.2	0.2	0.0	0.0	0.0	0.0	2.0	3.9	4.6	4.8	0.1	0.2	0.5	0.6
Number of strides	All numbers equal															

Hindlimb

	Stride length				Stride duration				Stance length				Stance duration			
	m				s				m				s			
MV	0.58	0.81	1.50	1.54	0.57	0.41	0.33	0.33	0.34	0.35	0.39	0.42	0.35	0.19	0.10	0.10
SD	0.1	0.1	0.1	0.2	0.1	0.0	0.0	0.0	0.0	0.0	0.0	0.0	0.0	0.0	0.0	0.0
Number of strides					128	168	134	134	All numbers equal							

	Swing length				Swing duration				Dutyfactor				Speed			
	m				s								m/s			
MV	0.24	0.46	1.11	1.12	0.22	0.22	0.24	0.23	61	45	29	30	1.02	1.95	4.48	4.62
SD	0.0	0.1	0.1	0.2	0.0	0.0	0.0	0.0	2.8	4.1	3.1	4.6	0.2	0.2	0.5	0.4
Number of strides	All numbers equal															

Kinematics

Forelimb

Forelimb	Shoulder blade				Shoulder joint				Upper arm				Elbow joint			
Touch down	64	63	60	66	143	140	143	150	78	78	83	84	143	140	145	139
Lift off	*92*	*94*	*102*	*104*	121	122	123	134	35	33	26	35	134	134	149	152
Mid-stance	-	-	-	-	134	132	138	142	-	-	-	-	136	136	140	136
Effective motion	*28*	*31*	*42*	*43*	21	19	20	16	43	44	56	50	9	6	5	13
Stance amplitude	*30*	*33*	*43*	*42*	21	19	23	19	43	45	59	54	18	18	32	32
Swing amplitude	*30*	*35*	*54*	*48*	30	34	39	40	48	53	72	64	49	62	73	76
Number of strides	132	160	116	128	Values in italics are marker-based values corrected to reflect the X-Ray video and position data.											

Forelimb	Lower arm				Carpal joint				Pastern			
Touch down	64	62	62	55	191	192	202	196	127	130	141	143
Lift off	99	101	123	118	141	140	165	153	42	39	41	35
Mid-stance	-	-	-	-	191	192	202	196	-	-	-	-
Effective motion	35	39	61	63	50	51	37	43	86	91	100	108
Stance amplitude	48	46	62	63	46	56	55	63	79	85	98	105
Swing amplitude	65	82	105	110	100	121	123	129	115	145	168	173

Hindlimb

Hindlimb	Pelvis				Hip joint				Thigh				Stifle joint			
Touch down	65	58	59	63	123	118	118	115	57	61	59	52	140	143	144	147
Lift off	57	53	46	41	154	155	156	153	97	102	110	112	130	136	141	146
Mid-stance	-	-	-	-	123	118	119	116	-	-	-	-	140	143	142	146
Effective motion	8	4	13	22	31	37	38	38	40	41	52	59	10	7	3	1
Stance amplitude	10	7	15	23	34	38	40	36	43	42	51	58	14	15	20	20
Swing amplitude	9	9	25	29	37	44	54	53	44	49	62	75	35	50	57	57
Number of strides	108	166	120	137	Values in italics are marker-based values corrected to reflect the X-Ray video and position data.											

Hindlimb	Lower leg				Ankle joint				Hind pastern			
Touch down	83	82	85	95	132	132	135	134	48	49	50	39
Lift off	33	34	31	35	146	149	160	155	112	115	129	121
Mid-stance	-	-	-	-	132	132	134	133	-	-	-	-
Effective motion	50	49	54	60	15	17	25	21	64	66	80	82
Stance amplitude	49	48	54	59	28	38	50	59	65	66	79	85
Swing amplitude	58	64	87	92	34	50	62	55	67	73	91	102

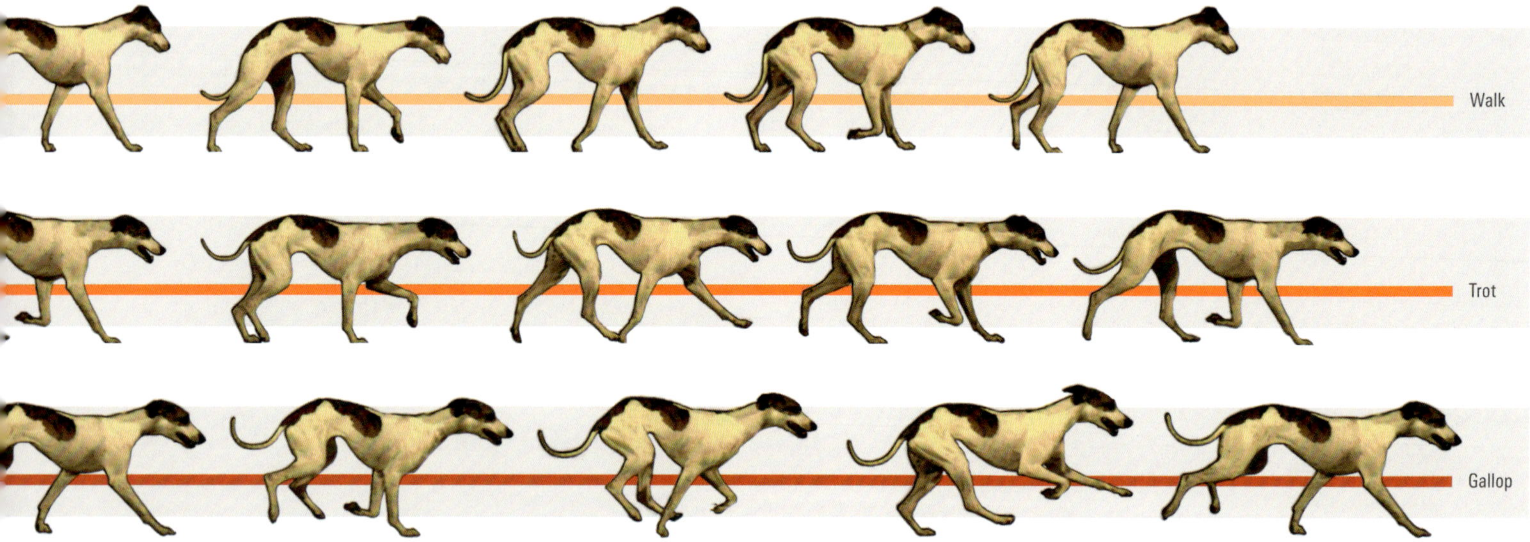

WHIPPET

DESCRIPTION OF BREED

Although the Whippet is regarded as one of the youngest breeds of sighthound, reliable information about its origin and emergence is thin on the ground. The breeding of Whippets as we know them today began between 1830 and 1840 among the working class in northern England. The dogs were used mainly to bait rabbits or for dog racing, which made speed the main criterion for selection. It appears that all kinds of breeds were crossed, but the only thing we know for certain is that long-legged terriers and, of course, sighthounds played an important role in the emergence of the Whippet. The breed was recognized by the Kennel Club in 1890. The first Whippets were introduced to Germany around 1900, where ten years later the group "Friends of the Sighthound" was established. Well-built and well-performing Whippets were imported from England and used as the basis of the German breeding program.
The standard requires a very deep chest and shoulder blades which reach to the spine, but otherwise the requirements are fairly general. The hindlimbs should reach well under the body during locomotion and the forelimbs should have a long, low reach. Height at the withers should be between 47 and 51 centimeters in dogs and between 44 and 47 centimeters in bitches.

PROPORTIONS

As the legginess index shows, the chest in Whippets is no deeper than in other breeds of a similar size. One of the males we measured was slightly larger than the standard allows, though this did not appear to affect his limb segment proportions. The only sex-specific differences we observed were a shorter femur and a slightly longer trunk in relation to withers height in bitches. The shoulder blade is slightly shorter in Whippets than in most other breeds, but the lower leg is longer. Limb segment variability is very low, especially in the forelimb.

GAIT PARAMETERS

Comparatively speaking, relative stride length is quite high in this breed, in almost all gaits. The relative stride length of the forelimb in the walk and the trot and that of the hindlimb in the trot is low, however.
In the forelimbs, forwards excursion remains constant between the gaits, only deviating in the leading limb in the gallop, where it is greater due to the movement of the forearm. Backwards excursion is more or less equal in the walk and the trot and greater in the gallop, where the values for the trailing limb exceed those for the leading limb. The difference between the symmetrical gaits and the gallop can be put down to the movement of the shoulder blade and forearm, while the difference between the trailing and leading limb is caused by a difference in the movement of the upper arm. In the hindlimbs too, forwards excursion is greatest in the leading limb thanks to the position of the femur and lower leg. When it comes to backwards excursion, a clear increase is seen from the walk to the trot to the gallop, where values for the trailing limb exceed those for the leading limb. The position of the pelvis and femur play the greatest role here.
In a cross-breed comparison, the relative excursion values for this breed differ significantly from gait to gait and limb to limb. In the forelimbs, forwards excursion is situated towards the top of the range in the walk and the trot, as is the backwards excursion of the hindlimbs in the trot and the gallop. The backwards excursion of the forelimbs is low in the trot and the gallop. The forwards excursion of the hindlimbs is always low, and their backwards excursion is low in the walk.

Proportions

Body	♀		♂	
Number of dogs	9		2	
	MV	SD	MV	SD
Withers height [cm]	34	1.4	37	0.4
Body length / Withers height	1.0	0.0	1.0	0.0
Withers height / croup height	1.0	0.0	1.1	0.1
Legginess index	1.1	0.1	1.0	0.0

Forelimb	♀		♂	
Number of dogs	9		2	
	Proportion of total length			
	%	SD	%	SD
Shoulder blade	28	1.8	27	1.1
Upper arm	26	1.0	27	1.1
Lower arm	29	1.3	31	0.8
Pastern	16	1.3	16	1.3

Hindlimb	♀		♂	
Number of dogs	9		2	
	Proportion of total length			
	%	SD	%	SD
Thigh	37	2.3	35	3.8
Lower leg	35	2.4	35	4.5
Hind pastern	28	1.2	30	0.7

Gait parameters

Forelimb

	Stride length				Stride duration				Stance length				Stance duration			
	m				s				m				s			
MV	0.35	0.52	-	-	0.52	0.37	-	-	0.20	0.23	-	-	0.31	0.17	-	-
SD	0.0	0.1	-	-	0.1	0.0	-	-	0.0	0.0	-	-	0.1	0.0	-	-
Number of strides					79	69	0	0	All numbers equal							

	Swing length				Swing duration				Dutyfactor				Speed			
	m				s								m/s			
MV	0.15	0.29	-	-	0.21	0.19	-	-	58	47	-	-	0.68	1.43	-	-
SD	0.0	0.1	-	-	0.0	0.0	-	-	6.6	3.8	-	-	0.1	0.3	-	-
Number of strides	All numbers equal															

Hindlimb

	Stride length				Stride duration				Stance length				Stance duration			
	m				s				m				s			
MV	0.38	0.57	-	-	0.53	0.35	-	-	0.23	0.22	-	-	0.33	0.15	-	-
SD	0.1	0.1	-	-	0.1	0.0	-	-	0.0	0.0	-	-	0.1	0.0	-	-
Number of strides					111	126	0	0	All numbers equal							

	Swing length				Swing duration				Dutyfactor				Speed			
	m				s								m/s			
MV	0.16	0.35	-	-	0.20	0.21	-	-	62	41	-	-	0.73	1.63	-	-
SD	0.0	0.1	-	-	0.0	0.0	-	-	6.4	6.3	-	-	0.1	0.4	-	-
Number of strides	All numbers equal															

Kinematics

Forelimb	Shoulder blade				Shoulder joint				Upper arm				Elbow joint			
Touch down	*60*	*61*	-	-	147	152	-	-	77	80	-	-	152	147	-	-
Lift off	*93*	*99*	-	-	132	140	-	-	44	41	-	-	158	155	-	-
Mid-stance	-	-	-	-	143	148	-	-	-	-	-	-	152	145	-	-
Effective motion	*33*	*38*	-	-	14	12	-	-	33	40	-	-	6	8	-	-
Stance amplitude	*36*	*40*	-	-	16	18	-	-	34	41	-	-	20	25	-	-
Swing amplitude	*36*	*40*	-	-	24	29	-	-	37	45	-	-	45	58	-	-
Number of strides	79	69	0	0	Values in italics are marker-based values corrected to reflect the X-Ray video and position data.											

Forelimb	Lower arm				Carpal joint				Pastern			
Touch down	75	66	-	-	198	197	-	-	123	130	-	-
Lift off	114	114	-	-	171	169	-	-	56	55	-	-
Mid-stance	-	-	-	-	198	197	-	-	-	-	-	-
Effective motion	39	47	-	-	28	27	-	-	67	75	-	-
Stance amplitude	45	50	-	-	34	43	-	-	65	73	-	-
Swing amplitude	60	73	-	-	80	92	-	-	104	118	-	-

Hindlimb	Pelvis				Hip joint				Thigh				Stifle joint			
Touch down	50	39	-	-	*109*	107	-	-	*64*	*68*	-	-	152	155	-	-
Lift off	44	36	-	-	*140*	137	-	-	*101*	*107*	-	-	138	145	-	-
Mid-stance	-	-	-	-	*114*	108	-	-	-	-	-	-	152	155	-	-
Effective motion	6	4	-	-	*31*	30	-	-	*37*	*39*	-	-	14	10	-	-
Stance amplitude	10	8	-	-	30	30	-	-	*40*	*39*	-	-	19	22	-	-
Swing amplitude	7	10	-	-	30	35	-	-	*39*	*45*	-	-	36	53	-	-
Number of strides	111	126	0	0	Values in italics are marker-based values corrected to reflect the X-Ray video and position data.											

Hindlimb	Lower leg				Ankle joint				Hind pastern			
Touch down	88	86	-	-	143	143	-	-	55	56	-	-
Lift off	42	43	-	-	162	164	-	-	120	121	-	-
Mid-stance	-	-	-	-	144	143	-	-	-	-	-	-
Effective motion	46	43	-	-	20	23	-	-	65	65	-	-
Stance amplitude	46	45	-	-	34	42	-	-	68	65	-	-
Swing amplitude	53	64	-	-	40	61	-	-	73	85	-	-

PROPORTIONS

Two males were slightly larger than specified by the standard, though there were no obvious differences in their limb segment proportions compared to other individuals of the breed. In all the animals we measured, trunk length and height at the withers were identical, giving them a perfectly square build. Not enough male dogs were involved in the study for us to be able to identify any sex-specific differences. The forelimb proportions of this breed are average compared to other breeds, while in the hindlimbs the lower leg is slightly shorter.

GAIT PARAMETERS

Comparatively speaking, relative stride length is short in the walk and middle-of-the-range in the trot. Relative stance phase length is always short, but in the trot it is compensated by the slightly longer relative swing phase.

Forelimb excursion is, in both directions, the same in the trot as it is in the walk, as can be seen from the limb angulation, which remains almost constant. The forwards excursion of the hindlimbs is greater in the walk, while backwards excursion is greater in the trot. These differences can be put down to differences in the positioning of the pelvis and femur.

Compared to that observed in other breeds, relative excursion in both directions is low in the walk and the trot in both the forelimbs and the hindlimbs. Our data do not confirm that this breed displays a particularly ground-covering gait.

MINIATURE SCHNAUZER

DESCRIPTION OF BREED

The Miniature Schnauzer originated in southern Germany. As mentioned in connection with the Schnauzer, the distinction between Pinschers and Schnauzers is a relatively new one, and the equivalent of the Miniature Schnauzer was the Affenpinscher, which at the end of the 19th century was, however, a fairly broad term for a whole ragbag of different dogs. The miniature form of the Pinscher has been around for a long time, with written sources which describe the animals as lap dogs or salon dogs dating from as early as 1880. Räber (1993) believes it to be unlikely that Miniature Schnauzers are descended from Silky Pinschers, however, as the latter were significantly smaller. The miniature form of the Schnauzer began to be bred officially in 1897 following the Erfurt Dog Show, with the aim of obtaining a smaller version of the Schnauzer. Starting in 1899, dog shows opened a separate class for rough-haired Miniature Pinschers and Affenpinschers, thus creating a breeding base. The dogs have been officially known as Miniature Schnauzers since 1910.

Height at the withers should be between 30 and 35 centimeters, dogs should be square-built and the gait should be free, balanced and vigorous.

LITERATURE

ACEVEDO L. M. & RIVERA J. L. (2006): New insights into skeletal muscle fibre types in the dog with particular focus towards hybrid myosin phenotypes. Cell Tissue Res. 323 (2), 283-303.

ADRIAN M. J., ROY W. E. & KARPOVICH P. V. (1966): Normal gait of the dog. An electrogoniometric study. Am. J. Vet. Res. 27, 90-95.

AFELT, Z. & KASICKI, S. (1975): Limb coordinations during locomotion in cats and dogs. Acta Neurobiol. Exp. 35: 369-378.

AFELT Z., BLASZCZYK J. & DOBRZECKA C. (1975): Speed control in animal locomotion: transitions between symmetrical andn nonsymmetrical gaits in the dog. Acta Neurobiol. Exp. 43: 235-250.

AGUERA E., DIZ, A., VAZQUEZ-AUTON J.M., VIVO, J. & MONTERDE J.G. (1990a): Muscle fibre morphometry in three dog muscles of different functional purpose in different breeds. Anat. Histol. Embryol. 19 (4), 289-293.

AGUERA E., VAZQUEZ-AUTON J. M., VIVO J. & MONTERDE J. G. (1990b): Differences in fibre population in dog muscles of different functional purpose. Anat. Histol. Embryol. 19 (2), 128-134.

AINSWORTH D. M., SMITH C. A., EICKER S. W., DUCHARME N. G., HENDERSON K. S. et al. (1997): Pulmonary-locoexercising in dogs and horses. Respir. Physiol. 110 (2-3), 287-294.

AINSWORTH D. M., SMITH C. A., HENDERSON K. S. & DEMPSEY J. A. (1996): Breathing during exercise in dogs - passive or active? J. Appl. Physiol. 81 (2), 586-595.

ALEXANDER R. M. (1976): Mechanics of bipedal locomotion. 493-504. In: P. S. Davies (ed.): Perspectives in Experimental Biology. Oxford: Pergamon Press.

ALEXANDER R. M. (1984): Elastic energy stores in running vertebrates. Am. Zool. 24 (1), 85-94.

ANDERSON T. M., VONHOLDT B. M., CANDILLE S. I., MUSIANI M. & GRECO C. (2009): Molecular and evolutionary history of melanism in North American gray wolves. Science 323, 1339-1343.

ANDERSONE Ž., LUCCHINI V., RANDI E. & OZOLI Š. J. (2002): Hybridisation between wolves and dogs in Latvia as documented using mitochondrial and microsatellite DNA markers. Mammal. Biol. 67 (2), 79-90.

ANSORGE H., HOLZAPFEL M., KLUTH G., REINHARDT I. & WAGNER C. (2010): Die Rückkehr der Wölfe. Das erste Jahrzehnt. BIUZ 40 (4), 244-253.

ARMSTRONG R., SAUBERT IV C., SEEHERMAN H. & TAYLOR C. (1982): Distribution of fiber types in locomotory muscles of dogs. Am. J. Anat. 163 (1), 87-98.

ARSHAVSKII Y. I., KOTS Y. M., ORLOVSKII G. N., RODIONOV I. M. & SHIK M. L. (1965): Investigation of the biomechanics of running by the dog. Biophysics 10, 737-746.

BADOUX D. M. & HOOGEVEEN P. (1976): Some notes on the biomechanics of the normal and dysplastic canine acetabulum. Proc. Konink. Nederl. Akad. Wet., C 79 (2), 97-111.

BARDET J. (1998): Diagnosis of shoulder instability in dogs and cats: a retrospective study. J. Am. Anim. Hosp. Assoc. 34 (1), 42-54.

BELYAEV D. K. (1979): Destabilizing selection as a factor in domestication. J. Hered. 70 (5), 301-308.

BENECKE N. (1994): Der Mensch und seine Haustiere: die Geschichte einer jahrtausendealten Beziehung. Stuttgart: Theiss Verlag.

BENNINGER M. I., SEILER G. S., ROBINSON L. E., FERGUSON S. J., BONEL H. M. et al. (2006): Effects of anatomic conformation on three-dimensional motion of the caudal lumbar and lumbosacral portions of the vertebral column of dogs. Am. J. Vet. Res. 67 (1), 43-50.

BERTRAM J. E., LEE D. V., CASE H. N. & TODHUNTER R. J. (2000): Comparison of the trotting gaits of Labrador Retrievers and Greyhounds. Am. J. Vet. Res. 61 (7), 832-838.

BERTRAM J. E. A., LEE D. V., TODHUNTER R. J., FOELS W. S., WILLIAMS A. J. et al. (1997): Multiple force platform analysis of the canine trot: a new approach to assessing basic characteristics of locomotion. Vet. Comp. Orthop. Traumatol. 10 (3), 160-169.

BESANCON M. F., CONZEMIUS M. G., DERRICK T. R. & RITTER M. J. (2003): Comparison of vertical forces in normal greyhounds between force platform and pressure walkway measurement systems. Vet. Comp. Orthop. Traumatol. 16 (3), 153-157.

BEYERSDORF P. (2007): 100 Jahre Spaniels in Deutschland. Mürlenbach/Eifel: Kynos Verlag.

BLASZCZYK J. W. (2001): Gait transitions during unrestrained locomotion in dogs. Equine Vet. J. (Suppl.) 33, 112-115.

BLICKHAN R. (1989): The spring-mass model for running and hopping. J. Biomech. 22 (11-12), 1217-1227.

BLICKHAN R., SEYFARTH A., GEYER H., GRIMMER S., WAGNER H. et al. (2007): Intelligence by mechanics. Phil. Trans. R. Soc. A 365, 199-220.

BOCKSTAHLER B. A., SKALICKY M., PEHAM C., MÜLLER M. & LORINSON D. (2007): Reliability of ground reaction forces measured on a treadmill system in healthy dogs. Vet. J. 173, 373-378.

BOYKO A. R., QUIGNON P., LI L., SCHOENEBECK J. J., DEGENHARDT J. D. et al. (2010): A Simple Genetic Architecture Underlies Morphological Variation in Dogs. PLoS Biol. 8 (8), 1-13.

BRAMBLE D. & JENKINS F. (1993): Mammalian locomotor-respiratory integration: implications for diaphragmatic and pulmonary design. Science 262 (5131), 235-240.

BRAUND K. G., HOFF E. J. & RICHARDSON K. E. (1978): Histochemical identification of fiber types in canine skeletal muscle. Am. J. Vet. Res. 39 (4), 561-565.

BREBNER N. S., MOENS N. M. M. & RUNCIMAN J. R. (2006): Evaluation of a treadmill with integrated force plates for kinetic gait analysis of sound and lame dogs at a trot. Vet. Comp. Orthop. Traumatol. 19 (4), 205-212.

BREIT S. & KUNZEL W. (2001): On biomechanical properties of the sacroiliac joint in purebred dogs. Ann. Anat. 183 (2), 145-150.

BROWN C. M. & DALZELL B. (1986): Dog locomotion and gait analysis. Wheat Ridge, Colorado: Hoflin Publishing.

BRYANT J. D., BENNETT M. B., BRUST J. & ALEXANDER R. M. (1987): Forces exerted on the ground by galloping dogs (*Canis familiaris*). J. Zool. 213 (2), 193-204.

BUDIANSKY S. (1992): The covenant of the wild: why animals chose domestication. New York: W. Morrow.

BUDSBERG S. C. & VERSTRAETE M. C. (1993): Canine center of pressure patterns during a trotting gait. Vet. Surg. 22 (5), 373-374 (Meeting Abstracts).

BUDSBERG S. C., VERSTRAETE M. C. & SOUTAS-LITTLE R. W. (1987): Force plate analysis of the walking gait in healthy dogs. Am. J. Vet. Res. 48, 915-918.

BUDSBERG S. C., VERSTRAETE M. C., BROWN J. & REECE L. (1995): Vertical loading rates in clinically normal dogs at trot. Am. J. Vet. Res. 56 (10), 1275-1280.

BÜCKNER H. J. (1971): Allometrische Untersuchungen an den Vorderextremitäten adulter Caniden. Ann. Anat. 186, 11-46.

BÜRGER R. & LANG J. (1993): Kinetische Studie über die Lendenwirbelsäule und den lumbosakralen Übergang beim Deutschen Schäferhund. Schweizer Arch. Tierheilk. 135 (2), 35-43.

BURKHOLDER B. L. (1959): Movements and behavior of a wolf pack in Alaska. J. Wildl. Man. 23, 1-11.

BUXTON D. F. & PECK D. (1990): Density of muscle spindle profiles in the intrinsic forelimb muscles of the dog. J. Morph. 203 (3), 345-359.

CAICOYA A. G., ILLERT M. & JÄNIKE R. (1999): Monosynaptic la pathways at the cat shoulder. J. Physiol. 518, 825-841.

CARRIER D. R. (1996): Function of the intercostal muscles in trotting dogs: ventilation or locomotion? J. Exp. Biol. 199 (7), 1455-1465.

CARRIER D. R., GREGERSEN C. & SILVERTON N. (1998): Dynamic gearing in running dogs. J. Exp. Biol. 201 (23), 3185-3195.

CARRIER D. R., CHASE K. & LARK K. (2005): Genetics of canid skeletal variation: Size and shape of the pelvis. Genome Res. 15 (12), 1825-1830.

CARRIER D. R., DEBAN S. M. & FISCHBEIN T. (2008): Locomotor function of the pectoral girdle 'muscular sling' in trotting dogs. J. Exp. Biol. 209, 2224-2237.

CARRIER D. R., DEBAN S. M. & FISCHBEIN T. (2008): Locomotor function of forelimb protractor and retractor muscles of dogs: evidence of strut-like behavior at the shoulder. J. Exp. Biol. 211 (1), 150-162.

CASINOS A., BOU J., CASTIELLA M. J. & VILADIU C. (1986): On the allometry of long bones in dogs (*Canis familiaris*). J. Morph. 190 (1), 73-79.

CAVAGNA G. A., HEGLUND N. C. & TAYLOR C. R. (1977): Mechanical work in terrestrial locomotion: two basic mechanisms for minimizing energy expenditure. Am. J. Physiol. 233 (5), R243-R261.

CERNY H. & CIZINAUSKAS S. (1995): The clavicle of newborn dogs. Acta Vet. Brno 64 (2), 139-145.

CHAIX L. (2000): A Preboreal dog from the Northern Alps (Savoie, France). 49-59. In: S. J. Crockford (Ed.): Dogs Through Time: An Archaeological Perspective, Proceedings of the 1st ICAZ Symposium on the History of the Domestic Dog. Oxford: Archaeopress, B.A.R..

CHASE K., CARRIER D. R., ADLER F., JARVIK T., OSTRANDER E. et al. (2002): Genetic basis for systems of skeletal quantitative traits: Principal component analysis of the canid skeleton. PNAS 99 (15), 9930-9935.

CIUCCI P., BOITANI L., FRANCISCI F. & ANDREOLI G. (1997): Home range, activity and movements of a wolf pack in central Italy. J. Zool. 243 (4), 803-819

CLUTTON-BROCK J. (1995): Origins of the Dog: domestication and early history. 7-20. In: J. Serpell (ed.): The domestic dog: its evolution, behaviour and interactions with people. New York: Cambridge University Press.

COLBORNE G. R., WALKER A. M., TATTERSALL A. J. & FULLER C. J. (2006): Effect of trotting velocity on work patterns of the hind limbs of Greyhounds. Am. J. Vet. Res. 67 (8), 1293-1298.

COPPINGER R. & COPPINGER L. (2001): Wolves evolve into dogs. 39-67. In: R. Coppinger, L. Coppinger (Eds.): Dogs: A Startling New Understanding of Canine Origin, Behavior and Evolution New York: Scribner.

CROCKFORD S. J. (2005): Native dog types in North America before arrival of European dogs. Proc. 30th world congress of the world small animal veterinary association, Mexico City.

DAVIS S. J. M. & VALLA F. R. (1978): Evidence for domestication of the dog 12,000 years ago in the Natufian of Israel. Nature 276, 608-610.

DEBAN S. & CARRIER D. R. (2002): Hypaxial muscle activity during running and breathing in dogs. J. Exp. Biol. 205 (13), 1953-1967.

DENNY, M. W. (2008): Limits to running speed in dogs, horses and humans J. Exp. Biol. 211, 3836-3849.

DETELS A. (1980): Das Gliedmaßenskelett des Dakkels (Dachshund, Dachshunds). Diss. med. vet. Veterinärmedizinische Fakultät, Ludwig-Maximilians-Universität München.

DICKOMEIT M. J. (2002): Anatomische und biomechanische Untersuchungen am Ellbogengelenk des Hundes (*Canis familiaris*). Diss. med. vet. Veterinärmedizinische Fakultät, Ludwig-Maximilians-Universität München.

DILL D. B., EDWARDS H. T. & TALBOTT J. H. (1932): Studies in muscular activity. VII. Factors limiting the capacity for work. J. Physiol. 77 (1), 49-62.

DOBSON G. P., PARKHOUSE W. S., WEBER J. M., STUTTARD E., HARMAN J. et al. (1988): Metabolic changes in skeletal muscle and blood of greyhounds during 800-m track sprint. Am. J. Physiol. 255 (3 Pt 2), R513-R519.

DONAT K. (1971): Fixation of the clavicle in the cat and dog. Ann. Anat. 128 (4), 365-374.

DRISCOLL C. A. & MACDONALD D. W. (2010): Top dogs: wolf domestication and wealth. J. Biol. 9 (2), 10.

ECKSTEIN F., LÖHE F., STEINLECHNER M., MÜLLER-GERBL M. & PUTZ R. (1993b): Kontaktflächen des menschlichen Humeroulnargelenks in Abhängigkeit von der Anpresskraft - ihr Zusammenhang mit subchondraler Mineralisierung und Gelenkmorphologie der Incisura trochlearis. Ann. Anat. 175, 545-552.

ECKSTEIN F., STEINLECHNER M., MÜLLER-GERBL M. & PUTZ R. (1993a): Mechanische Beanspruchung und subchondrale Mineralisierung des menschlichen Ellbogengelenks. Unfallchirurg 96, 99-104.

ELLER D. (2003): Anatomische und biomechanische Untersuchungen am Schultergelenk (Articulatio humeri) des Hundes (*Canis familiaris*). Diss. med. vet. Veterinärmedizinische Fakultät, Ludwig-Maximilians-Universität München.

FANCHON L., VALETTE J.-P., SANAA M. & GRANDJEAN D. (2006): The measurement of ground reaction force in dogs trotting on a treadmill: an investigation of habituation. Vet. Comp. Orthop. Traumatol. 19 (2), 81-86.

FIFE M., BAILEY C., LEE D. & CARRIER D. R. (2001): Function of the oblique hypaxial muscles in trotting dogs. J. Exp. Biol. 204 (13), 2371-2381.

FISCHER M. S. (1999): Kinematics, EMG, and inverse dynamics of the therian forelimb – a synthetical approach. Ann. Anat. 238, 41-54.

FISCHER M. S. (2001): Locomotory organs of Mammals: New mechanics and feed-back pathways but conservative central control. Zoology 103, 230-239.

FISCHER M. S. & BLICKHAN R. (2006): The tri-segmented limbs of therian mammals: kinematics, dynamics, and self-stabilization. A review. J. Exp. Zool. 305A, 935-952.

FISCHER M. S. & LEHMANN R. (1998): Application of cineradiography for the metric and kinematic study of in-phase gaits during locomotion of the pika (Ochotona rufescens, Mammalia: Lagomorpha). Zoology 101, 148-173.

FISCHER M. S., SCHILLING N., SCHMIDT M. & WITTE H. (2002): Basic limb kinematics of small therian mammals. J. Exp. Biol. 205 ((Pt 9)), 1315-1338.

FISCHER M. S. & WITTE H. (1998): The functional morphology of the three-segmented limb of mammals and its specialities in small and medium-sized mammals. Proc. Europ. Mechanics Coll. Euromech 375 Biology and Technology of Walking, 10-17.

FONDON III J. & GARNER H. (2004): Molecular origins of rapid and continuous morphological evolution. PNAS 101 (52), 18058-18063.

FRITTS S. H. (1983): Record dispersal by a wolf from Minnesota. J. Mammal. 64, 166-167.

FULLER T. K. (1991): Effect of snow depth on wolf activity and prey selection in North central Minnesota. Can. J. Zool. 69, 283-287.

GARCIA M. A. (2005): Ichnologie générale de la grotte Chauvet. Bull. Soc. Préhist. Franç. 102, 103-108.

GERMONPRÉ M., SABLIN M. V., STEVENS R. E., HEDGES R. E. M., HOFREITER M. et al. (2009): Fossil dogs and wolves from Palaeolithic sites in Belgium, the Ukraine and Russia: osteometry, ancient DNA and stable isotopes. J. Archaeol. Sci. 36 (2), 473-490.

GERTH N., SUM S., JACKSON S. & STARCK J. (2009): Muscle plasticity of Inuit sled dogs in Greenland. J. Exp. Biol. 212 (8), 1131-1139.

GOLDSPINK G. (1977): Mechanics and energetics of muscle in animals of different sizes with particular reference to the muscle fiber composition. 37-55. In: T. J. Pedley (Ed.): Scale Effects in Animals Locomotion. New York: Academic Press.

GOSLOW G., SEEHERMAN H., TAYLOR C., McCUTCHIN M. & HEGLUND N. (1981): Electrical activity and relative length changes of dog limb muscles as a function of speed and gait. J. Exp. Biol. 94 (1), 15-42.

GRAY M. M., SUTTER N. B., OSTRANDER E. A. & WAYNE R. K. (2010): The IGF1 small dog haplotype is derived from Middle Eastern gray wolves. BMC Biol. 8 (1), 16.

GREGERSEN C., SILVERTON N. & CARRIER D. R. (1998): External work and potential for elastic storage at the limb joints of running dogs. J. Exp. Biol. 201 (23), 3197-3210.

GREGORY C., CULLEN J., POOL R. & VASSEUR P. (1986): The canine sacroiliac joint: preliminary study of anatomy, histopathology, and biomechanics. Spine 11 (10), 1044-1048.

GRIFFIN T. M., MAIN R. P. & FARLEY C. T. (2004): Biomechanics of quadrupedal walking: how do four-legged animals achieve inverted pendulum-like movements? J. Exp. Biol. 207 (Pt 20), 3545-3558.

GRIFFON D. J. (1994): Vertical ground reaction force redistribution during experimentally induced shoulder lameness in dogs. Vet. Comp. Orthop. Traumatol. 7, 154-157.

GRILLNER S. (1975): Locomotion in vertebrates: central mechanisms and reflex interaction. Physiol. Rev. 55 (2), 247-303.

GROß M. (2008): Ansätze zur Modellierung der Gangart Trab eines Airedale Terriers (*Canis lupus* f. familiaris, Canidae) auf der Grundlage experimentell erhobener Daten. Diplomarbeit Biologisch-Pharmazeutische Fakultät, Friedrich-Schiller-Universität Jena.

GUNN H. (1978): Differences in the histochemical properties of skeletal muscles of different breeds of horses and dogs. J. Anat. 127 (Pt 3), 615-634.

GUREVICH L., BENNETT W. H., ANDERSON M. A., MANN F. A., JIANG B. et al. (1994): A force plate and display system for studying ground reaction forces in dogs. Biomed. Sci. Instrum. 30, 81-86.

GUY P. S. & SNOW D. H. (1981): Skeletal muscle fibre composition in the dog and its relationship to athletic ability. Res. Vet. Sci. 31 (2), 244-248.

HARE B. (2002): The domestication of social cognition in dogs. Science 298 (5598), 1634-1636.

HARE B., PLYUSNINA I., IGNACIO N. & SCHEPINA O. (2005): Social cognitive evolution in captive foxes is a correlated by-product of experimental domestication. Curr. Biol. 15 (3), 226-230.

HARE B. & TOMASELLO M. (2005): Human-like social skills in dogs? Trends Cogn. Sci. 9 (9), 439-444.

HARE W. C. D. (1959): Radiographic anatomy of the canine pectoral limb. Part 1: Fully developed limb. J. Am. Vet. Med. Assoc. 135, 264-271.

HARE W. C. D. (1960): The age at which epiphyseal union takes place in the limb bones of the dog. Wiener tierärztl. Mschr. 72, 224-245.

HARE W. C. D. (1961): The ages at which the centers of ossification appear roentgenographically in the limb bones of the dog. Am. J. Vet. Res. 22, 825-835.

HAYES R. D. (1995): Numerical and functional responses of wolves and regulation of moose in the Yukon. Master thesis Department of Biological Sciences, Simon-Fraser-University.

HILZHEIMER M. (1926): Natürliche Rassengeschichte der Haussäugetiere. Berlin/Leipzig: W. de Gruyter.

HITZ D. (1973): Untersuchungen über Skelett- und Gelenksveränderungen beim Bassethund. Diss. med. vet. Veterinärmedizinische Fakultät, Universität Zürich.

HO S., KOLOKOTRONIS S. & ALLABY R. (2007): Elevated substitution rates estimated from ancient DNA sequences. Biol. Lett. 3 (6), 702-705.

HO S., PHILLIPS M., COOPER A. & DRUMMOND A. (2005): Time dependency of molecular rate estimates and systematic overestimation of recent divergence times. Mol. Biol. Evol. 22 (7), 1561-1568.

HÖCHE C. (2008): Anatomischer und histochemischer Vergleich morphologisch verschiedener Züchtungen von *Canis lupus* f. familiaris. Diplomarbeit Biologisch-Pharmazeutische Fakultät, Friedrich-Schiller-Universität Jena.

HOLLER, P.J., BRAZDA, V., DAL-BIANCO, B., LEWY, D, MUELLER et al. (2010): Kinematic motion analysis of the joints of the forelimbs and hind limbs of dogs during walking exercise regimens. Am J Vet 71:734–740.

HONGO T., LUNDBERG A., PHILLIPS C. G. & THOMPSON R. F. (1984): The pattern of monosynaptic Ia-connections to hindlimb motor nuclei in the baboon: a comparison with the cat. Proc. R. Soc. Lond. B 221, 261-289.

HOPPELER H. & FLÜCK M. (2002): Normal mammalian skeletal muscle and its phenotypic plasticity. J. Exp. Biol. 205 (Pt 15, Review), 2143-2152.

HOROWITZ A. (2009): Attention to attention in domestic dog (*Canis familiaris*) dyadic play. Anim. Cogn. 12 (1), 107-118.

HOTTINGER H. A., DECAMP C. E., OLIVIER N. B., HAUPTMAN J. G. & SOUTAS-LITTLE R. W. (1996): Noninvasive kinematic analysis of the walk in healthy largebreed dogs. Am. J. Vet. Res. 57 (3), 381-388.

HUTTON W. C., FREEMAN M. A. & SWANSON S. A. (1969): The forces exerted by the pads of the walking dog. J. Small Anim. Pract. 10 (2), 71-77.

JAYES A. S. & ALEXANDER R. M. (1978): Mechanics of locomotion of dogs (*Canis familiaris*) and sheep (Ovis aries). J. Zool. 185, 289-308.

JAYES A. S. & ALEXANDER R. M. (1982): Estimates of mechanical stresses in leg muscles of galloping greyhounds (*Canis familiaris*). J. Zool. 198 (3), 315-328.

JEDRZEJEWSKI W., SCHMIDT K., THEUERKAUF J., JEDRZEJEWSKA B. & OKARMA H. (2001): Daily movements and territory use by radio-collared wolves (*Canis lupus*) in BiałowieÐa Primeval Forest in Poland. Can. J. Zool. 79 (11), 1993-2004.

JÐDRZEJEWSKI W., SCHMIDT K., THEUERKAUF J., JÐDRZEJEWSKA B., SELVA N. et al. (2002): Kill rates and predation by wolves on ungulate populations in BiałowieÐa Primeval Forest (Poland). Ecology 83 (5), 1341-1356.

JENKINS, F.A.J. & CAMAZINE, S.M. (1977): Hip structure and locomotion in ambulatory and cursorial carnivores. J. Zool. (Lond.) 181, 351-370.

JEVENS D. J., HAUPTMAN J. G., CAMP C. E. D., BUDSBERG S. C. & SOUTAS-LITTLE R. W. (1993): Contribution to variance in force-plate analysis of gait in dogs. Am. J. Vet. Res. 54, 612-615.

KAISER A. (2006): Morphologische und biomechanische Eigenschaften des Karpalgelenks (Articulatio carpi) des Hundes (*Canis familiaris*). Diss. med. vet. Veterinärmedizinische Fakultät, Ludwig-Maximilians-Universität München

KAPATKIN A. S., ARBITTIER G., KASS P. H., GILLEY R. S. & SMITH G. K. (2007): Kinetic gait analysis of healthy dogs on two different surfaces. Vet. Surg. 36 (6), 605-608.

KAVANAU J. L. & RAMOS J. (1975): Influences of light on activity and phasing of carnivores. Am. Nat. 109, 391-418.

KEMP T. (2005): Functional trade-offs in the limb bones of dogs selected for running versus fighting. J. Exp. Biol. 208 (18), 3475-3482.

KIMURA T. (1993): Correlations between feet posture and muscle fiber size in mammals - comparison of the anterior tibial muscle fiber among plantigrade type, digitigrade type and unguligrade type. J. Anat. 68 (6), 617-625.

KINZEL G. L., VAN SICKLE D. C. & HILLBERRY B. M. (1976): Preliminary study of the in vivo motion in the canine shoulder. Am. J. Vet. Res. 37 (12), 1505-1510.

KUBEIN-MEESENBURG D., NÄGERL H. & FANGHÄNEL J. (1990a): Elements of a general theory of joints. 1. Basic kinematic and static function of diarthrosis. Ann. Anat. 170, 301-308.

KUSAK J., SKRBINŠEK A. M. & HUBER D. (2005): Home ranges, movements, and activity of wolves (*Canis lupus*) in the Dalmatian part of Dinarids, Croatia. Eur. J. Wildl. Res. 51 (4), 254-262.

KUZON JR. W., ROSENBLATT J., PYNN B., MARCHETTI P., PLYLEY M. et al. (1989): A comparative histochemical and morphometric study of canine skeletal muscle. Can. J. Vet. Res. 53 (2), 125-132.

LARK K., CHASE K. & SUTTER N. B. (2006): Genetic architecture of the dog: sexual size dimorphism and functional morphology. Trends Genet. 22 (10), 537-544.

LATORRE R., GIL F., VAZQUEZ J. M., MORENO F., MASCARELLO F. et al. (1993): Morphological and histochemical characteristics of muscle fibre types in the flexor carpi radialis of the dog. J. Anat. 182 (Pt 3), 313-320.

LEE D., BERTRAM J. & TODHUNTER R. (1999): Acceleration and balance in trotting dogs. J. Exp. Biol. 202 (24), 3565-3573.

LEONARD J., WAYNE R., WHEELER J., VALADEZ R., GUILLÉN S. et al. (2002): Ancient DNA evidence for Old World origin of New World dogs. Science 298 (5598), 1613-1616.

LIESER B. (2003): Morphologische und biomechanische Eigenschaften des Hüftgelenks (Articulatio coxae) des Hundes (*Canis familiaris*). Diss. med. vet. Veterinärmedizinische Fakultät, Ludwig-Maximilians-Universität München.

LINDBLAD-TOH K., WADE C. M., MIKKELSEN T. S., KARLSSON E. K., JAFFE D. B. et al. (2005): Genome sequence, comparative analysis and haplotype structure of the domestic dog. Nature 438 (7069), 803-819.

LINDSTEDT S. L., HOPPELER H., BARD K. M. & THRONSON H. A. J. (1985): Estimate of muscle-shortening rate during locomotion. Am. J. Physiol. 249 (6 pt 2), R699-R703.

LOTT D., LOEFFLER K. & KLEINE-KUHLMANN R. (1990): Bestimmung der Gelenkwinkel an den Gliedmaßen von Hunden, 2. Mitteilung: Gelenkwinkel bei Deutschen Schäferhunden und Boxern. Kleintierpraxis 35.

LUMER H. (1940): Evolutionary allometry in the skeleton of the domesticated dog. Am. Nat., 439-467.

MADORE E., HUNEAULT L., MOREAU M. & DUPUIS J. (2007): Comparison of trot kinetics between dogs with stifle or hip arthrosis. Vet. Comp. Orthop. Traumatol. 20 (2), 102-107.

MAI M. (1995): Funktionell-anatomische Untersuchungen an den Articulationes membri thoracici von Deutschem Schäferhund, Dobermann, Deutscher Dogge, Rauhaar- und LanghaarDachshund. Diss. med. vet. Veterinärmedizinische Fakultät, Ludwig-Maximilians-Universität München.

MAIERL J. (2003): Zur funktionellen Anatomie und Biomechanik des Ellbogengelenks (Articulatio cubiti) des Hundes (*Canis familiaris*). Habilitationsschrift Veterinärmedizinische Fakultät, Ludwig-Maximilians-Universität München.

MAIERL J. & BÖTTCHER P. (1999): Dreidimensionale Visualisierung der subchondralen Knochendichte entlang der Oberflächennormalen am Schulter und Ellbogengelenk des Hundes. Ann. Anat. (Suppl.) 181, 288.

MALLONÉE J. (2008): Movements of radio collared wolves and their significance on pack assembly. J. Am. Sci. 4 (1), 53-58.

MARX J., OLSSON M. C. & LARSSON L. (2006): Scaling of skeletal muscle shortening velocity in mammals representing a 100.000-fold difference in body size. Pflügers Arch. - Eur. J. Physiol. 452 (2), 222-230.

MASON D. R., SCHULZ K. S., FUJITA Y. & KASS P. H. (2005): In vitro force mapping of normal canine humeroradial and humeroulnar joints. Am. J. Vet. Res. 66 (1), 132-135.

MAXWELL L. C., BARCLAY J. K., MOHRMAN D. E. & FAULKNER J. A. (1977): Physiological characteristics of skeletal muscles of dogs and cats. Am. J. Physiol. 233, C14-C18.

MAZZORIN J. D. G. & TAGLIACOZZO A. (2000): Morphological and osteological changes in the dog from the Neolithic to the Roman period in Italy. 141-161. In: S. J. Crockford (Ed.): Dogs Through Time: An Archaeological Perspective. Proceedings of the 1st ICAZ Symposium on the History of the Domestic Dog. Oxford: Archaeopress, B.A.R..

McCARTHY P. H. & WOOD A. K. (1988): Anatomic and radiologic observations of the clavicle of adult dogs. Am. J. Vet. Res. 49 (6), 956-959.

McLAUGHLIN R. M. (2001): Kinetic and kinematic gait analysis in dogs. Vet. Clin. North Am. Small Anim. Pract. 31, 193-201.

McLAUGHLIN R. M. & ROUSH J. K. (1995): Effects of increasing velocity on braking and propulsion times during force plate gait analysis in Greyhounds. Am. J. Vet. Res. 56, 159-161.

McLAUGHLIN R. M. J. & ROUSH J. K. (1994): Effects of subject stance time and velocity on ground reaction forces in clinically normal greyhounds at the trot. Am. J. Vet. Res. 55 (12), 1666-1671.

MECH L. D. (1994): Regular and homeward travel speeds of Arctic wolves. J. Mammal. 75 (3), 741-742.

MECH L. D., JR. L. D. F., REAM R. R. & WINSHIP J. W. (1971): Movements, behavior, and ecology of timber wolves in northeastern Minnesota. 35-51. In: L. D. Mech, L. D. Frenzel Jr. (Eds.): Ecological studies of the timber wolf in northeastern Minnesota. USDA Forest Service Research Paper NC-52. North Central Forest Experimental Station. St. Paul, Minnesota.

MÉNDEZ J. & KEYS A. (1960): Density and composition of mammalian muscle. Metabolism 9, 184-188.

MERRILL S. & MECH L. (2003): The usefulness of GPS telemetry to study wolf circadian and social activity. Wildl. Soc. Bull., 947-960.

MOREY D. F. (2005): Burying key evidence: the social bond between dogs and people. J. Arch. Sci. 33 (2), 158-175.

MÜLLER J. (2010): Zur Domestikation des Wolfs - wann, wie, wo? Unser Rassehund 10, 26-30.

MÜLLER-GERBL M. (1991): CT-Osteoabsorptiometrie (CT-OAM) und ihr Einsatz der Langzeitbeanspruchung der großen Gelenke in vivo. Habilitationsschrift Medizinische Fakultät, Ludwig-Maximilians-Universität München.

NANUA P. & WALDRON K. J. (1995): Energy comparison between trot, bound, and gallop using a simple model. J. Biomech. 117 (4), 466-473.

NAPIERLA H. & UERPMANN H.-P. (2010): A 'New' Palaeolithic Dog from Central Europe. Int. J. Osteoarchaeol 20, DOI: 10.1002/oa.1182.

NEWSHOLME S., LEXELL J. & DOWNHAM D. (1988): Distribution of fibre types and fibre sizes in the tibialis cranialis muscle of beagle dogs. J. Anat. 160, 1-8.

NEWTON C. D. (1985): Normal joint range of motion in the dog and cat. 1101-1106. In: C.D. Newton, D.M. Nunamaker (eds.): Textbook of Small Animal Orthopaedics, Philadelphia: Lippincott.

NIELSEN C., STOVER S. M., SCHULZ K. S., HUBBARD M. & HAWKINS D. A. (2003): Two-dimensional link-segment model of the forelimb of dogs at a walk. Am. J. Vet. Res. 64 (5), 609-617.

NOBIS G. (1979): Der älteste Haushund lebte vor 14.000 Jahren. Umschau 79, 19.

OFF W. & MATIS U. (1997a): Ganganalyse beim Hund. Teil 1: Dynamometrische und kinematische Meßverfahren und ihre Anwendung beim Tetrapoden. Tierärztl. Praxis 25, 8-14.

OFF W. & MATIS U. (1997b): Ganganalyse beim Hund. Teil 2: Aufbau eines Ganglabors und bewegungsanalytische Untersuchungen. Tierärztl. Praxis 25, 303-311.

OKARMA H., JĐDRZEJEWSKI W., SCHMIDT K., S.ĐNIEĐKO, BUNEVICH A. N. et al. (1998): Home ranges of wolves in BiałowieĐa Primeval Forest, Poland, compared with other Eurasian populations. J. Mammal. 79, 842-852.

OWEN M. R., RICHARDS J., CLEMENS D. N., DREW S. T., BENNET D. et al. (2004): Kinematics of the elbow and stifle joints in greyhounds during treadmill trotting - An investigation of familiarisation. Vet. Comp. Orthop. Traumatol. 17, 141-145.

PAGE A. E., ALLAN C., JASTY M., HARRIGAN T. P., BRAGDON C. R. et al. (1993): Determination of loading parameters in the canine hip in vivo. J. Biomech. 26 (4/5), 571-579.

PARKER H. G., KIM L., SUTTER N., CARLSON S., LORENTZEN T. et al. (2004): Genetic structure of the purebred domestic dog. Science 304 (5674), 1160-1164.

PARKER H. G. & OSTRANDER E. A. N. (2005): Canine genomics and genetics: running with the pack. PLoS Genet. 1 (5), 507-513.

PARKER H. G., VONHOLDT B. M., QUIGNON P., MARGULIES E. H., SHAO S. et al. (2009): An expressed Fgf4 retrogene is associated with breed-defining chondrodysplasia in domestic dogs. Science 325, 995-998.

PASI, B. M. & CARRIER, D. R. (2003): Functional trade-offs n the limb muscles of dogs selected for running vs. fighting. J. Evol. Biol. 16: 324-332.

PERRET, C. & BERTHOZ, A. (1973): Evidence of static and dynamic fusimotor actions on the spindle response to sinusoidal stretch during locomotor activity in the cat. Exp. Brain Res. 18 (2), 178-188.

PETERSEN, A.M.W., PEDERSEN, B.K. (2005): The anti-inflammatory effect of exercise. J Appl Physiol 98: 1154–1162

PETTE D. & STARON R. S. (1990): Cellular and molecular diversities of mammalian skeletal muscle fibers. Rev. Physiol. Biochem. Pharmacol. 116 (Review), 1-76 .

PRIEUR W. D. (1980): Coxarthrosis in the dog part I: Normal and abnormal biomechanics of the hip joint. Vet. Surg. 9, 145-149.

RÄBER H. (1993): Enzyklopädie der Rassehunde - Band1. Stuttgart: Kosmos Verlag.

RÄBER H. (1995): Enzyklopädie der Rassehunde - Band2. Stuttgart: Kosmos Verlag.

RÄBER H. (2008): Die Schweizer Hunderassen. Schweizerische Kynologische Gesellschaft .

RAUSCHER T. (1986): Metrische Untersuchungen an den Gelenken und Muskel der Schultergliedmaße bei zwei Hunderassen - ein Vergleich zwischen Dackel (Dachshund und Dachshund) und dem Deutschen Schäferhund. Diss. med. vet. Veterinärmedizinische Fakultät, Ludwig-Maximilians-Universität München.

RENBERG W. C., JOHNSTON S. A., YE K. & BUDSBERG S. C. (1999): Comparison of stance time and velocity as control variables in force plate analysis of dogs. Am. J. Vet. Res. 60 (7), 814-819.

RICHTER V. & LOEFFLER K. (1976): Rassespezifische Merkmale am Becken des Hundes. Dtsch. Tierärztl. Wschr. 83, 455-461.

RIEGERT S. (2004): Anatomische und biomechanische Untersuchungen am Kniegelenk (Articulatio genus) des Hundes (*Canis familiaris*). Diss. med. vet. Veterinärmedizinische Fakultät, Ludwig-Maximilians-Universität München.

RIGGS C. M., DECAMP C. E., SOUTAS-LITTLE R. W., BRADEN T. D. & RICHTER M. A. (1993): Effects of subject velocity on force plate-measured ground reaction forces in healthy Greyhounds at the trot. Am. J. Vet. Res. 54 (9), 1523-1526.

RINGROSE R. P. (1997): Self-stabilizing running. Ph.D. thesis. Boston: Massachusetts Institute of Technology (pp.131).

RISER W. H. & SHIRER J. F. (1965): Normal and abnormal growth of the distal foreleg in large and giant dogs. J. Amer. Vet. Radiol. Soc. 6, 50-64.

RITTER D., NASSAR P., FIFE M. & CARRIER D. (2001): Epaxial muscle function in trotting dogs. J. Exp. Biol. 204 (17), 3053-3064.

RODRIGUEZ-BARBUDO P., VAAMONDE R., AGUERA E., CARPIO M., MORENO F. et al. (1984): Estudio histoquimico y morfometrico del musculo tibial craneal en perros de diferentes aptitudes dinimicas (galgo, pastor alemin y fox terrier). Anat. Histol. Embryol. 13 (4), 300-312.

ROMERO-SUAREZ, S., SHEN, J., BROTTO, L., HALL, T., MO, C., et al. (2010): Muscle-specific inositide phosphatase (MIP/MTMR14) is reduced with age and its loss accelerates skeletal muscle aging process by altering calcium homeostasis. AGING 2 (8): 504-513.

ROOS H., BRUGGER S. & RAUSCHER T. (1992): Über die biologische Wertigkeit der Bewegung in den Radioulnargelenken bei Katze und Hund. Anat. Histol. Embryol. 21, 199-205.

ROOS H., RAUSCHER T. & BRUGGER S. (1993): Die biologische Wertigkeit der Bewegungen im Schultergelenk von Katze und Hund. Tierärztl. Praxis 21 (4), 323-327.

ROUSH J. K. & McLAUGHLIN M. J. (1994): Effects of subject stance time and velocity on ground reaction force in clinically normal Greyhounds at the walk. Am. J. Vet. Res. 55, 1672-1676.

RUCHAY S. (2008): Das Labrum glenoidale beim Hund. Eine anatomisch-histologische Studie unter besonderer Berücksichtigung des Kollagenfaserverlaufs. Diss. med. vet. Klinikum Veterinärmedizin, Justus-Liebig-Universität Gießen.

RUMPH P. F., KINCAID S. A. & BAIRD D. K. (1993): Vertical ground reaction force distribution during experimentally induced acute synovitis in dogs. Am. J. Vet. Res. 54, 365-369.

RUMPH P. F., LANDER J. E., KINCAID S. A., BAIRD D. K., KAMMERMANN J. R. et al. (1994): Ground reaction force profiles from force platform gait analyses of clinically normal mesomorphic dogs at the trot. Am. J. Vet. Res. 55 (6), 756-761.

RUMPH P. F., STEISS J. E. & MONTGOMERY R. D. (1997): Effects of selection and habituation on vertical ground reaction force in Greyhounds. Am. J. Vet. Res. 58 (11), 1206-1208.

RUMPH P. F., STEISS J. E. & WEST M. S. (1999): Interday variation in vertical ground reaction force in clinically normal Greyhounds at the trot. Am. J. Vet. Res. 60 (6), 679-683.

SAETRE P., LINDBERG J., LEONARD J., OLSSON K., PETTERSSON U. et al. (2004): From wild wolf to domestic dog: gene expression changes in the brain. Mol. Brain Res. 126 (2), 198-206.

SAMPSON J. & BINNS M. M. (2006): The Kennel Club and the early history of dog shows and breed clubs. 19-30. In: E. A. Ostrander, U. Giger, K. Lindblad-Toh (Eds.): The dog and its genome. New York: Cold Spring Harbor.

SCHENDEL M., DEKUTOSKI M., OGILVIE J., OLSEWSKI J. & WALLACE L. (1995): Kinematics of the canine lumbar intervertebral joints: an in vivo study before and after adjacent instrumentation. Spine 20 (23), 255.

SCHERER P. K., HILLBERRY B. M. & SICKLE D. V. (1979): Determining the in-vivo areas of contact in the canine schoulder. J. Biomech. 101, 271-278.

SCHILLING N. & CARRIER D. R. (2009): Function of the epaxial muscles during trotting. J. Exp. Biol. 212 (7), 1053-1063.

SCHILLING, N. & CARRIER, D. R. (2010): Function of the epaxial muscles in walking, trotting, and galloping dogs: Implications for the evolution of epaxial muscle funtion in tetrapods. J. Exp. Biol. 213, 1490-1502.

SCHILLING N., FISCHBEIN T., YANG E. & CARRIER D. R. (2009): Function of the extrinsic hindlimb muscles in trotting dogs. J. Exp. Biol. 212 (7), 1036-1052.

SCHLEGEL F. R. (1913): Die Extremitäten der Caniden, ihre Beziehungen zur Körpersymmetrie und die Verhältnisse ihrer relativen Proportionen. Diss. phil. Philosophisch-Naturwissenschaftliche Fakultät, Universität Bern.

Schleip, R. (2003): Faszien und Nervensystem. Z. osteopath. Medizin 4(1), 20-28.

SCHMIDT P. A. & MECH L. D. (1997): Wolf pack size and food acquisition. Am. Nat. 150 (4), 513-517.

SCHOLLE H.-C., SCHUMANN N. P., BIEDERMANN F., ROELEVELD K., STEGEMAN D. F. et al. (2001): Spatio-temporal surface EMG characteristics from rat triceps brachii muscle during treadmill locomotion indicate selective recruitment of functionally distinct muscle regions. Exp. Brain Res. 138 (1), 26-36.

SCHREIBER J. (1947): Beiträge zur vergleichenden Anatomie und zur Mechanik des Kniegelenkes. Wiener tierärztl. Mschr. 34, 725-744.

SCHRÖDER M. (1978): Beitrag zur Entwicklung des Skelettes der Vordergliedmaßen beim Deutschen Schäferhund. Diss. med. vet. Veterinärmedizinische Fakultät, Ludwig-Maximilians-Universität München.

SEYFARTH A., GEYER H., GÜNTHER M. & BLICKHAN R. (2002): A movement criterion for running. J. Biomech. 35 (5), 649-655.

SHAHAR R. & BANKS-SILLS L. (2002): Biomechanical analysis of the canine hind limb: calculation of forces during three-legged stance. Vet. J. 163 (3), 240-250.

SHAHAR R. & MILGRAM J. (2001): Morphometric and anatomic study of the hind limb of a dog. Am. J. Vet. Res. 62 (6), 928-933.

SHAHAR R. & MILGRAM J. (2005): Morphometric and anatomic study of the forelimb of the dog. J. Morph. 263 (1), 107-117.

SHARIR A., MILGRAM J. & SHAHAR R. (2006): Structural and functional anatomy of the neck musculature of the dog (*Canis familiaris*). J. Anat. 208 (3), 331.

SHARP J. W., BAILEY C. S., JOHNSON R. D. & KITCHELL R. L. (1991): Spinal root origin of the radial nerve and nerves innervating shoulder muscles of the dog. Anat. Histol. Embryol. 20 (3), 205-214.

SHEARIN A. L. & OSTRANDER E. A. (2010): Canine morphology: Hunting for Genes and tracking mutations. PLoS Biol. 8 (3), 1-6.

SIMMONS A., CARRIER D., FARMER C. & GREGERSEN C. (1997): Lack of locomotor-cardiac coupling in trotting dogs. Am. J. Physiol. Regul. Integr. Comp. Physiol. 273 (4), 1352-1360.

SMERDU V., CEHOVIN T., STRBENC M. & FAZARINC G. (2009): Enzyme- and immunohistochemical aspects of skeletal muscle fibers in Brown Bear (Ursus arctos). J. Morph. 270 (2), 154-161.

SNOW D., BILLETER R., MASCARELLO F., CARPENE E., ROWLERSON A. et al. (1982): No classical type IIB fibres in dog skeletal muscle. Histochem. Cell Biol. 75 (1), 53-65.

STARK H. (2007): Die 3D-Architektur der Muskelfaszikel in ausgewählten Muskeln und ihre Relevanz zur Kraftentwicklung. Dissertation Biologisch-Pharmazeutische Fakultät, Friedrich-Schiller-Universität Jena.

STEUERNAGEL M. (2009): Identifikation und Verteilung der Fasertypen in der perivertebralen Muskulatur des Beagles. Diplomarbeit Biologisch-Pharmazeutische Fakultät, Friedrich-Schiller-Universität Jena.

STRBENC M., SMERDU V., ZUPANC M., TOZON N. & FAZARINC G. (2004): Pattern of myosin heavy chain isoforms in different fibre types of canine trunk and limb skeletal muscles. Cells Tissues Organs 176 (4), 178-186.

STREITBERGER K., FISCHER M. S., DISTL O. & EPPLEN J. T. (2008): Molekulargenetische Untersuchung zur genetischen Variabilität in Hunderassen. Unser Rassehund 7, 30-33.

SUNDQVIST A., BJORNERFELDT S., LEONARD J., HAILER F., HEDHAMMAR A. et al. (2006): Unequal contribution of sexes in the origin of dog breeds. Genetics 172 (2), 1121-1128.

SUTTER N., BUSTAMANTE C., CHASE K., GRAY M., ZHAO K. et al. (2007): A single IGF1 allele is a major determinant of small size in dogs. Science 316 (5821), 112-117.

TASHMAN S. & ANDERST W. (2003): In-vivo measurement of dynamic joint motion using high speed biplane radiography and CT: application to canine ACL deficiency. J. Biomech. 125, 238-245.

TAYLOR C. R. (1994): Relating mechanics and energetics during exercise. Adv. Vet. Sci. Comp. Med. 38A, 181-215.

THEUERKAUF J. (2009): What Drives Wolves: Fear or Hunger? Humans, Diet, Climate and Wolf Activity Patterns. Ethology, 1-9.

THEUERKAUF J., GULA R., PIRGA B., TSUNODA H., EGGERMANN J. et al. (2007): Human impact on wolf activity in the Bieszczady Mountains, SE Poland. Ann. Zool. Fenn. 44, 225-231.

THURBER J. M. & PETERSON R. O. (1993): Effects of pupulation density and pack size on the foraging ecology of gray wolves. J. Mammal. 74, 879-889.

TOKURIKI M. (1973): Electromyographic and joint-mechanical studies in quadrupedal locomotion. I. Walk. Nippon Juigaku Zasshi. Japan. J. Vet. Sci. 35 (5), 433-436.

TOKURIKI M. (1973): Electromyographic and joint-mechanical studies in quadrupedal locomotion. II. Trot. Nippon Juigaku Zasshi. Japan. J. Vet. Sci. 35 (6), 525-533.

TOKURIKI M. (1974): Electromyographic and joint-mechanical studies in quadrupedal locomotion. III. Gallop. Nippon Juigaku Zasshi. Japan. J. Vet. Sci. 36 (2), 121-132.

TOKURIKI M. (1979): Cinematographic and electromyographic analysis of vertical standing jump in the dog. J. Exp. Biol. 83 (1), 271-282.

TOKURIKI M. & NAKATA A. (1991): Trunk muscle activity during locomotion in the newt, lizard, dog and cat. 221-227. In: M. Shimamura, S. Grillner, V.R. Edgerton (Eds.): Neurobiological basis for human locomotion. Tokyo: Japan Scientific Societies Press.

TRUT L. (1999): Early canid domestication: the farm-fox experiment. Am. Scient. 87 (2), 160-169.

TRUT L., OSKINA I. & KHARLAMOVA A. (2009): Animal evolution during domestication: the domesticated fox as a model. Bioessays 31 (3), 349-360.

TRUT L. N., KHARLAMOVA A. V., CARRIER D. R., CHASE K., KUKEKOVA A. V. et al. (2006): Morphology and behavior: are they coupled at the genome level? 81-93. In: E. A. Ostrander, U. Giger, K. Lindblad-Toh (Eds.): The genome of the domestic dog. New York: Cold Spring Harbor.

TRUT L. N., PLIUSNINA I. Z. & OSKINA I. N. (2004): An experiment on fox domestication and debatable issues of evolution of the dog. Genetika 40 (6), 794-807.

UNKEL-MOHRMANN F. (1999): Das ganganalytische Profil des Deutschen Schäferhundes. Diss. med. vet. Veterinärmedizinische Fakultät, Ludwig-Maximilians-Universität München.

USHERWOOD J., WILLIAMS S. & WILSON A. (2007): Mechanics of dog walking compared with a passive, stiff-limbed, 4-bar linkage model, and their collisional implications. J. Exp. Biol. 210 (3), 533-540.

USHERWOOD J. R. & WILSON A. M. (2005): Biomechanics: no force limit on greyhound sprint speed. Nature 438 (7069), 753-754.

VAN VALKENBURGH B. & RUFF C. B. (1987): Canine tooth strength and killing behaviour in large carnivores. J. Zool. 212 (3), 379-398.

VASSEUR P. & ARNOCZKY S. (1981): Collateral ligaments of the canine stifle joint: anatomic and functional analysis. Am. J. Vet. Res. 42 (7), 1133-1137.

VERARDI A., LUCCHINI V. & RANDI E. (2006): Detecting introgressive hybridization between free-ranging domestic dogs and wild wolves (*Canis lupus*) by admixture linkage disequilibrium analysis. Mol. Ecol. 15 (10), 2845-2855.

VERGINELLI F., CAPELLI C., COIA V. & MUSIANI M. (2005): Mitochondrial DNA from prehistoric canids highlights relationships between dogs and south-east european wolves. Mol. Biol. Evol. 22 (12), 2541-2551.

VILA C. & LEONARD J. (2007): Origin of Dog Breed Diversity. 38-58. In: P. Jensen (Ed.): The Behavioural Biology of Dogs. Portland (OR): Book News, Inc..

VILA C., WALKER C., SUNDQVIST A., FLAGSTAD Ø., ANDERSONE Z. et al. (2003): Combined use of maternal, paternal and bi-parental genetic markers for the identification of wolf–dog hybrids. Heredity 90 (1), 17-24.

VORNHOLT E. (2004): Der Große Münsterländer. Melsungen: Vlg. J. Neumann-Neudamm AG.

WADA, N., HORI, H., TOKURIKI, M. (1993): Electromyographic and kinematic studies of tail movements in dogs during treadmill locomotion. J. Morph. 217: 105-113.

WAGNER H. & BLICKHAN R. (1999): Stabilizing function of skeletal muscles: an analytical investigation. J. Theor. Biol. 199 (2), 163-179.

WALTER R. & CARRIER D. R. (2009): Rapid acceleration in dogs: ground forces and body posture dynamics. J. Exp. Biol. 212, 1930-1939.

WALTER R. M. & CARRIER D. R. (2007): Ground forces applied by galloping dogs. J. Exp. Biol. 210 (Pt 2), 208-216.

WAYNE R. K. (1986a): Developmental constraints on limb growth in domestic and some wild canids. J. Zool. 210 (3), 381-399.

WAYNE R. K. (1986b): Limb morphology of domestic and wild canids: The influence of development on on morphological change. J. Morph. 187 (3), 301-319.

WEISE G. (1965): Über das Wachstum verschiedener Haushundrassen. Z. Säugetierk. , 257-282.

WENTINK G. (1976): The action of the hind limb musculature of the dog in walking. Acta anat. 96, 70-80.

WENTINK G. (1977): Biokinetical analysis of hind limb movements of the dog. Anat. Embryol. 151, 171-181.

WENTINK G. (1979): Dynamics of the hind limb at walk in horse and dog. Anat. Histol. Embryol. 155 (2), 179-190.

WILLIAMS S. B. (2009): Exploring the mechanical basis for acceleration: pelvic limb locomotor function during accelerations in racing greyhounds (*Canis familiaris*). J. Exp. Biol. 212 (4), 550-565.

WILLIAMS S. B., WILSON A. M., DAYNES J., PECKHAM K. & PAYNE R. C. (2008b): Functional anatomy and muscle moment arms of the thoracic limb of an elite sprinting athlete: the racing greyhound (*Canis familiaris*). J. Anat. 213 (4), 373-382.

WILLIAMS S. B., WILSON A. M., RHODES L., ANDREWS J. & PAYNE R. C. (2008a): Functional anatomy and muscle moment arms of the pelvic limb of an elite sprinting athlete: the racing greyhound (*Canis familiaris*). J. Anat. 213 (4), 361-372.

WINGFIELD C., LAW H. T. & STEAD A. C. (1993): Canine kinematic hindleg gait analysis using a microcomputer. J. Small Anim. Pract. 34 (7), 319-324.

WINHARD F. E. (2007): Anatomische und computertomographische Untersuchungen am gesunden und degenerativ veränderten Schulter- und Ellbogengelenk des Hundes (*Canis familiaris*). Diss. med. vet. Veterinärmedizinische Fakultät, Ludwig-Maximilians-Universität München.

WITTE H. F., HACKERT R., SAEGLITZ J., VOGES D., MAIERL J. et al. (2004): Untersuchung zur Biomechanik großer Gelenke beim Schäferhund. Abschlußbericht GKF.

ZEDER M., EMSHWILLER E., SMITH B. & BRADLEY D. (2006): Documenting domestication: the intersection of genetics and archaeology. Trends Genet. 22 (3), 139-155.

ZIEGLER H. (1934): Weitere Untersuchungen über den M. glutaeobiceps von Hund (*Canis familiaris*) und Katze (Felis catus dom.). Morph. Jb. 73, 385-391.

ZUCKERKANDL E. & PAULING L. B. (1962): Molecular disease, evolution, and genetic heterogeneity. 189-225. In: M. Kasha, B. Pullman: Horizons in Biochemistry. New York: Academic Press.

ACKNOWLEDGEMENTS

We would like to thank everyone who was involved in the Jena Study and in the making of this book.

The greatest supporter of our project, from the initial idea to the final proofreading, was Dr. Helga Eichelberg, without whom these insights into the locomotion of dogs would never have been obtained. Thanks are also due to the German Kennel Club (Verband für das Deutsche Hundewesen – VDH), Prof. Dr. Peter Friedrich, Bernhard Meyer, Udo Kopernik and Leif Kopernik for exceptional, open and constructive collaboration. Many thanks go to the Society for Cynological Research (Gesellschaft zur Förderung Kynologischer Forschung - GKF), the breed clubs involved in the study and their presidents and breeding boards for their generous funding and support. The fact that breeders associations have been financing basic research for many years in Germany is something which is widely appreciated internationally. The study was granted in various ways by the German Research Council (DFG) and the Bundesministerium für Bildung und Forschung (BMBF).

That we were able to study over three hundred dogs over a period of months and years at a private medical and biotechnological research center (Forschungszentrum für Medizintechnik und Biotechnologie - fzmb) in Bad Langensalza is a reflection of the particularly close link between industry and science in Thuringia. Without the initial willingness of Prof. Dr. Manfred Günthert, and that of Dr. Peter Miether, this would not have been possible. We owe a huge vote of thanks to these two gentlemen, the many helpers we had at the fzmb and, in particular, to Mr Manfred Gillner, our constant companion at the treadmill, for his untiring help in taking measurements.

We were only able to investigate small breeds such as the Chihuahua with the kind support of PD Dr. André Seyfarth at the Locomotion Laboratory in Jena. Indeed, that Chihuahuas were included in the study in the first place is solely down to a personal initiative for which we are very grateful.

A number of young scientists and student assistants were involved in performing the investigations which took place at the fzmb and in evaluating the results in Jena, and these we would like to thank in alphabetical order: Dr. Dirk Arnold, Anne Ansorg, Franziska Buch, Julia Ehrt, Martin Groß, Markus Hergeth, Matthias Kozlik, Dr. Cornelia Krause, Marcus Krauß, Anke Kügler, Irina Mischewski, Sabine Moritz, Sarah Müller, Manuela Östreicher, Dr. André Schmidt, Katharina Spange, Gabriele Unterhitzenberger, Saskia Wutke.

Thanks are due too to Ms Romy Peterson and Ms Katja Felbel for their patience and skill in obtaining the X-ray recordings. In terms of X-ray and computer tomography technology, our close cooperation with the Institute for Diagnostic and Interventional Radiology proved once again to be worth its weight in gold. Our thanks go to its director, Prof. Dr. Werner Kaiser, and to all the staff involved. Although the CT recordings were indispensable, our animation was also based on a number of detailed dissections for which we are grateful to Mr Matthias Krüger. Many thanks go too to Dr. John Nyakatura and Dr. Heiko Stark for their help on difficult questions during the processing of animation-related data.

The manuscript was subject to a strict regime of proofreading and revision, and our particular thanks here go to Prof. Dr. Horst Bleckmann, Prof. Dr. Dave Carrier, Ms Gaby Unterhitzenberger, Dr. Barbara Happe, Dr. Stefan Hertwig, PD Dr. Johann Maierl, Dr. Nadja Schilling and especially Dr. Manuela Schmidt.

We are grateful to Susann Hempel for the intro of the media library, Prof. Dr. Dr. Michael Schünke (Kiel) and Prof. Dave Carrier (Salt Lake City) for granting us their generous permission to use the figures on page 58 and 33, respectively. Prof. Dr. Wolfgang A. Linke (Bochum) was kind enough to check the particularly complex figure 112-113. We also thank Prof. Reinhard Schulz-Schaeffer for his continous support.

Ms Lucy Cathrow's translation of the manuscript into English was so convincing that we were sometimes moved to rework the German text in response. Her father, Ian Cathrow, did the final proofreading of the English book..

Finally, our greatest thanks go to the hundreds of dog owners who spared no cost or effort to join us in Bad Langensalza. This study, the largest-scale investigation in the world into dog locomotion, was only made possible by their help and support. Their enthusiasm was a great and constant motivator and our encounters with them remain etched in our memories.